Visualizing Calculus
by
Way of Maple™ :

An Emphasis on Problem Solving

Arnavaz Taraporevala
New York City College of Technology

Nadia Benakli
New York City College of Technology

Satyanand Singh
New York City College of Technology

Connect
Learn
Succeed™

VISUALIZING CALCULUS BY WAY OF MAPLE™: AN EMPHASIS ON PROBLEM SOLVING

Published by McGraw-Hill, a business unit of The McGraw-Hill Companies, Inc., 1221 Avenue of the Americas, New York, NY 10020. Copyright © 2012 by The McGraw-Hill Companies, Inc. All rights reserved. No part of this publication may be reproduced or distributed in any form or by any means, or stored in a database or retrieval system, without the prior written consent of The McGraw-Hill Companies, Inc., including, but not limited to, in any network or other electronic storage or transmission, or broadcast for distance learning.

Some ancillaries, including electronic and print components, may not be available to customers outside the United States.

 This book is printed on recycled, acid-free paper containing 10% postconsumer waste.

1 2 3 4 5 6 7 8 9 0 QDB/QDB 1 0 9 8 7 6 5 4 3 2 1

ISBN 978–0–07–803598–2
MHID 0–07–803598–8

Vice President, Editor-in-Chief: Marty Lange
Vice President, EDP: Kimberly Meriwether David
Senior Director of Development: Kristine Tibbetts
Editorial Director: Michael Lange
Developmental Editor: Eve L. Lipton
Marketing Manager: Alexandra Coleman
Project Coordinator: Mary Jane Lampe
Senior Buyer: Kara Kudronowicz
Manager, Creative Services: Michelle D. Whitaker
Cover Image: © Arnavaz Taraporevala, NadiaBenakli, and Satyanand Singh (created in Maple™)
Compositor: Lachina Publishing Services
Typeface: 12/14 Times Roman
Printer: Quad/Graphics

All credits appearing on page or at the end of the book are considered to be an extension of the copyright page.

Maple is a trademark of Waterloo Maple Inc.

www.mhhe.com

Preface

Visualizing Calculus by Way of *Maple*: An Emphasis on Problem Solving is a book that can be used as a supplement to any Calculus text or in a Calculus Laboratory setting. Although this book states definitions and theorems used in each chapter, it also has the utility to work with any calculus text. The student can cross reference concepts when and if necessary.

The objective of this book is for students to develop a better understanding and appreciation of Calculus by using *Maple*^{TM}, a computer algebra system. *Maple* is seen as a tool that expands the possibilities in problem solving, accuracy and research, while it encourages a deeper understanding of concepts. Although version 14 is used in this book, a number of the commands are available in the earlier versions as well.

All the computations were performed by using *Maple*. We begin our book with an introduction to basic *Maple* commands. This includes the use of the right click feature, available in *Maple* from version 10 and beyond. This feature allows the user greater flexibility in some instances, but it should be used with caution in others. The use of the palettes (like the Expressions and Common Symbols Palettes) that are available on the left hand side of any *Maple* screen (from version 10 and beyond) is also included.

For each chapter of the book, several examples with varying levels of difficulty are presented. At the end of each chapter, there are extensive exercises some of which can be used as projects to highlight the concepts. Some of the problems are more difficult and open ended with the goal of challenging and stimulating the reader in the creative aspects of mathematics. Emphasis is placed on the subtle points of Calculus such as the ability to interpret answers, know when to adjust parameters and focus on the salient points.

Prior to doing Vector Calculus, Linear Algebra is introduced. The use of matrices is incorporated to illustrate vectors and geometry in space, directional derivatives, tangents and differentials, gradient vectors etc. Consistent with our title, we have used examples from $\mathbb{R}^2$ and $\mathbb{R}^3$ to allow the reader to get a visual perspective. An appendix of select *Maple* commands is added at the end of the book. A list of the *with*(*packages*), used throughout the book are also included.

Acknowledgements

We give special thanks to our families for their patience during these last few months. We would also like to thank all the people who supported us in this endeavor. In particular, we give special thanks to Dr. Rhona Noll for reading several chapters of the manuscript and providing us with valuable suggestions. We are also grateful to our students whose motivation to learn calculus and explore its concepts with *Maple* is an inspiration. We hope this work will continue to motivate and inspire new generations of students to enjoy and learn calculus.

Special thanks are also given to McGraw-Hill publishers and their editorial staff who were instrumental in making this publication possible. Last, but not least we would like to thank the Maplesoft staff for their software support and permission to use *Maple* in the manner presented in our book.

Arnavaz Taraporevala, Satyanand Singh and Nadia Benakli

Table of Contents

Table of Contents

Chapter 0   Getting Started with *Maple*

Maple^{TM} is a powerful computer algebra system. The object of this chapter is to introduce basic *Maple* commands. As one begins to use *Maple*, it is essential that one becomes familiar with performing some basic operations such as addition, subtraction, multiplication and division of numbers or functions. When dealing with functions one is also expected to be familiar with graphing, factoring, and finding roots. In this chapter, we will illustrate these operations with some examples. These illustrated examples can be adapted to other functions that you would encounter in your course of study.

It is also important to keep in mind that *Maple* can perform the same operations with different commands, which gives the user more options. Sometimes one way may be more advantageous than another. We will illustrate this in some of the examples that follow.

The Basics

To get started with *Maple* depends on the platform you are using. On the PC or MAC you will see the *Maple* icon. In other systems you need to speak to your systems administrator to establish the manner in which it can be launched. In our text we are using the PC version in which *Maple* is launched by double clicking on the *Maple* icon.

Here is a picture of what we see once *Maple* is launched.

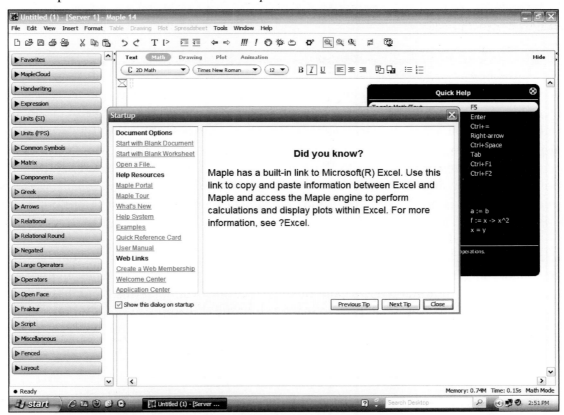

Note that each time you launch *Maple*; a different "*Did you know?*" window pops up. This window gives a number of helpful hints. Close this window. The next screen is the one given below.

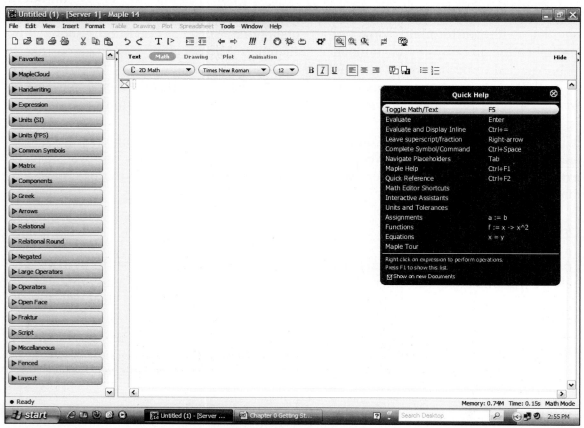

The *Quick Help* list on the right side gives a number of short cuts that can be used while executing *Maple* commands. Close this window or press the F1 key (that too closes this window). Now you are in the *Document* mode. This is the default setting for *Maple*. You can press on F1 to get the *Quick Help* list back again. One can open a *Maple* worksheet in either the *Document* mode or the *Worksheet* mode. To change the mode click on *File*, then click on *New* and select one of the options: *Worksheet Mode*, *Document Mode* or *Templates*. There is no (>) prompt in the "*Document* mode" whereas it is present in the "*Worksheet mode*" as shown below.

Each *Maple* worksheet contains a variety of built in characteristics that are not immediately apparent. Such characteristics are formatting, Document block boundaries, execution groups (earmarked for automatic execution), annotations, and bookmarks. Detailed explanations follow on how to make use of these features to one's advantage.

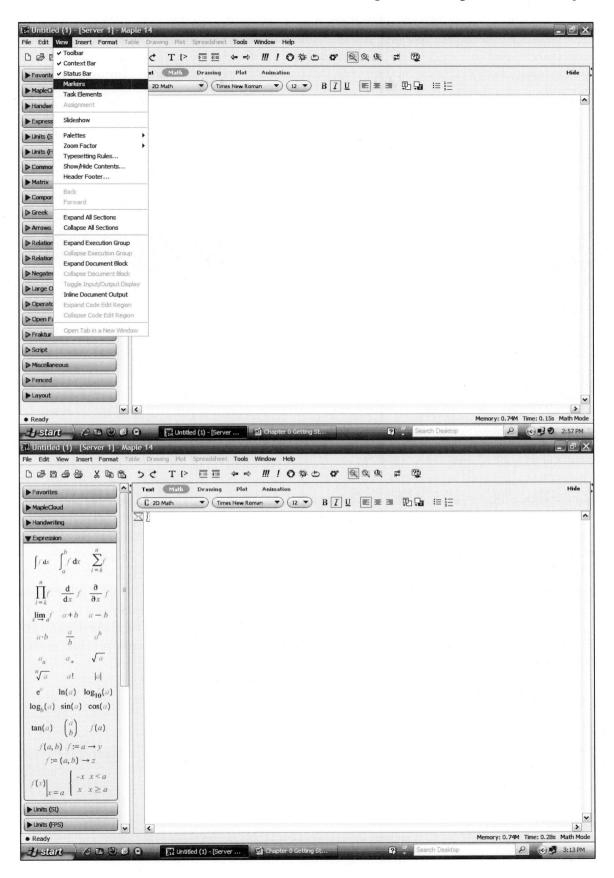

Unless otherwise stated, we will be in *Document* mode. On the left hand side of the screen there are several useful palettes such as the *Expression* palette and the *Common Symbols* palette. Click once on these palettes to see the following list of commands.

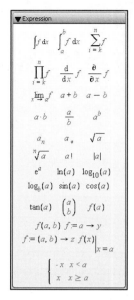
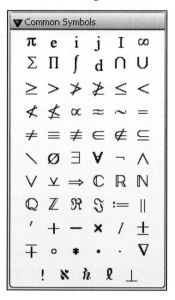

In the *Document* mode *2-D Math* is the default. In this chapter we will use *2-D Math*. This adds elegance to your work and the formulas match the typeset in most standard texts. It is also easy to spot errors. We will not use the *1-D* mode, however one can highlight text in the *2-D* mode, click on *Format* in the top pane of the worksheet, select *Convert To* and then the *1-D Math Input* option. This will be discussed in detail later.

On the top of the computer screen is a menu bar containing menus such as *File* and *Help*. In order to save the document, click on *File* and click on *Save As....* A *Save As* menu pops up. Under *File name* type drive:\filename.mw and then click on *Save*.

The labels (or equation numbers) are automatically of the form (1), (2), ... However, if you want to change it as we have throughout this text, one can do so by clicking on *Format* on top of the screen, selecting *Equation Labels* and then selecting *Label Display*. Enter 0 followed by the period symbol ⎪ (for chapter 0) for *Label Numbering Prefix* (as shown in the picture on the next page), then select *Sections Numeric* for *Label Numbering Scheme*, and click on OK.

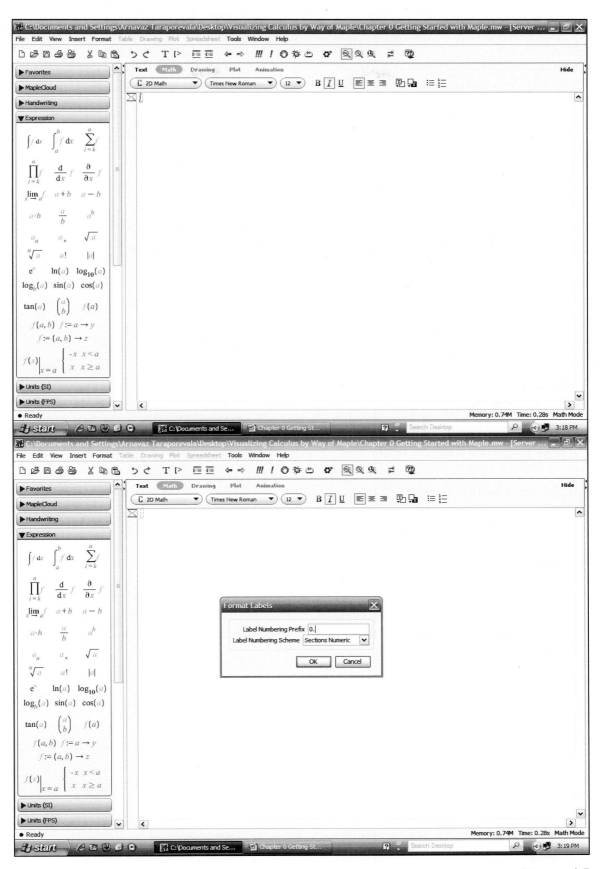

A *Maple* statement is executed only if it is entered in the *Math* mode (not the *Text* mode). Make sure that this icon is shaded when you want to execute a command (as seen in the first picture in this chapter). To toggle between the *Text* and *Math* modes, press F5 key. The variables entered as a part of the *Math* mode are automatically in italics. This is not the case in the *Text* mode. The colon : is used to suppress the output.

There are three basic commands in *Maple*. They are
Operational commands such as +, -, *, /.
Functional commands such as *factor, expand, plot*.
Programming commands such as *plot, if, end*.

You will discover that *Maple* is package driven and it will refuse to perform certain operations unless it is asked a priori to load a particular package. In addition to the frequently used commands, *Maple* has several specialized packages. These *"with"* packages activate different commands for certain related calculations. The *with(Student)* package gives us all the sub-packages that can be used. Make sure you are in the *Math* mode, type in *with(Student)* (notice that the italics come automatically) and then press the Enter key.

with(Student)

$$[Calculus1, LinearAlgebra, MultivariateCalculus, NumericalMethods, \qquad \textbf{(0.1)}$$
$$Precalculus, SetColors, VectorCalculus]$$

The *with(Student[Calculus1])* is an example of one such sub-package and activates commands related to Calculus I. For example, the *Tangent* command can be used to find the equation of a tangent line at a given point. A list of some common packages that will facilitate one's calculations is in Appendix A.

Example 1: Use the operational commands in *Maple* to evaluate
(a) $5+11$
(b) 851×98
(c) 69^{56}
(d) $5^8 - 34852$
(e) $70!$
(f) $|-132|$

Solution: *Maple* is a symbolic algebra system.
(a) To evaluate $5+11$ simply enter the number 5 followed by the + sign followed by the number 11 and press the Enter key.

$$5+11$$
$$16 \qquad\qquad \textbf{(0.2)}$$

(b)   For evaluating 851×98 simply enter the number 851 followed by the $\boxed{*}$ sign. Then enter the number 98 and press the $\boxed{Enter}$ key. Note that even though we entered $851 * 98$, we see $851 \cdot 98$ on the screen.

$851 \cdot 98$

$$83398 \tag{0.3}$$

(c)   In order to enter the number 69^{56}, enter 69, then use the $\boxed{\wedge}$ key for the exponent, enter the number 56 and press $\boxed{Enter}$.

69^{56}

$$9452552133242086160217597938992218547853669709340364610396839494642700730093767160420696522196845282081 \tag{0.4}$$

(d)   To enter $5^8 - 34852$, click on a^b in the *Expression* palette. Type 5 (as a is shaded), use the Tab key once (so b is shaded), and typing the number 8. Make sure that you use the right arrow key to bring the cursor down before typing the $\boxed{-}$ sign (otherwise everything you enter will be in the exponent).

$5^8 - 34852$

$$355773 \tag{0.5}$$

(e)   Click on the $a!$ in the *Expression* palette and enter the number 70.

$70!$

$$11978571669969891796072783721689098736458938142546425857555362864628009582789845319680000000000000000 \tag{0.6}$$

(f)   The absolute value of any number can be obtained by typing abs.

$\text{abs}(-132)$

$$132 \tag{0.7}$$

Remarks:

1.   If you want your answer in the same line press and hold the $\boxed{Ctrl}$ key and press the $\boxed{=}$ key.

$5 + 11 = 16$

You need to use the $\boxed{Enter}$ key to get to the next line. The output is suppressed when the colon $\boxed{:}$ is used at the end of the command.

$5 + 11 :$

2.   Another way to enter 69^{56} is by clicking on a^b in the *Expression* palette, typing 69 (when a is shaded), using the $\boxed{Tab}$ key once (now b is shaded), and typing the number 56 and press the $\boxed{Enter}$ key. The cursor is in the exponent after entering the number 56 and before using $\boxed{Enter}$ key. It is a good idea to use the right arrow key to bring the cursor down, before using $\boxed{Enter}$ key.

3.   Another way of inputting this is by clicking on the *Expression* palette on the left hand side and then clicking on the key marked $|a|$, entering the desired number, and pressing on the $\boxed{Enter}$ key.

$|-132|$

$$132 \tag{0.8}$$

4.   Try using your calculator to evaluate 69^{56} and $70!$.

Example 2: Use the operational commands in *Maple* to evaluate
(a) $3.68 - 6.5771$
(b) $\dfrac{9}{6001} - \dfrac{5}{2}$
(c) $\sqrt{24}$

Solution:
(a)   In order to obtain the difference $3.68 - 6.5771$ just enter this numerical expression and then press the $\boxed{Enter}$ key.

$3.68 - 6.5771$

$$-2.8971 \tag{0.9}$$

Observe that the answer is in the form of a decimal.

(b)   Now, suppose we want to subtract $\dfrac{9}{6001} - \dfrac{5}{2}$. Begin by typing the first fraction $\dfrac{9}{6001}$ as 9/6001. Since the *2-D Math* is the default setting, *Maple* automatically converts 9/6001 to a fraction that looks like $\dfrac{9}{6001}$ (make sure that you are in the *Math* mode). However, after entering the number 6001, the cursor is still in the denominator. Use the right arrow key before pressing the $\boxminus$ sign and entering the second fraction $\dfrac{5}{2}$. Otherwise, the $\boxminus$ sign and the fraction $\dfrac{5}{2}$ will appear in the denominator of the fraction $\dfrac{9}{6001}$ and will be seen as $\dfrac{9}{6001 - \dfrac{5}{2}}$.

After entering the difference $\dfrac{9}{6001} - \dfrac{5}{2}$, press the $\boxed{Enter}$ key.

$$\dfrac{9}{6001} - \dfrac{5}{2}$$

$$-\dfrac{29987}{12002} \tag{0.10}$$

(c)   The square root of a number can be obtained by typing *sqrt* or by clicking on the *Expression* palette on the left hand side and then clicking on the key marked $\sqrt{a}$, entering the desired number, and pressing on enter.

$$\sqrt{24}$$

$$2\sqrt{6} \tag{0.11}$$

Remarks:

1.   Notice that when the input is in the form of a fraction, the output will also be in the form of a fraction and when the input is in the form of a decimal the output is in the form of a decimal.

2.   Use your calculator to evaluate $\dfrac{9}{6001} - \dfrac{5}{2}$ as a fraction. What is your result?

3. $\dfrac{9}{6001} - \dfrac{5}{2}$ can also be evaluated by bringing the mouse anywhere on the difference, right clicking on it and then clicking on *Evaluate* to get the answer.

$$\dfrac{9}{6001} - \dfrac{5}{2}$$

$$-\dfrac{29987}{12002} \tag{0.12}$$

Example 3:

(a)   Express your answer for $\dfrac{9}{6001} - \dfrac{5}{2}$ as a decimal. Round your answer to the nearest integer.

(b)   Express your answer for $\sqrt{24}$ as a decimal with 9 significant digits.

Solution:

(a)   If we want the answer in the form of a decimal, then we will use the *evalf* command. *Maple* has a default setting of 10 significant digits.

$$evalf\left(\frac{9}{6001} - \frac{5}{2}\right)$$

$$-2.498500250 \tag{0.13}$$

To round the answer to the nearest integer, we use the *round* command.

$$round\left(\frac{9}{6001} - \frac{5}{2}\right)$$

$$-2 \tag{0.14}$$

(b)   In order to evaluate $\sqrt{24}$ as a decimal with 9 significant digits, one can use the *evalf* command to get the answer as a decimal. Type *evalf* $\lvert ($. Then select Insert, select Label and type 0.11, click on OK, then type $\rvert$ and then 9 (for the number of significant digits) and $\rvert)$ and then use the $\boxed{Enter}$ key.

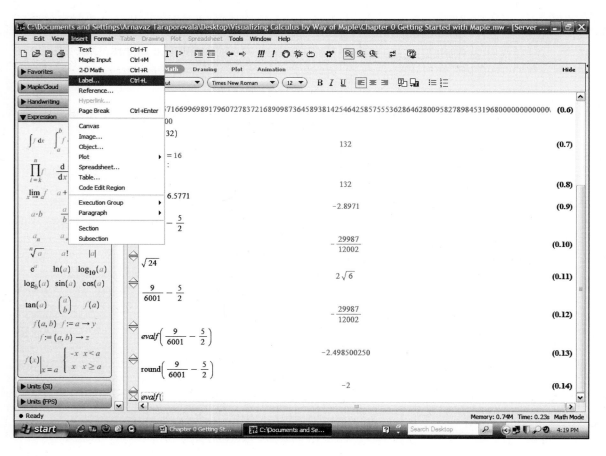

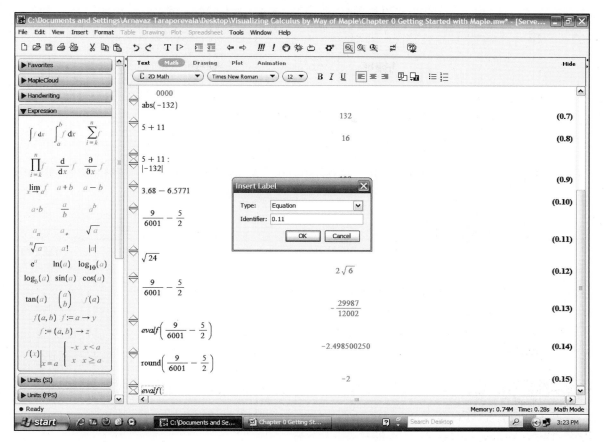

$$evalf\left((0.11), 9\right)$$

$$4.89897948 \qquad \textbf{(0.15)}$$

Maple remembers the label number and will automatically change it if the numbers for these labels change.

Remarks:

1. However, *Maple* can also compute as many digits of accuracy one assigns. For example if 6 significant digits are needed, we use the *evalf* command as in the equation above and include the number 6 as shown below.

$$evalf\left(\frac{9}{6001} - \frac{5}{2}, 6\right)$$

$$-2.49850 \qquad \textbf{(0.16)}$$

2. The difference between the *round* and *evalf* commands is seen below:

$$round\left(\frac{14}{3}\right)$$

$$5 \qquad \textbf{(0.17)}$$

$$evalf\left(\frac{14}{3}\right)$$

$$4.666666667 \tag{0.18}$$

3. The input for the difference in the *2-D Math* mode looked like $\frac{9}{6001} - \frac{5}{2}$. However, it looks different in the *1-D Math* mode. Highlight the difference $\frac{9}{6001} - \frac{5}{2}$, click on *Format*, select *Convert To*, select the *1-D Math Input* option, and press the $\boxed{Enter}$ key.

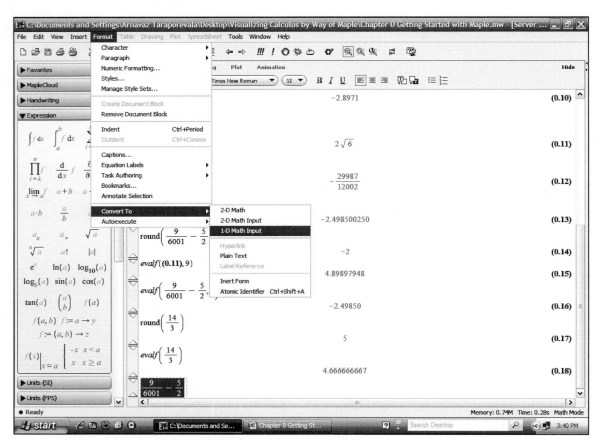

The result is shown below.

9/6001 – 5/2;

$$-\frac{29987}{12002} \tag{0.19}$$

Notice that the output looks the same as that in the *2-D Math* mode.

4. We can also set the number of digits globally for the document by giving the command digits equal to 5. An approximation to 5, 10, 20, 50 or 100 significant

digits can be obtained by first right clicking on the difference $\dfrac{9}{6001} - \dfrac{5}{2}$, clicking on the *Approximate* command and then scrolling down to the desired number of

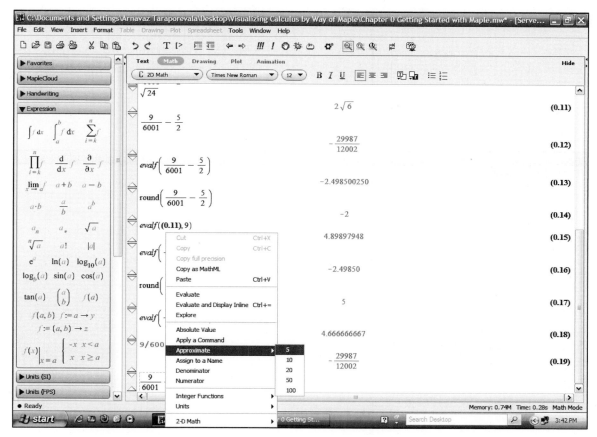

significant digits. This is seen in the picture and the command below.

$$\frac{9}{6001} - \frac{5}{2} \xrightarrow{\text{at 5 digits}} -2.4985$$

5.   Once again one can also obtain the answer by bringing the mouse anywhere on the square root, right clicking on it and then clicking on *Approximate* and then the desired number of significant digits.

$$\sqrt{24} \xrightarrow{\text{at 5 digits}} 4.8990$$

Example 4:   Evaluate

(a) $\cot\left(\dfrac{\pi}{3}\right)$

(b) $\sin^{-1}(1)$

(c) $\operatorname{arcsec}(-2)$

(d) $\arctan(0.8)$

Solution: *Maple* recognizes trigonometric functions if the angles are in parentheses.

(a)   To evaluate $\cot\left(\dfrac{\pi}{3}\right)$, we would enter it as *cot*, then open parenthesis, then use π from the *Common Symbols* palette, then / 3, use the right arrow before closing the parenthesis. If we do not use the right arrow after entering the number 3 the expression becomes $\cot\left(\dfrac{\pi}{3)}\right.$ rather than $\cot\left(\dfrac{\pi}{3}\right)$.

$$\cot\left(\frac{\pi}{3}\right)$$

$$\frac{1}{3}\sqrt{3} \tag{0.20}$$

(b)   The inverse trigonometric function for the sine function can be entered as $\arcsin(x)$ or as $\sin^{-1}(x)$.

$$\sin^{-1}(1)$$

$$\frac{1}{2}\pi \tag{0.21}$$

(c) $\operatorname{arc\,sec}(-2)$

$$\frac{2}{3}\pi \tag{0.22}$$

(d) $\arctan(0.8)$

$$0.6747409422 \tag{0.23}$$

Remarks:

1.   Instead of entering π from the *Common Symbols* palette, we can type Pi to get the same result. You must use the upper case for the letter p when typing Pi.

$$\cot\left(\frac{\text{Pi}}{3}\right)$$

$$\frac{1}{3}\sqrt{3} \tag{0.24}$$

2. Notice that as the number for $\arctan(0.8)$ was entered as a decimal, we got a floating-point solution. However, in the case of $\arctan\left(\dfrac{4}{5}\right)$ we needed to right click on it, select *Approximate* and then the desired number of significant digits (10 in our case).

$$\arctan\left(\dfrac{4}{5}\right)$$

$$\arctan\left(\dfrac{4}{5}\right) \tag{0.25}$$

$$\xrightarrow{\text{at 10 digits}}$$

$$0.6747409422 \tag{0.26}$$

This is because $\dfrac{4}{5}$ is entered as a fraction. This can be also achieved by using the *evalf* command.

Example 5: First evaluate the following numbers and then express each of the answers as a decimal

(a) e^{231}

(b) $\ln(9475)$

(c) $\log_{16}(1024)$

(d) $\log_{16}(9)$

Solution:

(a)   We will enter e^{231} by clicking on the *Expression* palette on the left hand side and then clicking on the key marked e^{a}, entering the number 231, and pressing on the $\boxed{Enter}$ key. Note that the answer as a decimal does not come right away. Right click on e^{231} (or e^{231} the *Maple* output) select *Approximate* and then select 10 as the desired number of significant digits.

$$e^{231}$$

$$e^{231} \tag{0.27}$$

$$\xrightarrow{\text{at 10 digits}}$$

$$2.099062257\ 10^{100} \tag{0.28}$$

(b)   We will enter $\ln(9475)$ by clicking on the *Expression* palette on the left hand side and then clicking on the key marked $\ln(a)$, entering the number 9475, and pressing on the $\boxed{Enter}$ key. Note that the answer as a decimal does not come right away. Right click on $\ln(9475)$, select *Approximate* and then select 10 as the desired number of significant digits.

$\ln(9475)$

$$\ln(9475) \tag{0.29}$$

$\xrightarrow{\text{at 10 digits}}$

$$9.156412030 \tag{0.30}$$

(c)   Enter $\log_{16}(1024)$ by clicking on the *Expression* palette on the left hand side and then clicking on the key marked $\log_{b}(a)$, entering the number 16 (as b is shaded), using the Tab key and entering the number 1024 (as a is shaded), and pressing on the $\boxed{Enter}$ key.

$\log_{16}(1024)$

$$\frac{5}{2} \tag{0.31}$$

$\xrightarrow{\text{at 10 digits}}$

$$2.500000000 \tag{0.32}$$

(d)   We will enter $\log_{16}(9)$ in a similar manner as we entered $\log_{16}(1024)$. Here we will need to right click on $\log_{16}(9)$, select *Approximate* and then select 10 as the desired number of significant digits.

$\log_{16}(9)$

$$\frac{1}{2}\frac{\ln(3)}{\ln(2)} \tag{0.33}$$

$\xrightarrow{\text{at 10 digits}}$

$$0.7924812505 \tag{0.34}$$

Remarks:

1.   However, if we enter 231 as 231.0 we automatically get the decimal representation for e^{231} as seen below.

$e^{231.0}$

$$2.099062257 \; 10^{100} \tag{0.35}$$

Another way of entering e^{231} is by typing $\exp(231)$. The letter e typed by using the keyboard key is not recognized by *Maple* as the base of the natural logarithm. Again as the answer is not a decimal right click on $\exp(231)$ or e^{231}, select *Approximate* and then select 10 as the desired number of significant digits.

$$\exp(231) \xrightarrow{\text{at 10 digits}} 2.099062257 \; 10^{100}$$

2. If we enter 9475 as 9475.0 we automatically get the decimal representation for $\ln(9475)$ as seen below.

$$\ln(9475.0)$$
$$9.156412030 \tag{0.36}$$

3. Again, if we enter 9 as 9.0 we automatically get the decimal representation for $\log_{16}(9)$ as seen below.

$$\log_{16}(9.0)$$
$$0.7924812504 \tag{0.37}$$

Example 6: Find the product $(3x-4)(x+5)^2$.

Solution: In order to perform the multiplication $(3x-4)(x+5)^2$ the *expand* command is used. This command is used to expand or multiply out the corresponding expression.

$$expand\left((3 \cdot x - 4) \cdot (x+5)^2\right)$$
$$3\,x^3 + 26\,x^2 + 35\,x - 100 \tag{0.38}$$

One could also right click on the expression and then click on *Expand*.

$$(3 \cdot x - 4) \cdot (x+5)^2 \xlongequal{\text{expand}} 3\,x^3 + 26\,x^2 + 35\,x - 100$$

Caution: In the *2-D input*, there is a subtle disadvantage. *Maple* sometimes assumes that a space between two expressions is a multiplication sign and proceeds in giving an answer in such a manner. However, this is not always the case. The authors of this book strongly urge the reader to use the * sign for multiplication and refrain from using the space bar for this purpose. The subtleties are illustrated in the examples below.

1. One could enter the expression "$3x$" as $3x$ (as *Maple* recognizes $3x$ as 3 times the variable x) or as $3 * x$, which is seen as $3 \cdot x$. However, care must be taken when entering the product $(3 \cdot x - 4) \cdot (x+5)^2$. In this case one has to use the multiplication sign between $(3x-4)$ and $(x+5)^2$. Otherwise, *Maple* will not

recognize it as a product as shown below by using both the *expand* command and using the right click feature and selecting *Expand*.

$$expand\left((3x-4)(x+5)^2\right)$$

$$9\,x(x+5)^2 - 24\,x(x+5) + 16 \tag{0.39}$$

$$(3x-4)(x+5)^2 \overset{expand}{=} 9\,x(x+5)^2 - 24\,x(x+5) + 16$$

Maple assumes that $x + 5$ is a variable, say y, and as a result $(3x-4)(x+5)^2$ is the square of the quantity $(3x-4)\,y$. Note that the expression $9\,x(x+5)^2$ means that the number 9 is multiplied by the square of the variable xy. However, if there is a space between $(3x-4)$ and $(x+5)^2$ rather than the multiplication sign *, then *Maple* recognizes the space as a multiplication sign. Right click on the expression and select *Expand* to get the following

$$(3x-4)\ (x+5)^2 \overset{expand}{=} 3x^3 + 26x^2 + 35x - 100$$

Use the * sign for multiplication and refrain from using the space bar for this purpose.

2.   Care must be taken when entering spaces between expressions. The answer for $\cos(2\cdot\pi)$ is 1.

$$\cos(2\cdot\pi)$$

$$1 \tag{0.40}$$

But if we leave a space between cos and $2\cdot\pi$ then we get a different answer. Right click on this answer, select *Approximate*, and 5 digits to get the answer below.

$$\cos\ (2\cdot\pi)$$

$$2\cos\pi \tag{0.41}$$

$$\xrightarrow{\text{at 5 digits}}$$

$$6.2832\cos \tag{0.42}$$

Basics of Graphing in Maple

Remark: When we perform *Maple* commands, it should be tempered with some knowledge from the user as to the domain of a function and some of the restrictions that we need. In other words we may plot a function over an interval that hides essential features. This will be illustrated in Examples 7 and 8 below.

Example 7: Graph the polynomial $f(x) = x^3 + 3x^2 - 4x - 12$ and find its domain, range, and zeros. Evaluate $f(5)$. Factor this polynomial. Solve the equation $f(x) = 12$. Obtain the floating-point approximation to the solution for the equation $f(x) = 12$.

Solution: We will enter $f(x) = x^3 + 3x^2 - 4x - 12$ as a *Maple* function by using the Expression palette and clicking on $f := a \rightarrow y$. Notice that the letter f is shaded. Enter the letter f and use the $\boxed{Tab}$ key. Now the letter a is shaded. Enter the letter x and use the $\boxed{Tab}$ key once more. Now enter the polynomial $x^3 + 3x^2 - 4x - 12$ and then press the $\boxed{Enter}$ key.

$f := x \rightarrow x^3 + 3 \cdot x^2 - 4 \cdot x - 12$

$$x \rightarrow x^3 + 3 \cdot x^2 - 4 \cdot x - 12 \qquad \textbf{(0.43)}$$

Another way of getting the symbol $\rightarrow$ is by typing $\boxed{}$ sign followed by $\boxed{>}$.

We can find the zeros of the polynomial by using the *solve* command.

$solve(f(x) = 0, x)$

$$2, \text{-}3, \text{-}2 \qquad \textbf{(0.44)}$$

The zeros of $f(x)$ are 2, -2, and –3. The graph of $f(x)$ can be sketched by using the *plot* command. If the end points of the values for x are not specified, the default value is x = -10 .. 10.

$plot(f(x), x)$

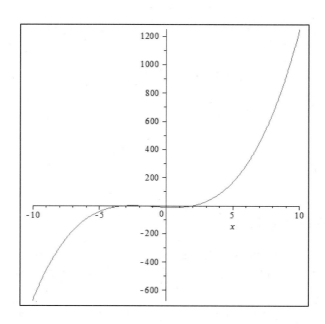

A better view is obtained by specifying the bounds for both x and y as shown below.

$$plot\left(f\left(x\right), x=-4\,..\,3,\ y=-14\,..\,25\right)$$

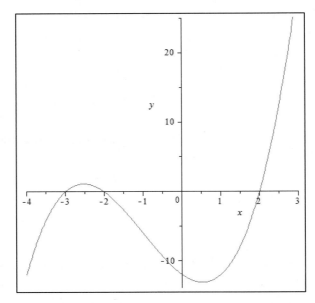

Since f is a polynomial function, the domain is (-∞, ∞) and the range is (-∞, ∞). Notice that if one were to right click on the function $f\left(x\right)$ the menu of options (for the *clickable Maple* feature) available is now limited. However, one can type the commands to factor $f\left(x\right)$ or evaluate $f\left(5\right)$ as shown below.

$$f\left(5\right)$$

$$168 \tag{0.45}$$

$$factor\left(f\left(x\right)\right)$$

$$\left(x-2\right)\left(x+3\right)\left(x+2\right) \tag{0.46}$$

Now let us find the solutions to the equation $f\left(x\right)=12$.

$$solve\left(f\left(x\right)=12,\ x\right)$$

$$\frac{1}{3}\left(243+6\sqrt{1383}\right)^{1/3}+\frac{7}{\left(243+6\sqrt{1383}\right)^{1/3}}-1,\ -\frac{1}{6}\left(243+6\sqrt{1383}\right)^{1/3} \tag{0.47}$$

$$-\frac{7}{2\left(243+6\sqrt{1383}\right)^{1/3}}-1+\frac{1}{2}\,I\,\sqrt{3}\left(\frac{1}{3}\left(243+6\sqrt{1383}\right)^{1/3}-\frac{7}{\left(243+6\sqrt{1383}\right)^{1/3}}\right),$$

$$-\frac{1}{6}\left(243+6\sqrt{1383}\right)^{1/3}-\frac{7}{2\left(243+6\sqrt{1383}\right)^{1/3}}-1-\frac{1}{2}\,\mathrm{I}\,\sqrt{3}\left(\frac{1}{3}\left(243+6\sqrt{1383}\right)^{1/3}-\frac{7}{\left(243+6\sqrt{1383}\right)^{1/3}}\right)$$

We will now obtain floating-point solutions to the equation $f(x)=12$ by using the *fsolve* command.

$$fsolve\left(f(x)=12,\,x\right)$$

$$2.487338444 \qquad\qquad\qquad\text{(0.48)}$$

Since two of the solutions are complex, we will need to use the *complex* option in the *fsolve* command to view all the solutions.

$$fsolve\left(f(x)=12,\,x,\,complex\right)$$

$$-2.743669222-1.456415830\,\mathrm{I},\ -2.743669222+1.456415830\,\mathrm{I},\ 2.487338444 \quad\text{(0.49)}$$

Hence the floating-point solutions to the equation $f(x)=12$ are –2.743669222 – 1.456415830 I, –2.743669222 + 1.456415830 I, and 2.487338444.

Remarks:
1.   The values for the floating-point solutions may not be exactly the same as in **(0.49)** but could be the same for six significant digits.

2.   The command $solve\left(f(x),x\right)$ automatically sets the expression equal to zero and solves for the indicated variable.

$$solve\left(f(x),x\right)$$

$$2,-3,-2 \qquad\qquad\qquad\text{(0.50)}$$

However, if we were interested in finding the solutions to the equation $f(x)=-12$ we would need the following command.

$$solve\left(f(x)=-12,x\right)$$

$$0,1,-4 \qquad\qquad\qquad\text{(0.51)}$$

The authors of this book strongly urge the reader to refrain from using the $solve\left(f(x),x\right)$ command and to continue using $solve\left(f(x)=0,x\right)$ command.

3.   We can enter the polynomial function as a *Maple* expression by using the assigned operator $:=$ as shown below. Note that we call this *Maple* expression, g,

so that we do not confuse it with the polynomial function $f(x)$ entered as a *Maple* function. Right click on the expression and select on *Factor* to get the factors of the polynomial.

$g := x^3 + 3 \cdot x^2 - 4 \cdot x - 12$

$$x^3 + 3 \cdot x^2 - 4 \cdot x - 12 \tag{0.52}$$

$\underset{=}{\overset{\text{factor}}{}}$

$$(x-2)(x+3)(x+2) \tag{0.53}$$

Alternately, the *factor* and *solve* commands can also be used to factor the polynomial and find the zeros.

$factor(g)$

$$(x-2)(x+3)(x+2) \tag{0.54}$$

$solve(g = 0, x)$

$$2, -3, -2 \tag{0.55}$$

We could use the right click feature to find the value of this polynomial when $x = 5$ by right clicking on the polynomial and clicking on *Evaluate at a Point*. The *Evaluate the expression at the point:* menu pops up. Use the $\boxed{Tab}$ key once, enter the number 5 and then click OK.

$$x^3 + 3 \cdot x^2 - 4 \cdot x - 12 \xrightarrow{\text{evaluate at a point}} 168$$

Hence we see $g(5) = 168$.

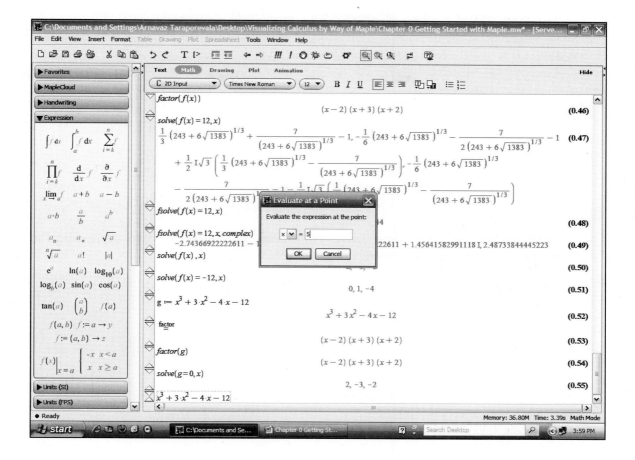

We could also use the *subs* command

$$subs\left(x = 5, \; g\right)$$

$$168 \tag{0.56}$$

We can use the plot command to sketch the graph of the polynomial function as we did in the example above. Instead of entering $g(x)$ we must just enter g.

4. We can plot the polynomial by right clicking on the expression. Scroll down to *Plots* and click on it. Then click *Plot Builder*.

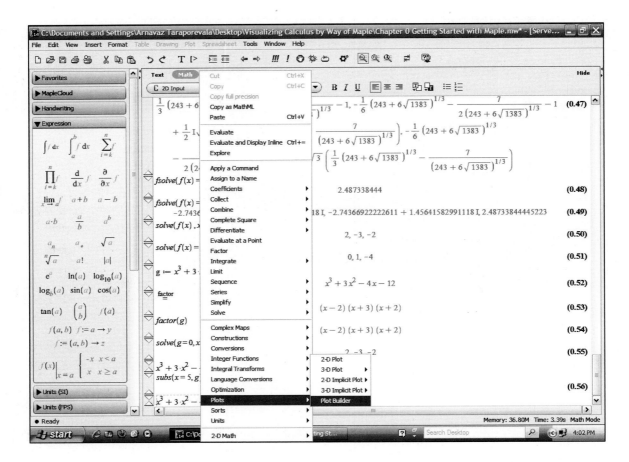

The *Interactive Plot Builder: Select Plot Type* menu pops up. Make sure that a *2-D plot* is selected. You will notice that the default settings for the *x*-axis are from -10 and 10. Set the left and right bounds for the *x Axis* at -4 and 3 (why?). Click on *Preview* to get a good look at the graph and find the limits for the y-axis. If you are satisfied, then click on *Options* to set the bounds for the *y*-axis. A *2-D Plot (plot)* menu pops up. Notice that under variables the *horizontal* axis is shaded in gray and that cannot be edited in this menu (try to edit and see what happens). However, there are boxes below this option. Enter -14 and 25 for the *Range from*, then change the default label *x* here to *y* and change the default orientation from *horizontal* to *vertical*. Click on *Preview* to see the graph. If you are satisfied with the shape of the graph, click on *Plot*.

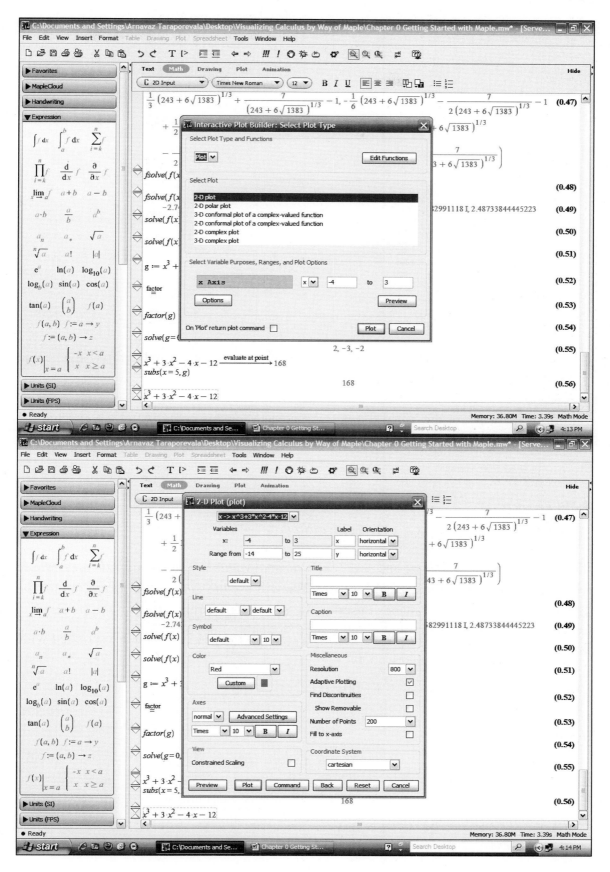

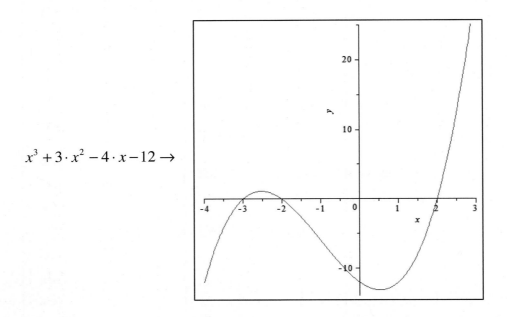

$$x^3 + 3 \cdot x^2 - 4 \cdot x - 12 \rightarrow$$

Example 8: Use *Maple* commands to identify, describe, and graph the conic $9x^2 + 4y^2 + 36x - 8y - 60 = 0$.

Solution: We will use the $with(geometry)$ package for this conic. After we execute the $with(geometry)$ package we will assign a name for the horizontal and vertical axes as the *x*-axis and the *y*-axis.

$with(geometry):$
$\_EnvHorizontalName := 'x' : \_EnvVerticalName := 'y' :$

If the $\_EnvHorizontalName$ and the $\_EnvVerticalName$ are not assigned, then *Maple* will ask for a name to be assigned for them. We will use the *conic* and *form* commands. The *conic* and *form* commands describe the output as a circle, an ellipse, a parabola, a hyperbola, cases of a point, two parallel lines, a double line, or a list of two intersecting lines.

$conic\left(c, 9 \cdot x^2 + 4 \cdot y^2 + 36 \cdot x - 8 \cdot y - 60 = 0, [x, y]\right)$

$$c \qquad\qquad\qquad\qquad \textbf{(0.57)}$$

$form(c)$

$$ellipse2d \qquad\qquad\qquad\qquad \textbf{(0.58)}$$

Hence $9x^2 + 4y^2 + 36x - 8y - 60 = 0$ is a (two-dimensional) ellipse. We will now use the *detail* command to get a complete description of this ellipse.

detail (c)

name of object	c
form of object	*ellipse2d*
center	$[-2,1]$
foci	$\left[\left[-2,1-\dfrac{5\sqrt{5}}{3}\right],\left[-2,1+\dfrac{5\sqrt{5}}{3}\right]\right]$
length of major axis	10
length of minor axis	$\dfrac{20}{3}$
equation of ellipse	$9x^2+4y^2+36x-8y-60=0$

(0.59)

Hence the center of this ellipse is $(-2,1)$. The foci are $\left(-2,1-\dfrac{5\sqrt{5}}{3}\right)$ and $\left(-2,1+\dfrac{5\sqrt{5}}{3}\right)$.

The length of the major axis is 10 and the length of the minor axis is $\dfrac{20}{3}$. We will now use the *draw* command to plot the ellipse.

draw (c)

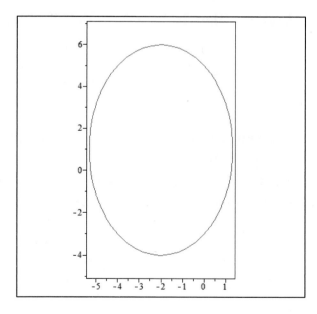

Without any additional commands we get a graph that is boxed. In order to get the *x*- and *y*-axes we use the option *axes = normal* and the option *thickness = 2*. This option allows us to make sure that the graph of the function is thicker than the default setting, which is 0. One can always increase the thickness of the plot.

draw $(c, axes = normal, thickness = 2)$

Chapter 0 Getting Started with *Maple*

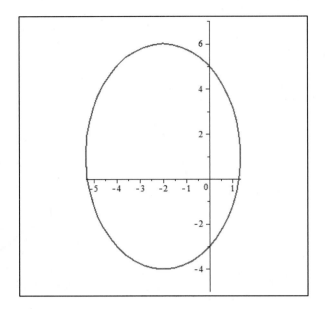

Remarks: We can complete squares using *Maple* commands. We will begin by activating the $with\left(Student\left[Precalculus\right]\right)$ package.

$with\left(Student\left[Precalculus\right]\right):$

$c1 := 9 \cdot x^2 + 4 \cdot y^2 + 36 \cdot x - 8 \cdot y - 60 = 0$

$$9x^2 + 4y^2 + 36x - 8y - 60 = 0 \qquad \textbf{(0.60)}$$

$CompleteSquare\left(CompleteSquare\left(c1, x\right), y\right)$

$$4\left(y-1\right)^2 - 100 + 9\left(x+2\right)^2 = 0 \qquad \textbf{(0.61)}$$

We can also achieve the same conclusion be completing the square using two steps. This is shown in the commands below.

$CompleteSquare\left(c1, x\right)$

$$9\left(x+2\right)^2 - 96 + 4y^2 - 8y = 0 \qquad \textbf{(0.62)}$$

$CompleteSquare\left(\left(0.62\right), y\right)$

$$4\left(y-1\right)^2 - 100 + 9\left(x+2\right)^2 = 0 \qquad \textbf{(0.63)}$$

Example 9: Consider the rounding function $f\left(x\right) = round\left(x\right)$, which rounds x to the nearest integer. Use this function to evaluate $f\left(-1.234\right)$, $f\left(-\dfrac{57}{14}\right)$, $f\left(\dfrac{57}{14}\right)$ and $f\left(63.59\right)$. Graph this function and find its domain and range.

Solution: Enter *f* as a *Maple* function.

$f := x \rightarrow \text{round}(x)$

$$x \rightarrow \text{round}(x) \qquad\qquad \textbf{(0.64)}$$

We will now evaluate this function when $x = -1.234$, $x = -\dfrac{57}{14}$, $x = \dfrac{57}{14}$, and $x = 63.59$

$f(-1.234)$

$$-1 \qquad\qquad \textbf{(0.65)}$$

$f\left(-\dfrac{57}{14}\right)$

$$-4 \qquad\qquad \textbf{(0.66)}$$

$f\left(\dfrac{57}{14}\right)$

$$4 \qquad\qquad \textbf{(0.67)}$$

$f(63.59)$

$$64 \qquad\qquad \textbf{(0.68)}$$

The graph of this function is sketched below.

$plot\left(\text{round}(x), x = -5 .. 5\right)$

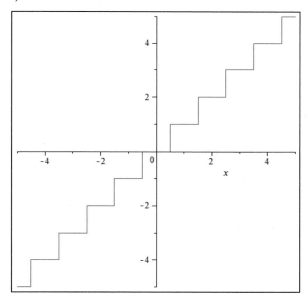

Notice that we cannot see the graph when *x* is between −1 and 1. We will also use the option *thickness* = 4. We will also use another option which is *discont* = *true*. This option finds the possible discontinuities (or jumps in this case) and breaks the *x*-axis into appropriate subintervals where the function has no breaks or holes. This will be discussed in further detail in Chapters 1 and 2.

$$plot\left(round\left(x\right), x = -5\mathinner{\ldotp\ldotp}5, discont = true, thickness = 4\right)$$

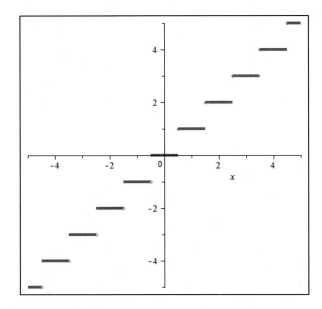

The domain of this function is the set of real numbers and the range is the set of integers.

Remark: Note that if we use the *evalf* command for $\dfrac{57}{14}$ we get the following:

$$evalf\left(\frac{57}{14}\right)$$

4.071428571 **(0.69)**

Example 9: Graph the functions $f(x) = x^5$ and $g(x) = \sin(x)$ on the same set of axes. Use this graph to find the number of intersection points of the two functions. Find all the points of intersection.

Solution: We will enter $f(x)$ and $g(x)$ as *Maple* functions.

$$f := x \rightarrow x^5$$

$x \rightarrow x^5$ **(0.70)**

$$g := x \rightarrow \sin(x)$$

$x \rightarrow \sin(x)$ **(0.71)**

We can plot both functions on the same set of axes using the *plot* command. However we must enter the two functions using braces. Since we are setting $x^5 = \sin(x)$ and it is known that $-1 \leq \sin(x) \leq 1$ for every value of x, we will get $-1 \leq x^5 \leq 1$. Hence we use the values of x from -1.1 to 1.1 in the graph below.

$$plot\left(\{f(x), g(x)\}, x = -1.1..1.1\right)$$

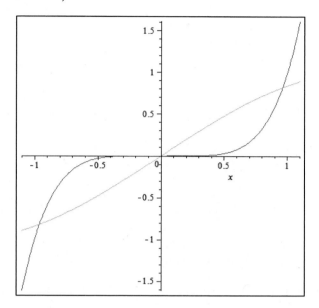

Is the function $f(x)$ in red and $g(x)$ in green or is it the other way around? When we plot two functions together on the same set of axes we use braces, { }. But *Maple* could graph the function $f(x)$ before the function $g(x)$ or vice versa. However, if we want to make sure they are plotted in a certain order we use brackets [ ]. That way we are sure that *Maple* is graphing the function $f(x)$ before the function $g(x)$. We could also specify the colors we want to use for the functions. We can now use the *color* option to specify the colors for the graphs. In this case the function $f(x)$ is in red and the function $g(x)$ is in blue.

$$plot\left(\left[f(x), g(x)\right], x = -1.1..1.1, color = \left[red, blue\right]\right)$$

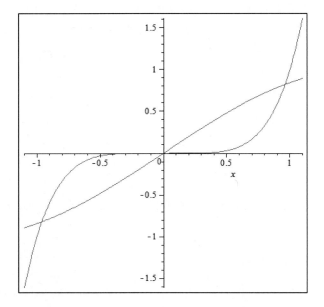

From the graph above we see that the two functions intersect at three points. We will now proceed to find the points of intersection by using the solve command.

$solve \left(f \left(x \right) = g \left(x \right), x \right)$
$$RootOf \left(-\sin \left(\_Z \right) + \_Z^5 \right) \tag{0.72}$$

This happened because *Maple* could not solve this equation and uses the expression *RootOf* as a placeholder. We will use the *fsolve* command to find numerical solutions of the equation.

$fsolve \left(f \left(x \right) = g \left(x \right), x \right)$
$$0. \tag{0.73}$$

This command only gives one solution to the equation $f(x) = g(x)$. We can force it to give all the solutions by inserting values for x in the vicinity of each root (for example, when we are looking for the solutions near $x = -1$ and near $x = 1$).

$solve \left(f \left(x \right) = g \left(x \right), x = -1 \right)$
$$-0.9610369415 \tag{0.74}$$

$solve \left(f \left(x \right) = g \left(x \right), x = 1 \right)$
$$0.9610369415 \tag{0.75}$$

The *fsolve* command gives all the solutions to polynomial equations without inserting any points. However, it does not do so when the equations are not polynomial equations as is seen in this example. Care must be taken when solving equations that are not polynomial equations. In order to get all the solutions, we will use the *Roots* command and use the *numeric* option. Before executing this command, we must activate the $with\big(Student[Calculus1]\big)$ package. From the graph above we know that the solutions are between –1.1 and 1.1. We will include this option in the command.

$$with\big(Student[Calculus1]\big)$$

$$Roots\big(f(x)=g(x),x=-1.1..1.1,numeric\big)$$

$$[-0.9610369415,0.,0.9610369415] \tag{0.76}$$

Hence the two curves intersect at –0.9610369415, 0 and 0.9610369415.

Remarks:

1. The *Roots* command gives only the real solutions to the equation. In order for *Maple* to compute floating-point solutions, the *numeric* option is used. Since the solutions in Example 9 were between –1.1 and 1.1 this was specified in the *Roots* command, when we used the *numeric* option. However, if we had not specified the range of values between –1.1 and 1.1, we would have obtained the same results as seen below.

$$Roots\big(f(x)=g(x),numeric\big)$$

$$[-0.9610369415,0.,0.9610369415] \tag{0.77}$$

This is because when the *numeric* option is used, the default setting in *Maple* is to find solutions in the interval from –10 to 10. If the roots that are outside the range from -10 to 10, then this must be specified in the command. The authors of this book strongly urge the reader to specify the range of values for *x* when using this option.

2. Note that if we did not use the *numeric* option then we would not get any results as shown below.

$$Roots\big(f(x)=g(x),x\big)$$

$$[\,] \tag{0.78}$$

3. It is essential that the reader be aware of the differences between *solve* and *fsolve* commands. Whenever you ask *Maple* to do a problem with the *solve* command, *Maple* resorts to solving the problem by using basic or high school algebra. The *fsolve* command on the other hand uses numerical methods such as a variant of an approximation technique known as Newton's Method (see Chapter 4 for more details).

4.   The *fsolve* command usually gives only one solution to the equation. However, in the case of solving equations involving only polynomials, this command gives all the solutions.

5.   We did not use the *Roots* command in Example 7 as it only gives real solutions. See what happens if we use this command with or without the *numeric* option. Before using this command we will execute the $with\big(Student[Calculus1]\big)$ package.

$f := x \rightarrow x^3 + 3 \cdot x^2 - 4 \cdot x - 12$
$$x \rightarrow x^3 + 3 \cdot x^2 - 4 \cdot x - 12 \tag{0.79}$$
$with\big(Student[Calculus1]\big):$
$Roots\big(f(x) = 12,\, x\big)$

$$\left[\frac{1}{3} \frac{\left(243 + 6\sqrt{1383}\right)^{2/3} + 21 - 3\left(243 + 6\sqrt{1383}\right)^{1/3}}{\left(243 + 6\sqrt{1383}\right)^{1/3}}\right] \tag{0.80}$$

$Roots\big(f(x) = 12,\, x,\, numeric\big)$
$$[2.487338444] \tag{0.81}$$

Exercises

1. Use the operational commands in *Maple* to evaluate $56487 - 798413$, $(-216)^{-7}$, $(-216)^{43}$, $96!$, and $314!$.

2. Use the operational commands to evaluate $(35 + 13^3)4 - 15 \div (3 + 12)$ and $35 + 13^3 \times 4 - 15 \div 3 + 12$. Are your answers different? Give reasons for your answer.

3. Use the operational commands in *Maple* to evaluate 5.9741×68.412645 and $201.9743 \div 51.6472$. Are both your answers expressed as a decimal? Give reasons for your answer.

4. Use the operational commands in *Maple* to evaluate $\dfrac{173}{56422} + \dfrac{91}{76}$ and $\dfrac{5}{2} \div \dfrac{6002}{9}$. Express these answers as decimals. Round each answer to the nearest integer.

5. Express $\dfrac{2}{5}$ as a decimal. Evaluate $(0.4)^{32}$ and $\left(\dfrac{2}{5}\right)^{32}$. Are your answers expressed as a decimal or a fraction? Why? Find the decimal representation of $\left(\dfrac{2}{5}\right)^{32}$.

6. Express $\dfrac{1}{4}$ as a decimal. Evaluate $(0.25)^{29}$ and $\left(\dfrac{1}{4}\right)^{29}$. Are your answers different? Why? Find the decimal representation of $\left(\dfrac{1}{4}\right)^{29}$.

6. Evaluate $\left(\dfrac{67}{66}\right)^{99}$ and $\dfrac{67^{99}}{66^{99}}$. Are your answers different? Why? Find the decimal representation for each of your answers.

8. Use the operational commands in *Maple* to evaluate $\sqrt{9}$, $26\sqrt{144}$, $\sqrt{12}$ and $4\sqrt{48}$. Express the numbers $\sqrt{12}$ and $4\sqrt{48}$ as decimals (with 8 significant digits).

9. Evaluate $\sin\left(\dfrac{\pi}{3}\right)$, $\sec\left(\dfrac{71\pi}{6}\right)$, $\tan\left(\dfrac{11\pi}{6}\right)$ and $\cos\left(\dfrac{25\pi}{4}\right)$.

10. Evaluate $\cos^{-1}\left(\dfrac{\sqrt{2}}{2}\right)$, $\operatorname{arccsc}\left(-\dfrac{2\sqrt{3}}{3}\right)$, $\tan^{-1}\left(-\dfrac{\sqrt{3}}{3}\right)$ and $\operatorname{arccot}(1)$.

11. Use the operational commands in *Maple* to evaluate e^{42}, e^{159}, and e^{7834}. Express your answers as decimals.

12. Use the operational commands in *Maple* to evaluate $\ln(531)$, $\ln(6451.297)$, and $\log_{15}(9784.12)$. Express your answers as decimals. Evaluate $\log_2(65536)$.

13. Expand $(2x-5y)^3(9x-4y)$ and $(11a+7)^5(7a-1)^2$.

14. Enter the function $f(x)=x^2-6x+5$ as a *Maple* expression. Evaluate $f(36)$. What are the factors of f?

15. Enter the function $f(x)=x^2-x-6$ as a *Maple* expression. Evaluate $f(2)$. What are the factors of f?

16. Enter the function $f(x)=x^2+4x-12$ as a *Maple* function. Evaluate $f(2)$. What are the factors of f? What are the zeros of f? Find the solutions of $f(x)=9$. Graph the function f.

17. Enter the function $f(x)=x^3+6x^2+11x+6$ as a *Maple* function. Evaluate $f(2)$. What are the factors of f? What are the zeros of f? Find the solutions of $f(x)=9$. Graph the function f.

18. Enter the following using the *2-D Mode* in *Maple*. Let $v:=3$ and $w:=6$. Find $v*w$, vw, and $v\ w$. Are all the answers the same? Give reasons for these answers.

19. Enter the following using the *2-D Mode* in *Maple*. Find $(x-y)*(x-y)$, $(x-y)(x-y)$, and $(x-y)\ (x-y)$. Right click each answer and select *Expand*. Are all the answers the same? Give reasons for these answers.

20. Consider the polynomial function $f(x)=x^3+3x^2-4x-12$ defined in Example 7. What happens when you use the command $plot(f(x),x=-\infty..\infty)$? Give reasons for your answer.

21 – 25 Use *Maple* commands to identify, describe, and graph the given curves.

21. $y^2+8x-8y+40=0$.

22. $25x^2 + 9y^2 - 50x + 36y - 164 = 0$.

23. $5x^2 + 5y^2 + 30x - 20y - 35 = 0$.

24. $y^2 - x^2 + 4x - 4 = 0$.

25.   Use *Maple* commands to identify, describe, and graph $8y^2 - 27x^2 + 4x - 6y - 4 = 0$.

26.   Consider the floor function $f(x) = \text{floor}(x)$, which is the largest integer less than or equal to x. Evaluate $f(-1.234)$, $f\left(-\dfrac{57}{14}\right)$, $f\left(\dfrac{57}{14}\right)$ and $f(63.59)$. Graph this function and find its domain and range.

27.   Let $g(x) = \text{ceil}(x)$. Find the domain of the functions f and g. Evaluate the values of $g(-1.234)$, $g\left(-\dfrac{57}{14}\right)$, $g\left(\dfrac{57}{14}\right)$, and $g(63.59)$. Graph this function (do not forget to use the *discont = true* option) and find its domain and range.

28.   Compare the graphs of the floor function, the ceiling function and the rounding function. Are they the same? Give reasons for your answer.

29.   Graph the functions $f(x) = x^2 - 4x + 1$ and $g(x) = \ln(x)$ on the same set of axes. Use this graph to find the number of intersection points of the two functions. Write the answer given when you use the command *solve(f(x) = g(x), x)*. Find all the points of intersection using the *Roots* command.

30.   Graph the functions $f(x) = x^2 - x - 1$ and $g(x) = e^{-x^2}$ on the same set of axes. Use this graph to find the number of intersection points of the two functions. Write the answer given when you use the command *solve(f(x) = g(x), x)*. Find all the points of intersection using the *Roots* command.

31.   Graph the functions $f(x) = \sin(x^2)$ and $g(x) = \cos(x^2)$ on the same set of axes in the interval $[0, 2\pi]$. Use this graph to find the number of intersection points of the two functions. Use some *Maple* commands to find the points of intersection on the interval $[0, 2\pi]$.

32.   Below is a list of some *Maple* commands used in this chapter. Describe the significance of each command. Can you find examples where each command is used?

(a) $abs(a)$

(b) $f := a \rightarrow y$

(c) $evalf(a)$

(d) $evalf(a, n)$

(e) $expand(expression)$

(f) $f(x)$

(g) $factor(expression)$

(h) $fsolve(f(x) = a, x)$

(i) $plot(f(x), x)$

(j) $plot(f(x), x = a..b, y = c..d)$

(k) $plot(\{f(x), g(x)\}, x = a..b, y = c..d)$

(l) $plot([f(x), g(x)], x = a..b, y = c..d, color = [color1, color2])$

(m) $plot(f(x), x = a..b, y = c..d, discont = true)$

(n) $Roots(f(x) = a, x, numeric)$

(o) $round(a)$

(p) $with(geometry)$

(q) $with(Student[Calculus1])$

33. Can you list some *Maple* commands that were used in this chapter but were not listed in exercise 32?

Chapter 1   An Introduction to *Maple* Commands

The object of this chapter is to introduce some more *Maple* commands while getting used to the ones introduced in Chapter 0. The right click feature will also be used here. In the examples below we will enter the functions as *Maple* functions.

More Graphs

Example 1:   Let $f(x) = 2x^3 - 3x^2 - 32x - 15$.

(a)   Find the domain, range, and zeros of the polynomial function f.

(b)   Graph the function f.

(c)   Use the *Drawing* menu to indicate the positions of the zeros of the polynomial along with its local maximum and minimum.

Solution: We will enter the polynomial function $f(x) = 2x^3 - 3x^2 - 32x - 15$ as a *Maple* function.

$$f := x \rightarrow 2 \cdot x^3 - 3 \cdot x^2 - 32 \cdot x - 15$$
$$x \rightarrow 2x^3 - 3x^2 - 32x - 15 \tag{1.1}$$

(a)   Since f is a polynomial of degree 3, the domain and range of this polynomial function is the set of all real numbers. We will find the zeros of the polynomial by using the *solve* command.

$$solve(f(x) = 0, x)$$
$$5, -\frac{1}{2}, -3 \tag{1.2}$$

The zeros of $f(x)$ are 5, $-\dfrac{1}{2}$, and -3.

(b)   The graph of f can be sketched by using the *plot* command.

$$plot(f(x), x = -4..6)$$

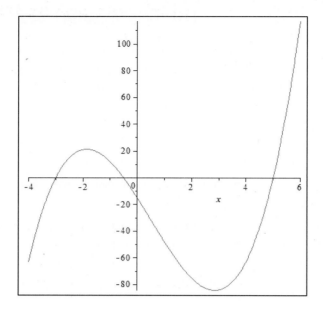

To indicate the positions of the zeros, local maximum and minimum of the polynomial, we will first insert grid lines as shown in the picture below. Click once on the graph. Click on the *Plot* menu and select *Toggle Gridlines* which is the right most in the *Plot* menu. The options in the *Plot* menu are *Plot style Polygon with Outline, Axis style Normal, Toggle Scaling Constrained, Point Probe, Rotate the plot, Scale plot axes, Translate plot axes, Execute click and drag code, Change Axis Properties, Change Gridlines Properties*, and *Toggle Gridlines*. Click on *Toggle Gridlines* to get a plot that has a grid.

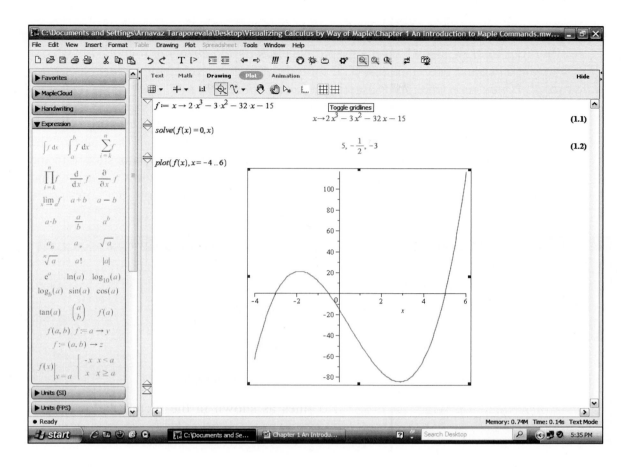

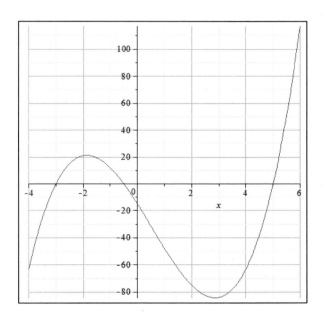

In order to draw arrows on the graph, we click on the *Drawing* menu. This is next to the *Text* and *Math* menus. The tools in the *Drawing* menu include *Selection tool*, *Pencil tool* (free style drawing), *Eraser tool*, *Text tool*, *Line tool*, *Rectangle tool*, *Round rectangle tool*, *Oval tool*, *Diamond tool*, *Drawing Alignment*, *Drawing Outline*, *Drawing Fill*, *Drawing Linestyle,* and *Drawing Canvas Properties*. To make the arrows select the option *Line tool* that is a part of the *Drawing* menu, then on *Drawing Linestyle*, select on the arrows as shown in the picture below. Place the cursor where you want to begin drawing the arrow and begin drawing. Double click once you have drawn the arrow an appropriate length.

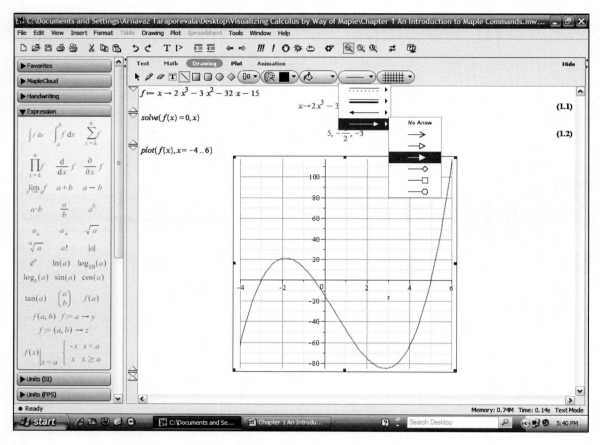

To insert text, select *Text tool* in the *Drawing* menu, click on the region where you want to insert the text, and then enter your text. Note that while you are actually inserting text with the graph, the *Text* menu is highlighted. The size of the region where the text is inserted can be increased or decreased by first clicking outside the plot, then clicking on the *Text tool* once again and then placing the cursor at the appropriate spot to increase or decrease the size. This is seen in the pictures below.

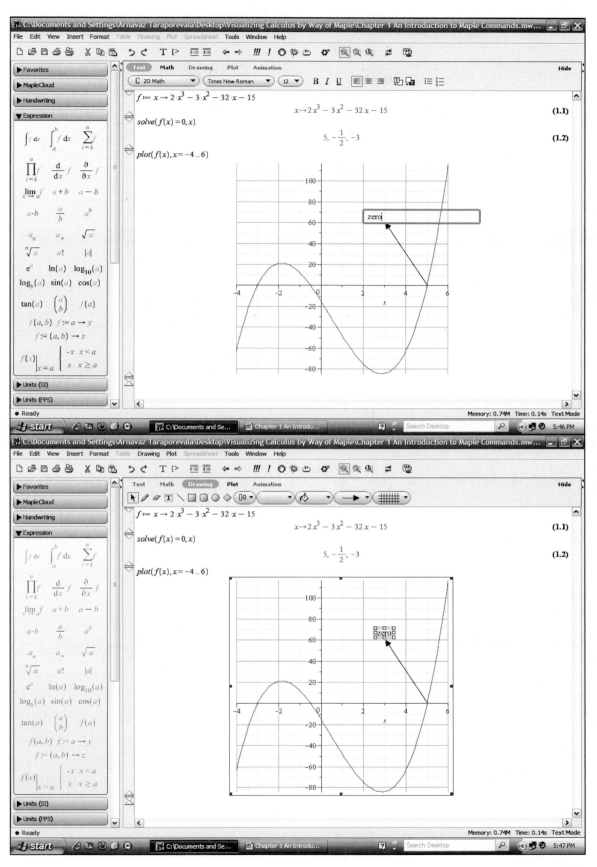

The final plot looks like the one below.

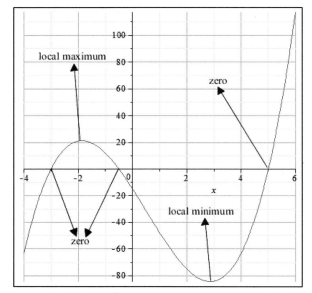

Remark: In order to use the zooming options of *Maple* one must create a *2-D smart plot object* and place the curve into this object. The *Pan* and *Scale* options of *Maple* then recalculated the curve. The curve is not recalculated when the function is graphed using the plot command or the plot builder. You can use the *Pan* and *Scale* options but you will only see the curve that has been set to the original set of axes.

Example 2: Graph the function $f(x) = \sqrt{8 - 2x - x^2}$ and find its domain, range and zeros.

Solution: Enter f as a *Maple* function.

$$f := x \rightarrow \sqrt{8 - 2 \cdot x - x^2}$$

$$x \rightarrow \sqrt{8 - 2 \cdot x - x^2} \qquad (1.3)$$

The domain of f is the set of all real numbers such that $8 - 2 \cdot x - x^2 \geq 0$. Utilize the *solve* command by setting the condition to be $8 - 2 \cdot x - x^2 \geq 0$. Notice that the $\geq$ symbol is in the *Common Symbols* palette. In the *Math* mode the $\geq$ can also be obtained by typing $\boxed{>}$ followed by $\boxed{=}$.

$$solve\left(8 - 2 \cdot x - x^2 \geq 0, \{x\}\right)$$

$$\{-4 \leq x, x \leq 2\} \qquad (1.4)$$

Hence the domain of f is the set of all real numbers x such that $-4 \le x \le 2$ or in interval notation [-4, 2]. By putting braces around x we get a solution in terms of inequalities. If we do not put braces, { }, around x in the solve commands we get the following.

$solve\left(8 - 2 \cdot x - x^2 \ge 0, x\right)$

$$RealRange\left(-4, 2\right) \tag{1.5}$$

Graph the function f using the interval [-4, 2] as the values of x. Let *Maple* show the range.

$plot\left(f\left(x\right), x = -4 .. 2\right)$

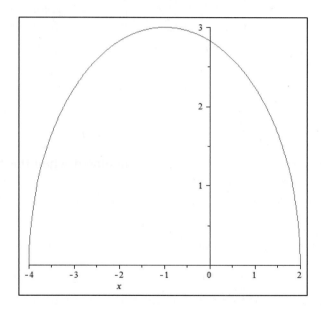

In this example the range is the interval [minimum value of $f\left(x\right)$, maximum value of $f\left(x\right)$]. Use the *minimize* and *maximize* commands to find the minimum and maximum values, respectively.

$minimize\left(f\left(x\right), x = -4 .. 2\right)$

$$0 \tag{1.6}$$

$maximize\left(f\left(x\right), x = -4 .. 2\right)$

$$\sqrt{9} \tag{1.7}$$

simplify radical
$=$

$$3 \tag{1.8}$$

The answer for the *maximize* command is $\sqrt{9}$. Right click on $\sqrt{9}$ and then click on *Simplify* and on *Radical* to simplify $\sqrt{9}$. Hence the range is [0, 3]. To find the zeros of the function we solve the equation $f(x) = 0$.

$solve\left(f(x) = 0, x\right)$

$$-4, 2 \qquad\qquad\qquad \textbf{(1.9)}$$

Therefore, the zeros are -4 and 2.

Example 3: Consider the rational function $f(x) = \dfrac{x^2 - 4}{x + 2}$. Find the domain of this function. Reduce this rational function to its lowest terms. Graph the rational function f and evaluate $f(-2)$.

Solution: We will enter $f(x)$ as a *Maple* function.

$f := x \rightarrow \dfrac{x^2 - 4}{x + 2}$

$$x \rightarrow \dfrac{x^2 - 4}{x + 2} \qquad\qquad\qquad \textbf{(1.10)}$$

Use the $\dfrac{a}{b}$ from the *Expression* palette to enter $\dfrac{x^2 - 4}{x + 2}$. The domain of this rational function is the set of all real numbers such that the denominator of f is not equal to zero. We will use the *solve* and *denom* commands (where *denom* refers to the denominator) to find the domain of the function.

$solve\left(denom\left(f(x)\right) = 0, x\right)$

$$-2 \qquad\qquad\qquad \textbf{(1.1)}$$

Hence the domain of f is the set of all real numbers except -2. The *normal* command will be used to reduce the rational expression to the normal form.

$normal\left(f(x)\right)$

$$x - 2 \qquad\qquad\qquad \textbf{(1.12)}$$

$plot\left(f(x), x = -4 .. 4\right)$

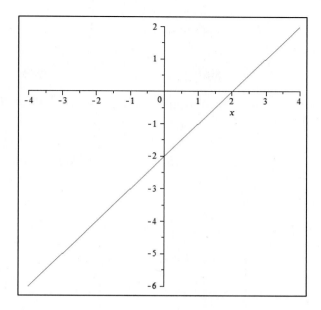

The graph of f appears to be the same as the line $y = x - 2$. However, -2 is not in the domain of f, the graph of f has a "hole" at $x = -2$. This hole is not visible using this plot command. It should be noted that *Maple* version 14 now identifies removable singularities on the graph with the command $discont = [showremovable]$ as shown below (this feature is not available in earlier versions of *Maple*).

$$plot\big(f(x), x = -4 .. 4, discont = [showremovable]\big)$$

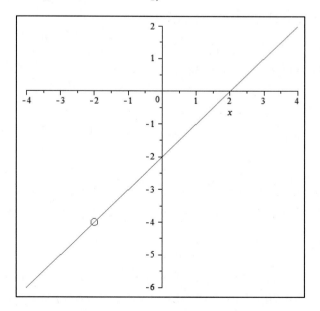

We finally evaluate $f(-2)$ to check our result.

$$f(-2)$$

Error, (in f) numeric exception; division by zero

Again $f(-2)$ is undefined as -2 is not in the domain of f.

Remark: We can use other plot options to get the graph below. This is explained in detail the upcoming chapters.

$$plot\left(f(x), x = -4 .. 4, thickness = 2, discont = \left[showremovable = \left[color = blue,\right.\right.\right.$$

$$\left.\left.\left.symbol = solidcircle\right]\right]\right)$$

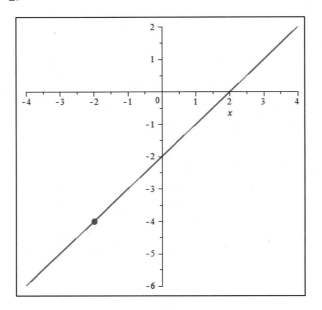

Example 4: Graph the function $f(x) = -\dfrac{1}{x^2 - 4x - 5}$ and find its domain, range and roots.

Solution: Enter $f(x)$ as a *Maple* function.

$$f := x \rightarrow -\frac{1}{x^2 - 4 \cdot x - 5}$$

$$x \rightarrow -\frac{1}{x^2 - 4 \cdot x - 5} \tag{1.13}$$

Since f is a rational function, the domain is the set of all real numbers such that the denominator is not equal to zero. As in the previous example we will use the *solve* and *denom* commands to find the domain of this function.

$$solve\left(denom(f(x)) = 0, x\right)$$

$$5, -1 \tag{1.14}$$

Therefore, the domain of f is the set of all real numbers except -1 and 5 or is the union of open intervals $(-\infty, -1) \cup (-1, 5) \cup (5, \infty)$. In order to find the range of f we will first plot f.

$plot\left(f(x), x = -4..8, y = -1.1..1\right)$

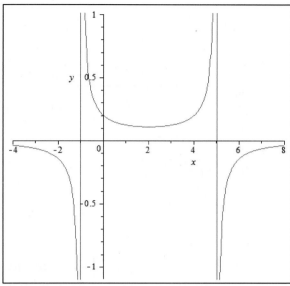

We will now proceed to find the range. For the intervals $(-\infty, -1)$ and $(5, \infty)$, the graph of $f(x)$ is below the x-axis. For the interval (-1, 5) we use the *minimize* command for x in [0, 4] and not for x in [-1, 5].

$minimize\left(f(x), x = 0..4\right)$

$$\frac{1}{9}$$ (1.15)

Thus the range of f is the union of open intervals $(-\infty, 0) \cup \left(\frac{1}{9}, \infty\right)$. The zeros of f are found by setting the numerator of $f(x)$ equal to zero. Both the *solve* and *numer* commands are used here (where *numer* refers to the numerator).

$solve\left(numer\left(f(x)\right) = 0, x\right)$

Since the numerator of this command did not give any output. Hence f has no zeros.

Remarks:

1. In the graph above, the vertical asymptotes are graphed along with the function. This happens because *Maple* plots points continuously. If one does not want to

see the vertical asymptotes, one should include the *discont = true* command as shown below.

$$plot\big(f(x), x = -4..8, y = -1.1..1, discont = true \big)$$

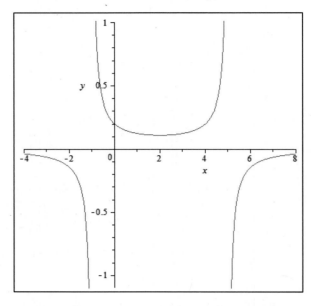

2. We can plot this function using the *RationalFunctionPlot* command in the *with*(*Student*[*Precalculus*]) package. Note this command could not be used while graphing the function in Example 3 since a vertical asymptote would be graphed at the removable discontinuity.

$$with\big(Student[Precalculus] \big):$$

$$RationalFunctionPlot\big(f(x), view = [-4..8, -1.1..1], functionoptions = [thickness = 2]$$

$$slantasymptoteoptions = [color = blue, linestyle = dash, thickness = 3],$$

$$verticalasymptoteoptions = [color = magenta, linestyle = dashdot, thickness = 2]\big)$$

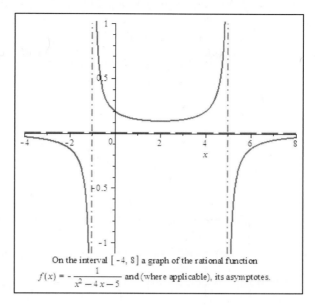

On the interval $[-4, 8]$ a graph of the rational function $f(x) = -\dfrac{1}{x^2 - 4x - 5}$ and (where applicable), its asymptotes.

Example 5: Find the domain, range, and zeros of the function $f(x) = x^{\frac{1}{3}}$. Evaluate $f(8)$, $f(0)$, and $f(-27)$. Graph the function $f(x)$.

Solution: Enter $f(x)$ as a *Maple* function.

$$f := x \rightarrow x^{\frac{1}{3}}$$

$$x \rightarrow x^{1/3} \tag{1.16}$$

From the definition of f, the domain and range are the set of all real numbers. To get the zeros of f we solve the equation $f(x) = 0$ using the *solve* command.

$$solve\left(f(x) = 0, x\right)$$

$$0 \tag{1.17}$$

We will now evaluate $f(8)$, $f(0)$, and $f(-27)$.

$$f(8)$$

$$8^{1/3} \tag{1.18}$$

simplify
$$=$$

$$2 \tag{1.19}$$

$$f(0)$$

$$0 \tag{1.20}$$

$f(-27)$

$$(-27)^{1/3} \tag{1.21}$$

simplify
$=$

$$\frac{3}{2} + \frac{3}{2} \, I \, \sqrt{3} \tag{1.22}$$

Notice the strange goings on. We used right click on $(8)^{\frac{1}{3}}$ and then click on *Simplify* and on *Simplify* to evaluate $f(8)$, and the same commands were used with $(-27)^{\frac{1}{3}}$ to evaluate $f(-27)$. When that was done, the answer for $f(-27)$ was a complex number. Let us now graph the function.

$plot\big(f(x), x = -3..3, y = -1.5..1.5\big)$

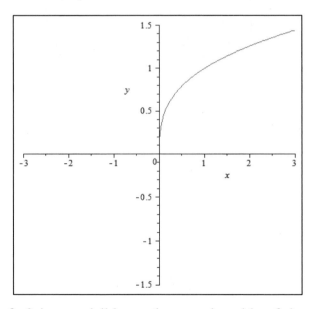

Observe that the graph of f is not visible on the negative side of the x-axis. We know that the domain of the function is the set of all real numbers, so the reason for this because the default setting for solving equations in *Maple* is that the underlying number system is the complex number system. We will now activate the $with(RealDomain)$ package. That way the basic underlying number system is now the set of real numbers. But before evaluating both $f(8)$ and $f(-27)$ and graphing the function f, we must first reenter the function f again (and not just execute the function).

$with(RealDomain):$

$f := x \rightarrow x^{\frac{1}{3}}$

$$x \rightarrow x^{1/3} \qquad \textbf{(1.23)}$$

$f(8)$

$$2 \qquad \textbf{(1.24)}$$

$f(-27)$

$$-3 \qquad \textbf{(1.25)}$$

Hence $f(8) = 2$, $f(0) = 0$, and $f(-27) = -3$. We will once again graph the function f.

$plot(f(x), x = -3..3, y = -1.5..1.5)$

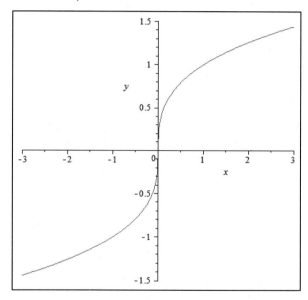

This is a more appropriate plot for the function.

Example 6: Let $f(x) = \sqrt{x+5}$ and $g(x) = \sqrt{x^2 - 9}$. Find the domain of each of the functions f and g. Find the rules of the functions $f + g$, $f - g$ and $\dfrac{f}{g}$ and identify their respective domains. Graph the functions $f + g$, $f - g$ and $\dfrac{f}{g}$ on the same set of axes.

Solution: Enter f and g as *Maple* functions.

$f := x \rightarrow \sqrt{x+5}$

$$x \rightarrow \sqrt{x+5} \qquad \textbf{(1.26)}$$

$$g := x \rightarrow \sqrt{x^2 - 9}$$

$$x \rightarrow \sqrt{x^2 - 9} \tag{1.27}$$

The domains of f and g are evaluated using the same procedure as in Example 2.

$$solve\left(x + 5 \geq 0, \{x\}\right)$$

$$\{-5 \leq x\} \tag{1.28}$$

$$solve\left(x^2 - 9 \geq 0, \{x\}\right)$$

$$\{x \leq -3\}, \{3 \leq x\} \tag{1.29}$$

Hence the domain of f is $[-5, \infty)$ and the domain of g is $(-\infty, -3] \cup [3, \infty)$. The rules of the functions $f + g$, $f - g$ and $\dfrac{f}{g}$ are given below.

$$\left(f + g\right)(x)$$

$$\sqrt{x+5} + \sqrt{x^2 - 9} \tag{1.30}$$

$$\left(f - g\right)(x)$$

$$\sqrt{x+5} - \sqrt{x^2 - 9} \tag{1.31}$$

$$\left(\frac{f}{g}\right)(x)$$

$$\frac{\sqrt{x+5}}{\sqrt{x^2 - 9}} \tag{1.32}$$

Therefore, $\left(f + g\right)(x) = \sqrt{x+5} + \sqrt{x^2 - 9}$, $\left(f - g\right)(x) = \sqrt{x+5} - \sqrt{x^2 - 9}$, and $\left(\dfrac{f}{g}\right)(x) = \dfrac{\sqrt{x+5}}{\sqrt{x^2 - 9}}$. Recall that the domains of both $f + g$ and $f - g$ is the intersection of the domains of f and g. Therefore the domain of $f + g$ and $f - g$ is $[-5, -3] \cup [3, \infty)$. Recall that the domain of $\dfrac{f}{g}$ is set of all real numbers x in the intersection of the domains of f and g such that $g(x) \neq 0$. So to find the domain of $\dfrac{f}{g}$ we will solve the equation $g(x) = 0$ using the *solve* command.

$$solve\left(g(x) = 0, x\right)$$

$$-3, 3 \tag{1.33}$$

Since 3 and –3 are not in the domain of $\dfrac{f}{g}$, the domain of $\dfrac{f}{g}$ is $[-5,-3)\cup(3,\infty)$. As we pointed out in chapter 0, in order to graph the functions $f+g$, $f-g$ and $\dfrac{f}{g}$ on the same set of axes, we will assign colors to each function and use brackets when using the *plot* command as shown below.

$$plot\left(\left[(f+g)(x),(f-g)(x),\left(\dfrac{f}{g}\right)(x)\right],x=-5..10,y=-6..6,color=[red,blue,brown]\right)$$

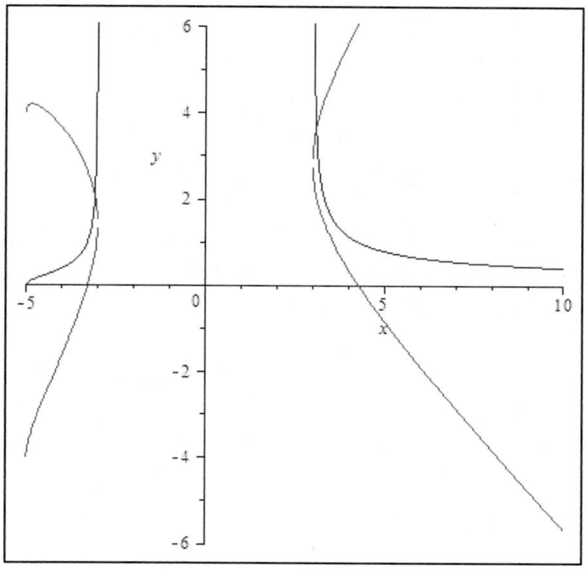

Notice even though -3 and 3 are in the domains of $(f+g)(x)$ and $(f-g)(x)$, the graph above does not show that information. This occurs because *Maple* is not plotting enough points. We can modify the code to fix this problem by using the *numpoints* command. This command specifies the minimum number of points used to generate the graph. In the next graph we are asking *Maple* to plot 5000 points (the default minimum number of points is at least 50).

$$plot\left(\left[(f+g)(x),(f-g)(x),\left(\dfrac{f}{g}\right)(x)\right],x=-5..10,y=-6..6,color=[red,blue,brown],\right.$$

$$\left.numpoints=5000\right)$$

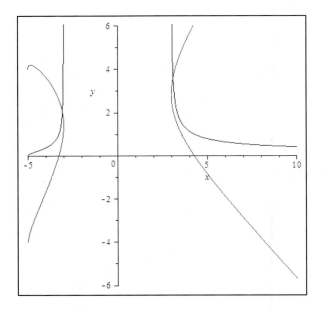

To make the distinction between the graphs even more pronounced, we could choose to represent each function with a different line style (e.g. solid, dot, dash, dashdot etc) as shown below. The *thickness = 3* command will also be used so that the dotted graph will be more visible.

$$plot\left(\left[(f+g)(x),(f-g)(x),\left(\frac{f}{g}\right)(x)\right],x=-5..10,y=-6..6,color=[red,blue,brown],\right.$$

$$\left.linestyle=[solid,dot,dash],thickness=3,numpoints=5000\right)$$

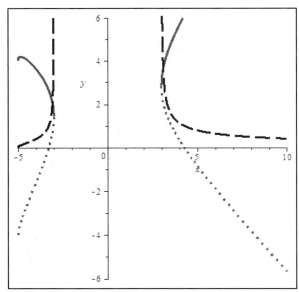

Another way to obtain a similar picture is first click on the plot, make sure you are on the *Plot* menu, select one curve by clicking on it, then using the right click, go to *Line* and represent each curve differently, e.g. solid, dash, etc.

Example 7: Let $f(x) = x^2 - 1$ and $g(x) = 3x - 5$. Find the domain of the functions f and g. Find the rule of the functions $f \circ g$ and $g \circ f$, and identify their respective domains. Graph the functions $f \circ g$ and $g \circ f$ on the same set of axes.

Solution: Enter $f(x)$ as a *Maple* function.

$f := x \rightarrow x^2 - 1$

$$x \rightarrow x^2 - 1 \qquad \qquad \textbf{(1.34)}$$

$g := x \rightarrow 3 \cdot x - 5$

$$x \rightarrow 3\,x - 5 \qquad \qquad \textbf{(1.35)}$$

Since f and g are polynomial function their domains are the set of all real numbers. The rules of the functions $f \circ g$ and $g \circ f$ are given below. We will use the symbol $\boxed{@}$ for the composition.

$(f \, @ \, g)(x)$

$$(3\,x - 5)^2 - 1 \qquad \qquad \textbf{(1.36)}$$

$(g \, @ \, f)(x)$

$$3\,x^2 - 8 \qquad \qquad \textbf{(1.37)}$$

From the above outputs we see that $(f \circ g)(x) = (3\,x - 5)^2 - 1$ and $(g \circ f)(x) = 3\,x^2 - 8$. Furthermore, the domains of $f \circ g$ and $g \circ f$ are also the set of all real numbers. The two graphs are plotted using the same command.

$plot\left(\left[(f \, @ \, g)(x), (g \, @ \, f)(x)\right], x = -3 .. 3, y = -10 .. 10, color = \left[red, blue\right]\right)$

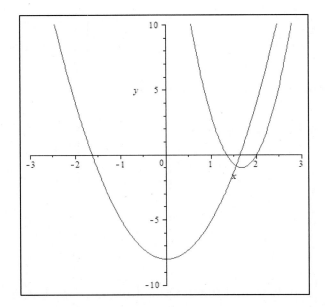

Remark: From the *Expression* Palette we could also use ∘ to denote the composition of two functions. However, this symbol will not be used as frequently because of some domain conditions that will be illustrated in the following example. Using the ∘ symbol, the rules of the functions $f \circ g$ and $g \circ f$ are given below.

$(f \circ g)(x)$

$$(3x-5)^2 - 1 \qquad\qquad \textbf{(1.38)}$$

$(g \circ f)(x)$

$$3x^2 - 8 \qquad\qquad \textbf{(1.39)}$$

We will now use the composition of functions to find the inverse of a function. Again, in this case we will activate the *with*(*RealDomain*) command. This is to make sure that the underlying number system is the set of all real numbers.

Example 8: Let $f(x) = \ln(x)$. Find the domain and range of the function $f(x)$. Find the rule for the inverse function $g(x)$. What are the domain and range of $g(x)$? Let $h(x) = x$. Plot the functions f, g, and $h(x)$ on the same set of axes. Find the rule of the functions $f \circ g$ and $g \circ f$, and identify their respective domains. Graph the functions $f \circ g$ and $g \circ f$ on the same set of axes. Does the graph of these functions reflect the domains of $f \circ g$ and $g \circ f$? Give reasons for your answer.

Solution: First enter the *with*(*RealDomain*) package and then enter $f(x)$ as a *Maple* function.

restart

$with(RealDomain):$

$f := x \rightarrow \ln(x)$

$$x \rightarrow \ln(x) \qquad \qquad \textbf{(1.40)}$$

The domain of f is the set of positive real numbers and the range is the set of all real numbers. In order to find the function g, we will solve the equation $x = f(y)$ for y and use the *unapply* command to express the answer as a function of x.

$solve(x = f(y), y)$

$$e^x \qquad \qquad \textbf{(1.41)}$$

$g := unapply((\textbf{1.41}), x)$

$$x \rightarrow e^x \qquad \qquad \textbf{(1.42)}$$

Hence the inverse function is $g(x) = e^x$ with a domain of the set of all real numbers and a range of the set of positive real numbers. We will now graph the functions f and g, and the line $y = x$ on the same set of axes.

$plot([f(x), g(x), x], x = -2..5, y = -3..5, color = [red, blue, black], linestyle = [dash, dot, solid],$
$\qquad thickness = 3)$

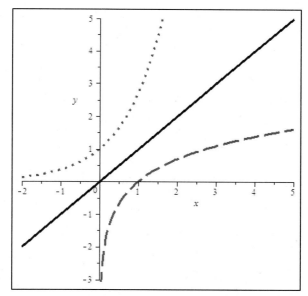

The rules of the functions $f \circ g$ and $g \circ f$ are given below.

$(f @ g)(x)$

$$x \qquad \qquad \textbf{(1.43)}$$

$$(g @ f)(x)$$

$$x \qquad\qquad \textbf{(1.44)}$$

Note that the domain of the function $f \circ g$ is the set of all real numbers and the domain of $g \circ f$ is the set of positive real numbers. We will now graph these functions on the same graph. Note that we will use different line styles so that the two graphs can be differentiated.

$$plot\left(\left[(f @ g)(x),(g @ f)(x)\right], x = -5 .. 5, color = \left[red, blue\right], linestyle = \left[dash, dot\right]\right.$$
$$\left. thickness = 3\right)$$

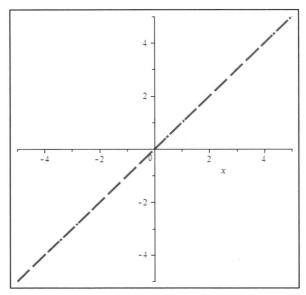

Note that even though the domain of $g \circ f$ is the set of positive real numbers, it is graphed for negative values of x. This is a "flaw" that can be corrected by using the following plot command.

$$plot\left(\left[`@`(f,g), `@`(g,f)\right], -5 .. 5, color = \left[red, blue\right], linestyle = \left[dash, dot\right], thickness = 3\right)$$

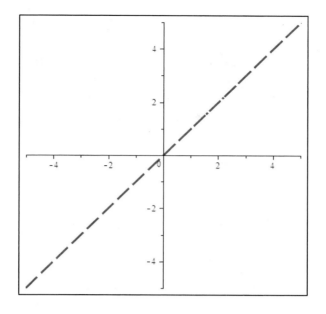

Remarks:

1. The advantage of using the command `` `@`(f,g) `` while graphing is that *Maple* does not simplify or evaluate the composition until the different conditions of finding the domain are satisfied. Note that we cannot evaluate $(g \circ f)(x)$ for negative values of x as the output is *undefined* as seen below.

    $$(f @ g)(-1)$$

    $$-1 \qquad\qquad\qquad\qquad\qquad \textbf{(1.45)}$$

    $$(g @ f)(-1)$$

    $$\textit{undefined} \qquad\qquad\qquad\qquad\qquad \textbf{(1.46)}$$

2. One can construct the graph of the functions f, g, and $h(x) = x$ on the same set of axes by using the function f and the *InversePlot* command in the $with(Student[Calculus1])$ package.

    $$with(Student[Calculus1]):$$
    $$InversePlot(f(x), x = -2..5)$$

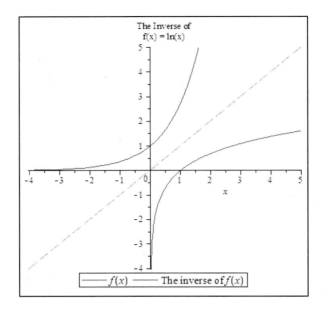

Notice that *Maple* is plotting the graphs on a different interval than the one specified in the *InversePlot* command. Indeed, *Maple* will choose the smallest interval containing the interval specified in the command in order to graph the function f and its inverse. If no interval is specified in the *InversePlot* command, the default interval is then $[-10,10]$. To control the horizontal and vertical ranges use the $view = [-2..5, -3..5]$ option. The problem does not give the actual inverse of the function. The graph below also uses options for the function, its inverse and the line $y = x$. Notice that the line $y = x$ is dashed – you may change the line style if you wish to do so.

$$InversePlot\left(\ln\left(x\right), view = [-2..5, -3..5], functionoptions = [color = black, thickness = 2],\right.$$
$$\left. inverseoptions = [color = red, thickness = 2], lineoptions = [color = brown, thickness = 2]\right)$$

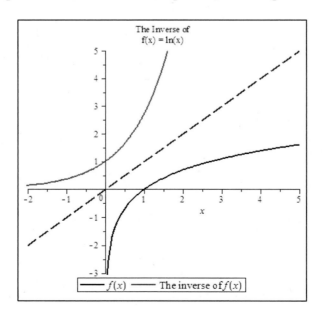

<u>Arrays</u>

Note: It is an extremely important tool to be able to generate tables with input and output values. This will be used in later chapters. Here is a procedure to create a table of values.

Example 9: Construct an array of 10 rows and 2 columns that has the numbers 1, 2, .. *n* in the first column and their squares in the second column.

Solution: The procedure below can be used to create an array or a table of values with 10 rows and 2 columns. Note that the values of *n* are from 1 through 10. We will call our array A. Note that $A[i, j]$ is the entry in the ith row and j^{th} column in the array.

$A := array(1..10, 1..2)$

for n **from** 1 **to** 10 **do** $A[n,1] := n$: $A[n,2] := A[n,1]^2$: **end do**:

$print(A)$

$$
\begin{bmatrix}
1 & 1 \\
2 & 4 \\
3 & 9 \\
4 & 16 \\
5 & 25 \\
6 & 36 \\
7 & 49 \\
8 & 64 \\
9 & 81 \\
10 & 100
\end{bmatrix}
\qquad \textbf{(1.47)}
$$

Remark: We can modify this procedure to have the title n for $A[1, 1]$ and n^2 for $A[2, 1]$ as shown below. In order for *Maple* to recognize that these are just titles it is important to put them in single quotes. Note that for n from 2 to 10 we see that $A[n, 2] = n - 1$ (otherwise we would then have the squares of the numbers from 2 .. 11).

$A := array(1..11, 1..2):$

$A[1,1] := 'n':$ $A[1,2] := 'n^2':$

for n **from** 2 **to** 11 **do** $A[n,1] := n-1$: $A[n,2] := A[n,1]^2$: **end do**:

$print(A)$

$$\begin{bmatrix} n & n^2 \\ 1 & 1 \\ 2 & 4 \\ 3 & 9 \\ 4 & 16 \\ 5 & 25 \\ 6 & 36 \\ 7 & 49 \\ 8 & 64 \\ 9 & 81 \\ 10 & 100 \end{bmatrix} \qquad\qquad (1.48)$$

Piecewise Functions

Example 10: Write a procedure for the piecewise function $f(x) = \begin{cases} -4 & \text{if } 0 \le x \le 3 \\ 10 & \text{if } 3 < x \le 5 \end{cases}$.

Graph this function and find its domain and range. Evaluate $f(-1)$, $f(1)$, $f(3)$, $f(5)$, and $f(7)$ and give reasons for the answers given.

Solution: We will write a procedure to create this piecewise function. It begins with *proc( )* and ends with the word *end*. To end an *if* statement we use *fi* or *end if*. Similarly to end a *do* loop we use *od* or *end do*.

f := **proc**(x) **if** $0 \le x$ **and** $x \le 3$ **then** -4 **elif** $3 < x$ **and** $x \le 5$ **then** 10 **else** *undefined* **end if end**

      proc(x) **if** $0 <= x$ **and** $x <= 3$ **then** -4 **elif** $3 < x$ **and** $x <= 5$ **then** 10 **else** *undefined* **end if**

         end proc   (1.49)

By the definition of the piecewise function the domain is $[0,5]$ and the range is $\{-4, 10\}$. Notice a slight variation in the *plot* command used, as f is the name of a procedure here not a *Maple* function and the input is from 0 to 5. As in chapter 0, we will use the *discont = true* command in the *plot* command. This ensures that *Maple* identify points of jump discontinuities and graphically displays them.

$plot(f, 0..5, discont = true)$

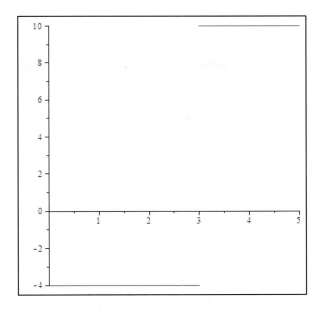

$f(-1);\ f(1);\ f(3);\ f(5);f(7);$

$$undefined$$
$$-4$$
$$-4$$
$$10$$
$$undefined \qquad\qquad\qquad\textbf{(1.50)}$$

The results show that $f(-1)$ and $f(7)$ are undefined as -1 and 7 are not in the domain of f. Since $0 \le 1 \le 3$ and $0 \le 3 \le 3$ we see that $f(1) = -4 = f(3)$. Furthermore, $f(5) = 10$ as $3 < 5 \le 5$.

Remarks: Alternatively, a built-in *Maple* command for piecewise functions is shown below.

$f := x \rightarrow piecewise(0 \le x$ **and** $x \le 3, -4, 3 < x$ **and** $x \le 5, 10, undefined)$
$$x \rightarrow piecewise(0 \le x \textbf{ and } x \le 3, -4, 3 < x \textbf{ and } x \le 5, 10, undefined) \textbf{ (1.51)}$$

This function can now be plotted as shown below.

$plot\left(f(x), x = 0 .. 5, discont = true\right)$

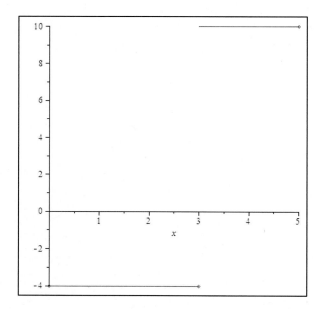

We can also evaluate the values of the functions at the points mentioned in Example 10.

$$f(-1); \; f(1); \; f(3); \; f(5); f(7);$$

$$undefined$$
$$-4$$
$$-4$$
$$10$$
$$undefined$$

(1.52)

Care must be taken while using this command. Since the domain of this function is [0, 5], one must include undefined as a part of the command. Failure to do so will give a default option, which is 0. This is shown below.

$$f := x \to piecewise(0 \le x \text{ and } x \le 3, -4, 3 < x \text{ and } x \le 5, 10)$$
$$x \to piecewise(0 \le x \text{ and } x \le 3, -4, 3 < x \text{ and } x \le 5, 10)$$

(1.53)

$$f(-1); \; f(1); \; f(3); \; f(5); f(7);$$

$$0$$
$$-4$$
$$-4$$
$$10$$
$$0$$

(1.54)

The plot command below also displays this feature.

$$plot\big(f(x), x = -1..6, discont = true, thickness = 3\big)$$

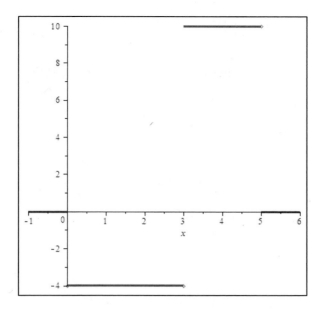

One would get a similar result if we clicked on $\begin{cases} -x & x < a \\ x & x \geq a \end{cases}$ in the *Expression Palatte* and edited it as seen below.

$$f := x \rightarrow \begin{cases} -4 & \text{if } 0 \leq x \leq 3 \\ 10 & \text{if } 3 < x \leq 5 \end{cases}$$

$$x \rightarrow piecewise(0 \leq x \textbf{ and } x \leq 3, -4, 3 < x \textbf{ and } x \leq 5, 10) \tag{1.55}$$

$$f(-1); \ f(1); \ f(3); \ f(5); f(7);$$

$$0$$
$$-4$$
$$-4$$
$$10$$
$$0 \tag{1.56}$$

The round, floor and ceiling functions are examples of particular types of piecewise functions. This will be discussed in further detail in the next chapters.

Chapter 1 An Introduction to *Maple* Commands

Exercises

1. Enter the function $f(x) = \dfrac{x^2 + 2x - 15}{x - 3}$ as a *Maple* function. What is the domain of *f*? Graph the function *f*.

2 – 17 Graph these function and find their domain, range, and zeros (if they exist).

2. $f(x) = x^3 + 2x^2 - 11x - 12$.

3. $f(x) = 6x^3 - 53x^2 + 67x + 70$.

4. $f(x) = -4x^4 + 9x^3 + 208x^2 - 84x - 864$.

5. $f(x) = |7x + 15|$.

6. $f(x) = |x^2 + 3x - 10|$.

7. $f(x) = |4x^2 - 7x - 102|$.

8. $f(x) = \sqrt{5x + 28}$.

9. $f(x) = \sqrt{16 - x^2}$.

10. $f(x) = -\sqrt{135 - 57x - 8x^2}$.

11. $f(x) = \dfrac{x^2 - x - 6}{x + 2}$.

12. $f(x) = \dfrac{x + 5}{x^2 + 7x - 18}$.

13. $f(x) = \dfrac{14x - 56}{6x^3 - 23x^2 - 334x + 1320}$.

14. $f(x) = e^{\sqrt{16 - x^2}}$.

15. $f(x) = \ln\left(\sqrt{16 - x^2}\right)$.

16. $f(x) = \ln\left|1 - x^2\right|$.

17. $f(x) = \ln\left|\sin(x)\right|$.

18. $f(x) = \dfrac{14x - 56}{6x^3 - 23x^2 - 334x + 1320}$.

19.   Let $f(x) = \begin{cases} 3x - 6 & \text{if} \quad 0 \le x \le 2 \\ 2x^2 - 6 & \text{if} \quad 2 < x \le 4 \end{cases}$. Graph this function and find its domain, range, and zero(s). Evaluate the values of $f(-1)$, $f(1)$, $f(2)$, $f(3)$, and $f(8)$.

20.   Let $f(x) = \begin{cases} 3 & \text{if} \quad -6 \le x < -2 \\ 5 & \text{if} \quad -2 \le x \le 4 \end{cases}$. Graph this function and find its domain, range, and zero(s). Evaluate the values of $f(-7)$, $f(-5)$, $f(-2)$, $f(4)$, and $f(6)$.

21.   Let $f(x) = \begin{cases} -5x + 10 & \text{if} \quad -2 \le x < 2 \\ -x^2 - 1 & \text{if} \quad 2 \le x \le 4 \end{cases}$. Graph this function and find its domain, range, and zero(s). Evaluate the values of $f(-2)$, $f(1)$, $f(2)$, $f(4)$, and $f(12)$.

22.   Let $f(x) = \begin{cases} -8 & \text{if} \quad -10 \le x \le -6 \\ -2 & \text{if} \quad -6 < x < 0 \\ 6 & \text{if} \quad 0 \le x \le 8 \end{cases}$. Graph this function and find its domain, range, and zero(s). Evaluate the values of $f(-10)$, $f(-6)$, $f(-1)$, $f(0)$, and $f(8)$.

23.   Let $f(x) = \sqrt{16 - 6x - x^2}$ and $g(x) = \sqrt{x + 7}$. Find the domain of the functions f and g. Find the rule of the functions $f + g$, $f - g$ and $\dfrac{f}{g}$ and identify their respective domains. Graph the functions $f + g$, $f - g$ and $\dfrac{f}{g}$ on the same set of axes.

24.   Let $f(x) = x^2 + 5x - 14$ and $g(x) = x^2 + 2x - 16$. Find the domain of the functions f and g. Find the rule of the functions $f + g$, $f - g$ and $\dfrac{f}{g}$ and identify

their respective domains. Graph the functions $f+g$, $f-g$ and $\dfrac{f}{g}$ on the same set of axes.

25. Let $f(x)=x^3$ and $g(x)=x^2$. Find the domain of the functions f and g. Find the rule of the functions $f+g$, $f-g$ and $\dfrac{f}{g}$ and identify their respective domains. Graph the functions $f+g$, $f-g$ and $\dfrac{f}{g}$ on the same set of axes.

26. Let $f(x)=x^2$ and $g(x)=\sqrt{x}$. Find the domain of the functions f and g. Find the rule of the functions $f\circ g$ and $g\circ f$, and identify their respective domains. Graph the functions $f\circ g$ and $g\circ f$ on the same set of axes. Does the graph of the functions reflect the domains of $f\circ g$ and $g\circ f$? Give reasons for your answer.

27. Let $f(x)=\dfrac{1}{4}x^2-3x-1$ and $g(x)=7x+15$. Find the domain of the functions f and g. Find the rule of the functions $f\circ g$ and $g\circ f$, and identify their respective domains. Graph the functions $f\circ g$ and $g\circ f$ on the same set of axes.

28. Let $f(x)=-x^2$ and $g(x)=\sqrt{133+9x-4x^2}$. Find the domain of the functions f and g. Find the rule of the functions $f\circ g$ and $g\circ f$, and identify their respective domains. Graph the functions $f\circ g$ and $g\circ f$ on the same set of axes. Does the graph of the functions reflect the domains of $f\circ g$ and $g\circ f$? Give reasons for your answer.

29. Let $f(x)=\dfrac{1}{x+3}$ and $g(x)=\dfrac{1}{x}-3$. Find the domain of the functions f and g. Find the rule of the functions $f\circ g$ and $g\circ f$, and identify their respective domains. Graph the functions $f\circ g$ and $g\circ f$ on the same set of axes. Does the graph of the functions reflect the domains of $f\circ g$ and $g\circ f$? Give reasons for your answer.

30. Let $f(x)=e^{-x^2}$ and $g(x)=-\sqrt{\ln(x)}$. Find the domain of the functions f and g. Find the rule of the functions $f\circ g$ and $g\circ f$, and identify their respective domains. Graph the functions $f\circ g$ and $g\circ f$ on the same set of axes. Does the graph of the functions reflect the domains of $f\circ g$ and $g\circ f$? Give reasons for your answer.

31. Let $f(x) = e^{-x^2}$ and $g(x) = -\sqrt{\ln(x)}$. Are $f(x)$ and $g(x)$ inverses of each another? Explain. What adjustments can you make to ensure that $f(x)$ and $g(x)$ inverses of one another?

32. Let $f(x) = \sqrt{x}$ and $g(x) = x^2 \sin^2(x)$. Find the domain of the functions f and g. Find the rule of the functions $f \circ g$ and $g \circ f$, and identify their respective domains. Graph the functions f, g and the line $y = x$ on the same set of axes. Does the graph of the functions reflect the domains of $(f \circ g)(x)$ and $(g \circ f)(x)$? Give reasons for your answer.

33. Let $f(x) = \dfrac{1}{x}$. Find the domain of the function f and graph it. Find the rule of the function $(f \circ f)(x)$ and identify its domain. Graph the function $f \circ f$. Explain how you can restrict the domain of f to make it invertible. Based on your result state the rule for the inverse of the function f.

34 - 46 Find the rule for the inverse function $g(x)$. What are the domain and range of $g(x)$? Let $h(x) = x$. Plot the functions f, g, and $h(x)$ on the same set of axes. Find the rule of the functions $f \circ g$ and $g \circ f$, and identify their respective domains. Graph the functions $f \circ g$ and $g \circ f$ on the same set of axes. Does the graph of these functions reflect the domains of $f \circ g$ and $g \circ f$? Give reasons for your answer.

34. $f(x) = \dfrac{e^{-x}}{1 + e^{-x}}$.

35. $f(x) = \tan(x)$.

36. $f(x) = \sqrt{e^x + 5}$.

37. $f(x) = \ln(11 + \ln(x))$.

38. $f(x) = \dfrac{5x - 9}{1 + x}$.

39. $f(x) = \ln\left(x - \sqrt{x^2 + 1}\right)$.

40. $f(x) = \cos^{-1}(5x - 1)$.

41. $f(x) = \cos\left(\tan^{-1}(x)\right)$.

42. $f(x) = \tan\left(\cos^{-1}(x)\right)$.

43. $f(x) = \sqrt{-2010 - x}$.

44. $f(x) = \ln\left(-2010 - x\right)$.

45. $f(x) = e^{e^x}$.

46. $f(x) = e^{e^{x^2 + 2x}}$ (Hint: Use completion of squares on the exponent and natural logarithm twice).

47. Let $f(x) = \dfrac{\sqrt{x-1}}{\sqrt{x+5}}$ and $g(x) = \sqrt{\dfrac{x-1}{x+5}}$. Find the domain of the functions f and g and graph them on the same set of axes. When is $f(x) = g(x)$? Does this violate any algebraic rules?

48. Construct an array of 20 rows and 2 columns that has the numbers 1, 2, …n in the first column and $1^1 - 1$, $2^2 - 1$, $3^3 - 1$, …, $n^n - 1$ in the second column.

49. Construct an array of 10 rows and 2 columns that has the first 10 numbers of the sequence $\{3^n\}$ in the first column and the corresponding terms for $\left\{ evalf\left(\left(1 + \dfrac{1}{3^n}\right)^{3^n}\right)\right\}$ in the second column.

50. Below is a list of some commands used in this chapter. Describe the significance of each command. Can you find examples where each command is used?

    (a) $A[i, j]$

    (b) $array(1..m, 1..n)$

    (c) $denom(f(x))$

    (d) $elif$

    (e) **end do**

    (f) $(f @ g)(x)$

    (g) $InversePlot(f(x), x = a..b)$

(h) $maximize\left(f\left(x\right), x = a..b\right)$

(i) $minimize\left(f\left(x\right), x = a..b\right)$

(j) $normal\left(f\left(x\right)\right)$

(k) $plot\left(f, a..b, discont = true\right)$

(l) $plot\left(f\left(x\right), x = a..b, y = c..d, discont = \left[showremovable\right]\right)$

(m) $plot\left(f\left(x\right), x = a..b, y = c..d, numpoints = n\right)$

(n) $plot\left(\left[`@`\left(f, g\right), `@`\left(g, f\right)\right], a..b, linestyle = \left[linestyle1, linestyle2\right]\right)$

(o) $proc\left(input\right)$

(p) $solve\left(f\left(x\right) \geq a, \left\{x\right\}\right)$

(p) $with\left(Student\left[Precalculus\right]\right)$

51. Can you list new *Maple* commands that were used in this chapter but were not listed in exercise 50?

Chapter 2 Limits

The object of this chapter is to introduce the limit commands. Limits will be calculated using a step-by-step procedure. Calculation of limits using lists will be explored. We will also include an example and exercises using the epsilon-delta definition of a limit that allows students to understand the precise definition of limits. The Intermediate Value Theorem will be introduced in this chapter.

Basic Definitions

***Definition*:** Let f be a function. Let c and L be real numbers. We say that the *limit of* $f(x)$ *as* x *approaches* c is L, denoted by $\lim_{x \to c} f(x) = L$, if for every $\varepsilon > 0$ there is a corresponding $\delta > 0$ such that if $0 < |x - c| < \delta$ then $|f(x) - L| < \varepsilon$.

***Definition*:** Let f be a function. Let c and L be real numbers. We say that the *limit of* $f(x)$ *as* x *approaches* c *from the left* is L, denoted by $\lim_{x \to c-} f(x) = L$, if for every $\varepsilon > 0$ there is a corresponding $\delta > 0$ such that if $c - \delta < x < c$ then $|f(x) - L| < \varepsilon$.

***Definition*:** Let f be a function. Let c and L be real numbers. We say that the *limit of* $f(x)$ *as* x *approaches* c *from the right* is L, denoted by $\lim_{x \to c+} f(x) = L$, if for every $\varepsilon > 0$ there is a corresponding $\delta > 0$ such that if $c < x < c + \delta$ then $|f(x) - L| < \varepsilon$.

***Theorem*:** $\lim_{x \to c} f(x) = L$ if and only if $\lim_{x \to c-} f(x) = L = \lim_{x \to c+} f(x)$.

***Theorem*:** Let f and g be functions such that $f(x) = g(x)$ for every value of x near c (except possibly at the point c). Suppose that $\lim_{x \to c} g(x)$ exists. Then $\lim_{x \to c} f(x)$ exists and $\lim_{x \to c} f(x) = \lim_{x \to c} g(x)$.

***Theorem*:** Let f and g be functions such that $f(x) \le g(x)$ for every value of x near c (except possibly at the point c). Suppose that $\lim_{x \to c} f(x)$ and $\lim_{x \to c} g(x)$ exist. Then $\lim_{x \to c} f(x) \le \lim_{x \to c} g(x)$.

***The Squeeze Theorem*:** Let f, g and h be functions such that $f(x) \le g(x) \le h(x)$ for every value of x near c (except possibly at the point c). Suppose that $\lim_{x \to c} f(x) = L = \lim_{x \to c} h(x)$. Then $\lim_{x \to c} g(x) = L$.

Definition of a continuous function: Let f be a function. Let c be a real number in the domain of f. We say that f is *continuous at c* if $\lim_{x \to c} f(x) = f(c)$. The function f is said to be a *continuous function* if it is continuous at every point in its domain.

Definition of a removable discontinuity: Let f be a function. Let c be a real number such that f is not continuous at c. We say that f has a *removable discontinuity* at c if $\lim_{x \to c} f(x)$ exists.

Theorem: Every polynomial function is continuous.

Limits using Maple Commands

In our first example we will use $\lim_{x \to a} f$ in the *Expression* palette to evaluate different limits using *Maple* commands.

Example 1: Graph the function $f(x) = \dfrac{x^2 + 2}{x + 2}$. Find $\lim_{x \to 2} f(x)$ if it exists. Is f continuous at $x = 2$? Give reasons for your answer.

Solution: Note that the domain of f is the set of all real numbers except -2. We will begin by entering f as a *Maple* function and graphing it.

$$f := x \to \frac{x^2 + 2}{x + 2}$$

$$x \to \frac{x^2 + 2}{x + 2} \qquad\qquad (2.1)$$

$plot\left(f(x), x = -5..3, y = -40..30\right)$

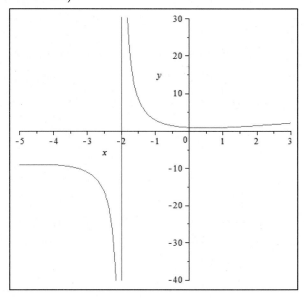

From the graph above it looks as though $\lim_{x \to 2} f(x)$ exists. We will now use *Maple* commands to find $\lim_{x \to 2} f(x)$, if it exists. Click on the expression $\lim_{x \to a} f$ in the *Expression Palette* on the left side of the screen. Notice that the letter x is highlighted. Type in x and use the $\boxed{Tab}$ key. Now the letter a is highlighted. Type in the number 2 and use the $\boxed{Tab}$ key once more. This time f is highlighted. Type in $f(x)$ and press the $\boxed{Enter}$ key to get the answer shown below.

$$\lim_{x \to 2} f(x)$$

$$\frac{3}{2} \qquad\qquad (2.2)$$

Hence $\lim_{x \to 2} f(x) = \frac{3}{2}$. Furthermore, as

$$f(2)$$

$$\frac{3}{2} \qquad\qquad (2.3)$$

we see that $\lim_{x \to 2} f(x) = \frac{3}{2} = f(2)$. Hence f is continuous at $x = 2$.

Remarks: The right click feature can be used to calculate the limit. After typing the expression $\dfrac{x^2 + 2}{x + 2}$, right click on it. Select Limit, type in the number 2 for Limit point, make sure that the circle *from both sides* is selected and then click on OK.

$$\frac{x^2 + 2}{x + 2} \xrightarrow{\text{limit}} \frac{3}{2}$$

The *limit* command can also be used and is shown below.

$$limit\left(\frac{x^2 + 2}{x + 2}, x = 2\right)$$

$$\frac{3}{2} \qquad\qquad (2.4)$$

Example 2: Graph the function $f(x) = \begin{cases} 3x + 2 & \text{if } -4 \le x < 1 \\ 4x^2 + 1 & \text{if } 1 \le x \le 2 \end{cases}$. Find $\lim_{x \to 1} f(x)$ if it exists. Is f continuous at $x = 1$? Give reasons for your answer.

Solution: Note that the domain of f is the interval $[-4, 2]$. We will begin by entering f as a *Maple* function using the piecewise command and graphing it.

$$f := x \rightarrow piecewise\left(-4 \le x \textbf{ and } x < 1, 3 \cdot x + 2, 1 \le x \textbf{ and } x \le 2, 4 \cdot x^2 + 1, undefined\right)$$

$$x \rightarrow piecewise\left(-4 \le x \textbf{ and } x < 1, 3 \, x + 2, 1 \le x \textbf{ and } x \le 2, 4 \, x^2 + 1, undefined\right) \qquad \textbf{(2.5)}$$

$$plot\left(f(x), x = -4 .. 2, y = -11 .. 18, discont = true\right)$$

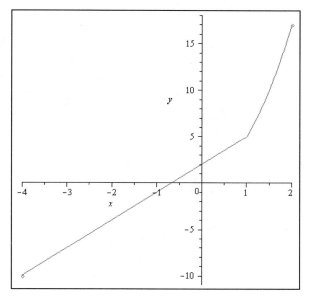

From the graph above it seems that $\lim_{x \to 1} f(x)$ exists. We will now use *Maple* commands to find $\lim_{x \to 1} f(x)$, if it exists.

$$\lim_{x \to 1} f(x)$$

$$5 \qquad \textbf{(2.6)}$$

Hence $\lim_{x \to 1} f(x) = 5$. Furthermore, as

$$f(1)$$

$$5 \qquad \textbf{(2.7)}$$

we see that $\lim_{x \to 1} f(x) = 5 = f(1)$. Hence f continuous at $x = 1$.

Remark: Do not forget to add *undefined* in the command for the piecewise function. If this is not added, then the default value for the function evaluated at any point that is not in its domain is 0. See the remark after Example 10 in Chapter 1.

Example 3: Graph the function $f(x) = \begin{cases} 4x^2 - 1 & \textbf{if } -3 \leq x \leq -1 \\ -7x - 9 & \textbf{if } -1 < x \leq 2 \end{cases}$. Find $\lim\limits_{x \to -1} f(x)$ if it exists. Is f continuous at $x = -1$? Give reasons for your answer.

Solution: Note that the domain of f is $[-3, 2]$. We will begin by entering f as a *Maple* function using the piecewise command and graphing it.

$f := x \to piecewise\left(-3 \leq x \text{ and } x \leq -1, 4 \cdot x^2 - 1, -1 < x \text{ and } x < 2, -7 \cdot x - 9, undefined\right)$

$\quad x \to piecewise\left(-3 \leq x \text{ and } x \leq -1, 4\, x^2 - 1, -1 < x \text{ and } x < 2, -7\, x - 9, undefined\right)$ **(2.8)**

$plot\left(f(x), x = -3 .. 2, y = -23 .. 35, discont = true\right)$

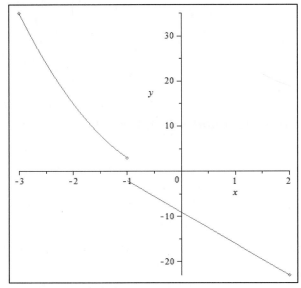

From the graph above we see that $\lim\limits_{x \to -1} f(x)$ does not exist. We will now verify this using *Maple* commands by first finding the left limit $\lim\limits_{x \to -1-} f(x)$ and the right limit $\lim\limits_{x \to -1+} f(x)$, if they exist.

$\lim\limits_{x \to -1-} f(x)$

$$3 \hspace{6cm} \textbf{(2.9)}$$

$\lim\limits_{x \to -1+} f(x)$

$$-2 \hspace{6cm} \textbf{(2.10)}$$

Hence the left limit $\lim\limits_{x \to -1-} f(x) = 3$ and the right limit $\lim\limits_{x \to -1+} f(x) = -2$. Since these values are not the same $\lim\limits_{x \to -1} f(x)$ does not exist. This can be shown using *Maple* commands.

$$\lim_{x \to -1} f(x)$$

undefined

(2.11)

Since $\lim_{x \to -1} f(x)$ does not exist, f is not continuous at $x = -1$.

Remark: The limit command can also be used as shown below.

$limit\left(f(x), x = -1, left\right)$

3

(2.12)

$limit\left(f(x), x = -1, right\right)$

−2

(2.13)

$limit\left(f(x), x = -1\right)$

undefined

(2.14)

Alternately, the right click feature can be used to calculate the limit. After typing the expression $piecewise\left(-3 \le x \text{ and } x \le -1, 4 \cdot x^2 - 1, -1 < x \text{ and } x < 2, -7 \cdot x - 9, undefined\right)$, right click on it. Select *Limit*, type in the number -1 for *Limit point*, make sure that the circle *from the left* is first selected and then click on OK.

$$piecewise\left(-3 \le x \text{ and } x \le -1, 4 \cdot x^2 - 1, -1 < x \text{ and } x < 2, -7 \cdot x - 9, undefined\right) \xrightarrow{\text{limit}} 3$$

To find the right limit repeat the procedure and make sure that the circle *from the right* is selected.

$$piecewise\left(-3 \le x \text{ and } x \le -1, 4 \cdot x^2 - 1, -1 < x \text{ and } x < 2, -7 \cdot x - 9, undefined\right) \xrightarrow{\text{limit}} -2$$

If the circle *from both sides* is selected then we get the following answer.

$$piecewise\left(-3 \le x \text{ and } x \le -1, 4 \cdot x^2 - 1, -1 < x \text{ and } x < 2, -7 \cdot x - 9, undefined\right) \xrightarrow{\text{limit}}$$
undefined

Note: The graph of a continuous function has no "holes" or "jumps". Care must be taken while using *Maple* since "holes" are not automatically visible on the graph unless additional commands are given.

Example 4: Let $f(x) = \dfrac{x^2 + 2x - 8}{x - 2}$. Find the left-hand limit $\lim_{x \to 2^-} f(x)$ and the right-hand limit $\lim_{x \to 2^+} f(x)$. Use these to find $\lim_{x \to 2} f(x)$ if it exists. Give reasons for your answer. Graph f to illustrate your answer.

Solution: Note that the domain of f is the set of all real numbers except 2. Since 2 is not in the domain of the function f, the function is not continuous at 2. We will begin by entering f as a *Maple* function.

$$f := x \rightarrow \frac{x^2 + 2 \cdot x - 8}{x - 2}$$

$$x \rightarrow \frac{x^2 + 2\,x - 8}{x - 2} \tag{2.15}$$

We will find the left-hand and right-hand limits.

$$\lim_{x \to 2-} f(x)$$

$$6 \tag{2.16}$$

Hence the left-hand limit is $\lim\limits_{x \to 2-} f(x) = 6$.

$$\lim_{x \to 2+} f(x)$$

$$6 \tag{2.17}$$

Thus the right-hand limit is $\lim\limits_{x \to 2+} f(x) = 6$. Since $\lim\limits_{x \to 2-} f(x) = 6 = \lim\limits_{x \to 2+} f(x)$, $\lim\limits_{x \to 2} f(x)$ exists and $\lim\limits_{x \to 2} f(x) = 6$. This is seen below using *Maple* commands.

$$\lim_{x \to 2} f(x)$$

$$6 \tag{2.18}$$

Notice what happens when we evaluate $f(2)$.

$f(2)$
Error, (in f) numeric exception: division by zero

Therefore, f has a removable discontinuity at 2. This will be illustrated in the following graph.

$$plot\left(f(x), x = -5..3, y = -1..7, discont = [showremovable]\right)$$

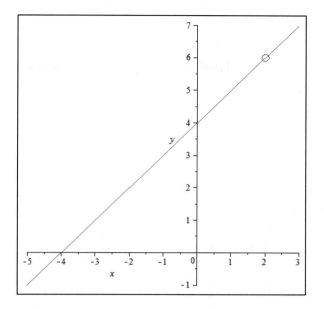

Remark: The *discont* command in *Maple* can be used to find possible real numbers where discontinuities may occur.

$$discont\big(f(x),x\big)$$

$$\{2\} \qquad\qquad\qquad \textbf{(2.19)}$$

We can use the same command to find the possible discontinuities of functions such as the round function and the secant function. Note that _*Z1* ~

$$discont\big(\mathrm{round}(x),x\big)$$

$$\{\_Z1\sim\} \qquad\qquad\qquad \textbf{(2.20)}$$

$$discont\big(\sec(x),x\big)$$

$$\left\{\pi\_Z2\sim+\frac{1}{2}\pi\right\} \qquad\qquad\qquad \textbf{(2.21)}$$

Note that as _*Zn* ~ denotes integers, the possible discontinuities of the round function are the integers and the possible discontinuities of the secant function are the numbers of the form $\pi x+\dfrac{\pi}{2}$ where x is any integer.

Step-by-step Limits

We will solve limit problems by using *Maple* commands intended for rules for limits. We will execute one step of calculation for each rule. To do so, we will activate the $with\big(Student[Calculus1]\big)$ package which will enable the *Maple Rule* routines. We will also need to set $infolevel\big[Student[Calculus1]\big] := 1$, for *Maple* to provide useful messages on the problem that needs to be solved. Each *Maple* routine involves two parameters, the rule to be applied in the step and the expression it is applied to. We will also use the inert form of the limit which is the *Limit* command. This command will return the limit unevaluated.

We will execute the command *restart* at the beginning of each problem.

Below is a list of different rules of limits recognized by *Maple*. We will illustrate these rules, when performing a step-by-step approach, with numerous examples throughout this chapter.

<u>Rules for evaluating limits using a step-by-step approach</u>

The tables below give the rules for limits and the *Maple* commands used for step-by-step evaluation of limits.

1.   Let k be a real number and let f and g be two functions such that $\lim\limits_{x \to c} f(x)$ and

    $\lim\limits_{x \to c} g(x)$ exist. Let k (a constant) be any real number. Then

Rule for limits	*Maple* rule for evaluating limits using a step-by-step approach
$\lim\limits_{x \to c}\big[k \cdot f(x)\big] = k \cdot \lim\limits_{x \to c} f(x)$	*constantmultiple*
$\lim\limits_{x \to c}\big[f(x) + g(x)\big] = \lim\limits_{x \to c} f(x) + \lim\limits_{x \to c} g(x)$	*sum*
$\lim\limits_{x \to c}\big[f(x) - g(x)\big] = \lim\limits_{x \to c} f(x) - \lim\limits_{x \to c} g(x)$	*difference*
$\lim\limits_{x \to c}\big[f(x) g(x)\big] = \lim\limits_{x \to c} f(x) \cdot \lim\limits_{x \to c} g(x)$	*product*
$\lim\limits_{x \to c}\big[f(x)^{g(x)}\big] = \Big[\lim\limits_{x \to c} f(x)\Big]^{\big[\lim\limits_{x \to c} g(x)\big]}$	*power*
$\lim\limits_{x \to c}\left[\dfrac{f(x)}{g(x)}\right] = \dfrac{\lim\limits_{x \to c} f(x)}{\lim\limits_{x \to c} g(x)}$	*quotient*

2.   The table below gives the limits of certain functions. Let c be a real number in the domain of the function that is considered. Let f be a function such that at $\lim\limits_{x \to c} f(x)$ exists. Let a be a positive real number such that $a \neq 1$.

Rule for limits	*Maple* rule for evaluating limits using a step-by-step approach
$\lim\limits_{x\to c} k = k$	*constant*
$\lim\limits_{x\to c} x = c$	*identity*
$\lim\limits_{x\to c} \sin(x) = \sin(c)$	sin
$\lim\limits_{x\to c} \cos(x) = \cos(c)$	cos
$\lim\limits_{x\to c} \tan(x) = \tan(c)$	tan
$\lim\limits_{x\to c} \cot(x) = \cot(c)$	cot
$\lim\limits_{x\to c} \sec(x) = \sec(c)$	sec
$\lim\limits_{x\to c} \csc(x) = \csc(c)$	csc
$\lim\limits_{x\to c} e^{f(x)} = e^{\lim\limits_{x\to c} f(x)}$	exp
$\lim\limits_{x\to c} a^{f(x)} = a^{\lim\limits_{x\to c} f(x)}$	*power*
$\lim\limits_{x\to c} \ln(x) = \ln(c)$	ln
$\lim\limits_{x\to c} \sin^{-1}(x) = \sin^{-1}(c)$	arcsin
$\lim\limits_{x\to c} \cos^{-1}(x) = \cos^{-1}(c)$	arccos
$\lim\limits_{x\to c} \tan^{-1}(x) = \tan^{-1}(c)$	arctan
$\lim\limits_{x\to c} \cot^{-1}(x) = \cot^{-1}(c)$	arccot
$\lim\limits_{x\to c} \sec^{-1}(x) = \sec^{-1}(c)$	arcsec
$\lim\limits_{x\to c} \csc^{-1}(x) = \csc^{-1}(c)$	arccsc

Remark: Note that the *power* rule is not used to find $\lim\limits_{x\to c} e^{f(x)} = e^{\lim\limits_{x\to c} f(x)}$. In this case one uses the *exp* rule. The *power* rule used to find limits is different from the *power* rule stated in chapters 3 and 6.

Example 5: Use the step-by-step procedure in *Maple* to evaluate the limit $\lim\limits_{x\to 1}\left(3x - x^2\right)$.

Solution: We will begin by executing the *restart* command. Activate the $with\left(Student\left[Calculus1\right]\right)$ package and set $infolevel\left[Student\left[Calculus1\right]\right] := 1$. We will also use the inert form of the limit which is the *Limit* command. This command will return the limit unevaluated.

restart

$with\big(Student[Calculus1]\big):$

$infolevel\big[Student[Calculus1]\big]:=1:$

$Rule[\;]\big(Limit\big(3\cdot x-x^2,x=1\big)\big)$

```
Creating problem #1
```

$$\lim_{x\to 1}\big(3\,x-x^2\big)=\lim_{x\to 1}\big(3\,x-x^2\big) \tag{2.22}$$

We use the *difference* command to find the limit of the difference.

$Rule[difference]\big((2.22)\big)$

$$\lim_{x\to 1}\big(3\,x-x^2\big)=\lim_{x\to 1}3\,x-\Big(\lim_{x\to 1}x^2\Big) \tag{2.23}$$

Use the *constantmultiple* command to factor 3 from the first limit.

$Rule[constantmultiple]\big((2.23)\big)$

$$\lim_{x\to 1}\big(3\,x-x^2\big)=3\Big(\lim_{x\to 1}x\Big)-\Big(\lim_{x\to 1}x^2\Big) \tag{2.24}$$

The *identity* rule will be used to find $\lim\limits_{x\to 1}x$. Note that $\lim\limits_{x\to 1}x=1$.

$Rule[identity]\big((2.24)\big)$

$$\lim_{x\to 1}\big(3\,x-x^2\big)=3-\Big(\lim_{x\to 1}x^2\Big) \tag{2.25}$$

The *power* rule is now used to find $\lim\limits_{x\to 1}x^2$.

$Rule[power]\big((2.25)\big)$

$$\lim_{x\to 1}\big(3\,x-x^2\big)=3-\Big(\lim_{x\to 1}x\Big)^2 \tag{2.26}$$

The *identity* rule is now used to find $\lim\limits_{x\to 1}x$.

$Rule[identity]\big((2.26)\big)$

$$\lim_{x\to 1}\big(3\,x-x^2\big)=2 \tag{2.27}$$

Hence $\lim\limits_{x\to 1}\left(3x-x^2\right)=2$. Finally we can show the step-by-step computation of this limit using the *ShowSteps* ( ) command. Note that description of the steps that occur in blue on the right side are visible in versions 13 and 14 but not in the earlier versions. However the *ShowSteps* ( ) command is available in the earlier versions.

ShowSteps ( )

$$\lim_{x\to 1}\left(3\,x-x^2\right)$$

$$
\begin{aligned}
&= \lim_{x\to 1} 3\,x - \lim_{x\to 1} x^2 && [\textit{difference}] \\
&= 3\lim_{x\to 1} x - \lim_{x\to 1} x^2 && [\textit{constantmultiple}] \\
&= 3 - \lim_{x\to 1} x^2 && [\textit{identity}] && \textbf{(2.28)} \\
&= 3 - \left(\lim_{x\to 1} x\right)^2 && [\textit{power}] \\
&= 2 && [\textit{identity}]
\end{aligned}
$$

Remarks: The following command shows the algorithm used by *Maple* while solving a limit problem. This feature is available in version 13 and 14 but not in the earlier versions.

ShowSolution $\left(Limit\left(3\cdot x-x^2,x=1\right)\right)$

Creating problem #2

$$\lim_{x\to 1}\left(3\,x-x^2\right)$$

$$
\begin{aligned}
&= \lim_{x\to 1} 3\,x + \lim_{x\to 1} -x^2 && [\textit{sum}] \\
&= 3\lim_{x\to 1} x + \lim_{x\to 1} -x^2 && [\textit{constantmultiple}] \\
&= 3 + \lim_{x\to 1} -x^2 && [\textit{identity}] && \textbf{(2.29)} \\
&= 3 - \lim_{x\to 1} x^2 && [\textit{constantmultiple}] \\
&= 3 - \left(\lim_{x\to 1} x\right)^2 && [\textit{power}] \\
&= 2 && [\textit{identity}]
\end{aligned}
$$

Note that the algorithm we used to find the limit in example 5 was one step shorter than the one used by *Maple* (since we used the *difference* command rather than the *sum* command). Thus we can come up with a "more efficient" algorithm. This difference is more evident in chapter 7.

Example 6: Use the step-by-step procedure in *Maple* to evaluate the limit $\lim\limits_{x \to 0}\left(3x^2 - 4\sec(x) - 5\right)$.

Solution: We will begin by executing the *restart* command. Do not forget to activate the $with\left(Student\left[Calculus1\right]\right)$ package and set $infolevel\left[Student\left[Calculus1\right]\right] := 1$.

restart

$with\left(Student\left[Calculus1\right]\right):$

$infolevel\left[Student\left[Calculus1\right]\right] := 1:$

$Rule\left[\ \right]\left(Limit\left(3 \cdot x^2 - 4 \cdot \sec(x) - 5, x = 0\right)\right)$

`Creating problem #1`

$$\lim_{x \to 0}\left(3\,x^2 - 4\sec(x) - 5\right) = \lim_{x \to 0}\left(3\,x^2 - 4\sec(x) - 5\right) \tag{2.30}$$

We use the *difference* command to find the limit of the difference.

$Rule\left[difference\right]\left((2.30)\right)$

$$\lim_{x \to 0}\left(3\,x^2 - 4\sec(x) - 5\right) = \lim_{x \to 0}3\,x^2 - \lim_{x \to 0}\left(4\sec(x)\right) + \lim_{x \to 0}(-5) \tag{2.31}$$

Note that the *difference* command here turns the difference into a sum for the constant term. Use the *constantmultiple* command to factor 3 from the first limit.

$Rule\left[constantmultiple\right]\left((2.31)\right)$

$$\lim_{x \to 0}\left(3\,x^2 - 4\sec(x) - 5\right) = 3\left(\lim_{x \to 0}x^2\right) - \lim_{x \to 0}\left(4\sec(x)\right) + \lim_{x \to 0}(-5) \tag{2.32}$$

The *power* rule will be used to find $\lim\limits_{x \to 0}x^2$.

$Rule\left[power\right]\left((2.32)\right)$

$$\lim_{x \to 0}\left(3\,x^2 - 4\sec(x) - 5\right) = 3\left(\lim_{x \to 0}x\right)^2 - \lim_{x \to 0}\left(4\sec(x)\right) + \lim_{x \to 0}(-5) \tag{2.33}$$

The *identity* rule is now used to find $\lim\limits_{x \to 0}x$.

$Rule\left[identity\right]\left((2.33)\right)$

$$\lim_{x \to 0}\left(3\,x^2 - 4\sec(x) - 5\right) = -\lim_{x \to 0}\left(4\sec(x)\right) + \lim_{x \to 0}(-5) \tag{2.34}$$

Use the *constantmultiple* command once more to factor 4 from the first limit.

$Rule[constantmultiple]((\textbf{2.34}))$

$$\lim_{x\to0}\left(3\,x^2-4\sec\left(x\right)-5\right)=-4\left(\lim_{x\to0}\sec\left(x\right)\right)+\lim_{x\to0}\left(-5\right) \tag{2.35}$$

The *sec* rule will be used to find $\lim_{x\to0}\sec\left(x\right)$.

$Rule[sec]((\textbf{2.35}))$

$$\lim_{x\to0}\left(3\,x^2-4\sec\left(x\right)-5\right)=-4+\lim_{x\to0}\left(-5\right) \tag{2.36}$$

The *constant* rule is now used to find $\lim_{x\to0}\left(-5\right)$.

$Rule[constant]((\textbf{2.36}))$

$$\lim_{x\to0}\left(3\,x^2-4\sec\left(x\right)-5\right)=-9 \tag{2.37}$$

Therefore, $\lim_{x\to0}\left(3x^2-4\sec\left(x\right)-5\right)=-9$.

Example 7: Use the step-by-step procedure in *Maple* to evaluate the limit $\lim_{x\to-\frac{\pi}{6}}\left(e^{\sin(x)+1}+8\cot^{3\cdot\csc(x)-4}\left(x\right)-1\right)$.

Solution: We will begin by executing the *restart* command. Do not forget to activate the $with\left(Student\left[Calculus1\right]\right)$ package and set $infolevel\left[Student\left[Calculus1\right]\right]:=1$.

$restart$
$with\left(Student\left[Calculus1\right]\right):$
$infolevel\left[Student\left[Calculus1\right]\right]:=1:$
$Rule[\ ]\left(Limit\left(e^{\sin(x)+1}+8\cdot\cot^{3\cdot\csc(x)-4}\left(x\right)-1,x=-\frac{\pi}{6}\right)\right)$

```
Creating problem #1
```

$$\lim_{x\to-\frac{1}{6}\pi}\left(e^{\sin(x)+1}+8\cot^{3\csc(x)-4}\left(x\right)-1\right)=\lim_{x\to-\frac{1}{6}\pi}\left(e^{\sin(x)+1}+8\cot^{3\csc(x)-4}\left(x\right)-1\right) \tag{2.38}$$

We use the *sum* command to find the limit of the sum.

Rule[*sum*]((**2.38**))

$$\lim_{x \to -\frac{1}{6}\pi} \left(e^{\sin(x)+1} + 8 \cot^{3\csc(x)-4}(x) - 1 \right) = \lim_{x \to -\frac{1}{6}\pi} e^{\sin(x)+1} + \lim_{x \to -\frac{1}{6}\pi} 8 \cot^{3\csc(x)-4}(x) + \lim_{x \to -\frac{1}{6}\pi} (-1)\,(\textbf{2.39})$$

Note here, as in the case of example 6, the *sum* command here turns the difference into a sum for the constant term. Use the *exp* command to find the limit of the first limit.

Rule[*exp*]((**2.39**))

$$\lim_{x \to -\frac{1}{6}\pi} \left(e^{\sin(x)+1} + 8 \cot^{3\csc(x)-4}(x) - 1 \right) = e^{\lim\limits_{x \to -\frac{1}{6}\pi}(\sin(x)+1)} + \lim_{x \to -\frac{1}{6}\pi} 8 \cot^{3\csc(x)-4}(x) + \lim_{x \to -\frac{1}{6}\pi} (-1) \;(\textbf{2.40})$$

The *sum* rule will be used to find the limit of the exponent of the first term.

Rule[*sum*]((**2.40**))

$$\lim_{x \to -\frac{1}{6}\pi} \left(e^{\sin(x)+1} + 8 \cot^{3\csc(x)-4}(x) - 1 \right) = e^{\lim\limits_{x \to -\frac{1}{6}\pi}\sin(x) + \lim\limits_{x \to -\frac{1}{6}\pi} 1} + \lim_{x \to -\frac{1}{6}\pi} 8 \cot^{3\csc(x)-4}(x) + \lim_{x \to -\frac{1}{6}\pi} (-1)$$

$$(\textbf{2.41})$$

The *sin* rule will be used to find the limit $\lim\limits_{x \to -\frac{1}{6}\pi} \sin(x)$.

Rule[*sin*]((**2.41**))

$$\lim_{x \to -\frac{1}{6}\pi} \left(e^{\sin(x)+1} + 8 \cot^{3\csc(x)-4}(x) - 1 \right) = e^{-\frac{1}{2} + \lim\limits_{x \to -\frac{1}{6}\pi} 1} + \lim_{x \to -\frac{1}{6}\pi} 8 \cot^{3\csc(x)-4}(x) + \lim_{x \to -\frac{1}{6}\pi} (-1) \quad (\textbf{2.42})$$

The *constant* rule is used to find $\lim\limits_{x \to -\frac{1}{6}\pi} 1$.

Rule[*constant*]((**2.42**))

$$\lim_{x \to -\frac{1}{6}\pi} \left(e^{\sin(x)+1} + 8 \cot^{3\csc(x)-4}(x) - 1 \right) = e^{\frac{1}{2}} + \lim_{x \to -\frac{1}{6}\pi} 8 \cot^{3\csc(x)-4}(x) + \lim_{x \to -\frac{1}{6}\pi} (-1) \;(\textbf{2.43})$$

The *constantmultiple* rule is used to find the limit of the second term.

Rule[*constantmultiple*]((**2.43**))

$$\lim_{x \to -\frac{1}{6}\pi} \left(e^{\sin(x)+1} + 8 \cot^{3\csc(x)-4}(x) - 1 \right) = e^{\frac{1}{2}} + 8 \left(\lim_{x \to -\frac{1}{6}\pi} \cot^{3\csc(x)-4}(x) \right) + \lim_{x \to -\frac{1}{6}\pi} (-1) \qquad (\textbf{2.44})$$

The *power* rule is used to find $\lim\limits_{x \to -\frac{1}{6}\pi} \cot^{3\csc(x)-4}(x)$.

Rule$[power]((\textbf{2.44}))$

$$\lim_{x \to -\frac{1}{6}\pi}\left(e^{\sin(x)+1} + 8\cot^{3\csc(x)-4}(x) - 1\right) = e^{\frac{1}{2}} + 8\left(\lim_{x \to -\frac{1}{6}\pi}\cot(x)\right)^{\lim\limits_{x \to -\frac{1}{6}\pi}(3\csc(x)-4)} + \lim_{x \to -\frac{1}{6}\pi}(-1) \quad \textbf{(2.45)}$$

The *difference* rule is used to find the limit of the exponent of the second term.

Rule$[difference]((\textbf{2.45}))$

$$\lim_{x \to -\frac{1}{6}\pi}\left(e^{\sin(x)+1} + 8\cot^{3\csc(x)-4}(x) - 1\right) = e^{\frac{1}{2}} + 8\left(\lim_{x \to -\frac{1}{6}\pi}\cot(x)\right)^{\lim\limits_{x \to -\frac{1}{6}\pi}3\csc(x) + \lim\limits_{x \to -\frac{1}{6}\pi}(-4)} + \lim_{x \to -\frac{1}{6}\pi}(-1) \quad \textbf{(2.46)}$$

The *constantmultiple* rule is used to find $\lim\limits_{x \to -\frac{1}{6}\pi} 3\csc(x)$.

Rule$[constantmultiple]((\textbf{2.46}))$

$$\lim_{x \to -\frac{1}{6}\pi}\left(e^{\sin(x)+1} + 6\cot^{3\csc(x)-4}(x) - 1\right) = e^{\frac{1}{2}} + 6\left(\lim_{x \to -\frac{1}{6}\pi}\cot(x)\right)^{3\left(\lim\limits_{x \to -\frac{1}{6}\pi}\csc(x)\right) + \lim\limits_{x \to -\frac{1}{6}\pi}(-4)} + \lim_{x \to -\frac{1}{6}\pi}(-1) \quad \textbf{(2.47)}$$

The *csc* rule is used to find $\lim\limits_{x \to -\frac{1}{6}\pi} \csc(x)$.

Rule$[csc]((\textbf{2.47}))$

$$\lim_{x \to -\frac{1}{6}\pi}\left(e^{\sin(x)+1} + 8\cot^{3\csc(x)-4}(x) - 1\right) = e^{\frac{1}{2}} + 8\left(\lim_{x \to -\frac{1}{6}\pi}\cot(x)\right)^{-6 + \lim\limits_{x \to -\frac{1}{6}\pi}(-4)} + \lim_{x \to -\frac{1}{6}\pi}(-1) \quad \textbf{(2.48)}$$

The *constant* rule is used to find $\lim\limits_{x \to -\frac{1}{6}\pi}(-4)$.

Rule$[constant]((\textbf{2.48}))$

$$\lim_{x \to -\frac{1}{6}\pi}\left(e^{\sin(x)+1} + 8\cot^{3\csc(x)-4}(x) - 1\right) = e^{\frac{1}{2}} + \frac{8}{\left(\lim\limits_{x \to -\frac{1}{6}\pi}\cot(x)\right)^{10}} + \lim_{x \to -\frac{1}{6}\pi}(-1) \quad \textbf{(2.49)}$$

Chapter 2 Limits

The *cot* rule is used to find $\lim\limits_{x \to -\frac{1}{6}\pi} \cot(x)$.

Rule[cot]((**2.49**))

$$\lim_{x \to -\frac{1}{6}\pi} \left(e^{\sin(x)+1} + 8\cot^{3\csc(x)-4}(x) - 1\right) = e^{\frac{1}{2}} + \frac{8}{243} + \lim_{x \to -\frac{1}{6}\pi}(-1) \qquad (2.50)$$

Note that the *cot* rule is used in the step above to find $\left(\lim\limits_{x \to -\frac{1}{6}\pi} \cot(x)\right)^{10}$. The *constant* rule is used to find $\lim\limits_{x \to -\frac{1}{6}\pi}(-4)$.

Rule[constant]((**2.50**))

$$\lim_{x \to -\frac{1}{6}\pi} \left(e^{\sin(x)+1} + 8\cot^{3\csc(x)-4}(x) - 1\right) = e^{\frac{1}{2}} - \frac{235}{243} \qquad (2.51)$$

Therefore, $\lim\limits_{x \to -\frac{1}{6}\pi} \left(e^{\sin(x)+1} + 8\cot^{3\csc(x)-4}(x) - 1\right) = e^{\frac{1}{2}} - \frac{235}{243}$.

Intermediate Value Theorem

***Intermediate Value Theorem*:** Let f be a continuous function on a closed interval $[a,b]$. Let C be a real number between $f(a)$ and $f(b)$. Then there is a point c in the open interval (a,b) such that $f(c) = C$.

Example 8: Let $f(x) = x\cos(x)$. Can the Intermediate Value Theorem be used to find a number c in the open interval $(-4,3)$ such that $f(c) = 0$? Give reasons for your answer.

Solution: Since both x and $\cos(x)$ are continuous functions of x, so is their product. Hence f is a continuous function. We will begin by entering f as a *Maple* function and graphing it in the interval $[-4,3]$.

$f(x) := x \to x \cdot \cos(x)$

$$x \to x\cos(x) \qquad (2.52)$$

$plot(f(x), x = -4..3)$

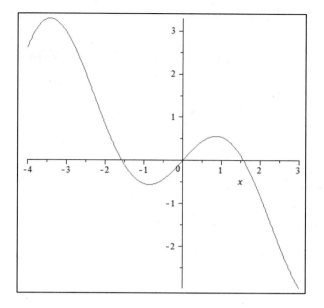

From the graph above we see that $f(-4) > 0$ and $f(3) < 0$. Hence by the Intermediate Value Theorem we can find a number c in the open interval $(-4,3)$ such that $f(c) = 0$. In fact from the graph above there are three values of c in $(-4,3)$ that satisfy the condition $f(c) = 0$. We will find these values by using the Roots command. In order to do that we will need to activate the $with(Student[Calculus1])$ package.

$with(Student[Calculus1])$:
$Roots(f(x) = 0, x = -4..3)$

$$\left[-\frac{1}{2}\pi, 0, \frac{1}{2}\pi \right]$$

(2.53)

Hence the solutions to $f(c) = 0$ in the interval $(-4,3)$ are $-\frac{1}{2}\pi$, 0, and $\frac{1}{2}\pi$. To obtain a floating point approximation we add the *numeric* option in the *Roots* command.

$Roots(f(x) = 0, x = -4..3, numeric)$

$$[-1.570796327, 0., 1.570796327]$$

(2.54)

Remarks:
1.   One can use the *fsolve* command to find solutions.

$$fsolve(f(x) = 0, x = -4..3)$$

$$0.$$

(2.55)

This gives only one solution. However if we narrow our interval of interest we can get all the solutions as shown below.

$$fsolve(f(x)=0, x=-4..-1)$$

$$-1.570796327 \tag{2.56}$$

$$fsolve(f(x)=0, x=-1..1)$$

$$0. \tag{2.57}$$

$$fsolve(f(x)=0, x=1..3)$$

$$1.570796327 \tag{2.58}$$

2. From the graph above we also see that $f(-4)>2$ and $f(3)<2$. Hence by the Intermediate Value Theorem there is at least one value of c in the open interval $(-4,3)$ such that $f(c)=2$. We can use the Roots command to find this value.

$$Roots(f(x)=2, x=-4..3)$$

$$[\;] \tag{2.59}$$

This does not get an exact value. However, we can obtain a floating point approximation using the numeric option as shown below.

$$Roots(f(x)=2, x=-4..3, numeric)$$

$$[-2.498755763] \tag{2.60}$$

Arrays and Limits

Definition of a limit of a sequence: Let $\{a_n\}$ be a sequence. Let a be a real number. We say that the limit of a_n as n approaches infinity is a, denoted by $\lim_{n\to\infty} a_n = a$, if for every $\varepsilon > 0$ there is a corresponding positive integer N such that if $n \ge N$ then $|a_n - a| < \varepsilon$. In this case we say that the sequence $\{a_n\}$ *converges to* a.

We will revisit sequences and limits in chapter 10. The example below shows applications of using arrays to find approximate values of limits. It gives a procedure to write an array of values for $\dfrac{n}{n+1}$ and use this to find the direction of the convergence of $\lim_{n\to\infty} \dfrac{n}{n+1}$.

Example 9: Show that the sequence $\dfrac{1}{2}, \dfrac{2}{3}, \dfrac{3}{4}, \ldots$ converges to 1. Construct an array for $\left\{\dfrac{n}{n+1}\right\}$ using the values of n from 101 till 120 to indicate the direction of convergence of the sequence $\left\{\dfrac{n}{n+1}\right\}$.

Solution: We will use *Maple* to find $\lim\limits_{n\to\infty}\dfrac{n}{n+1}$.

$$\lim_{n\to\infty}\frac{n}{n+1}$$

$$1 \hspace{8cm} \textbf{(2.61)}$$

We will now create an array or table of values for the sequence $\left\{\dfrac{n}{n+1}\right\}$ for the values of n from 101 till 120. The procedure below creates an array of 21 rows and 3 columns.

$A := array(1..21, 1..3)$

$A[1,1] := {}'n'\!: \;\; A[1,2] := {}'\dfrac{n}{n+1}'\!: \;\; A[1,3] := {}'evalf\left(\dfrac{n}{n+1}\right)'\!:$

for $\quad n \quad$ **from** $\quad 2 \quad$ **to** $\quad 21 \quad$ **do** $\quad A[n,1] := n+99 : \quad A[n,2] := \dfrac{A[n,1]}{A[n,1]+1} :$

$A[n,3] := evalf\left(\dfrac{A[n,1]}{A[n,1]+1}\right)$ **end do:**

$print(A)$

$$
\begin{bmatrix}
n & \dfrac{n}{n+1} & evalf\left(\dfrac{n}{n+1}\right) \\
\end{bmatrix}
$$

n	$\dfrac{n}{n+1}$	$evalf\left(\dfrac{n}{n+1}\right)$
101	$\dfrac{101}{102}$	0.99019607840
102	$\dfrac{102}{103}$	0.9902912621
103	$\dfrac{103}{104}$	0.9903846154
104	$\dfrac{104}{105}$	0.9904761905
105	$\dfrac{105}{106}$	0.9905660377
106	$\dfrac{106}{107}$	0.9906452056
107	$\dfrac{107}{108}$	0.9907407407
108	$\dfrac{108}{109}$	0.9908256881
109	$\dfrac{109}{110}$	0.99009090909
110	$\dfrac{110}{111}$	0.9909909910
111	$\dfrac{111}{112}$	0.9910714286
112	$\dfrac{112}{113}$	0.9911504425
113	$\dfrac{113}{114}$	0.9912280702
114	$\dfrac{114}{115}$	0.9913043478
115	$\dfrac{115}{116}$	0.9913793103
116	$\dfrac{116}{117}$	0.9914529915
117	$\dfrac{117}{118}$	0.9915254237
118	$\dfrac{118}{119}$	0.9915966387
129	$\dfrac{119}{120}$	0.9916666667
120	$\dfrac{120}{121}$	0.9917355372

(2.62)

Since the values of n are from 101 through 120 and the first row consists of labels for the columns, $A[n,1] := n + 99$. From the answers in the table above it looks as though the sequence converges to 1.

Remarks: One should exercise some caution while using sequences and arrays to determine the value of a limit. We recall that the limit $\lim_{x \to a} f(x) = L$ if $\lim_{n \to \infty} f(x_n) = L$ for every sequence $\{x_n\}$ such that $\lim_{n \to \infty} x_n = a$. Refer to exercise 30.

Example 10: Graph the function $f(x) = \dfrac{\sin(x)}{x}$. Find the domain of f. Construct an array of 21 rows and 5 columns in such a way that the first row consists of the labels of the columns, the first column is the value of n (where n takes values from 1 through 20), the second column is the floating point approximation of $-\dfrac{1}{n^3}$, the third column is the floating point approximation of $f\left(-\dfrac{1}{n^3}\right)$, the fourth column is the floating point approximation of $\dfrac{1}{n^3}$, and the fifth column is the floating point approximation of $f\left(\dfrac{1}{n^3}\right)$. Using the values in the array and the graph of f, can you estimate the values of the left-hand limit $\lim\limits_{x \to 0-} f(x)$ and the right-hand limit $\lim\limits_{x \to 0+} f(x)$, if they exist? Use these numbers to determine an estimate of $\lim\limits_{x \to 0} f(x)$, if it exists. Give reasons for your answer. Verify your answers using *Maple*.

Solution: We will enter f as a *Maple* function. Observe that the domain of f is the set of all real numbers except 0.

$$f(x) := x \to \frac{\sin(x)}{x}$$

$$x \to \frac{\sin(x)}{x} \qquad\qquad\qquad (2.63)$$

We will now modify the procedure in example 9 and create a table of values that indicate the direction of convergence. Since the values of n are from 1 through 20, we set the values of $A[n,1] := n - 1$ for the values of n are from 2 through 21.

$$A := array(1..21, 1..5)$$

$$A[1,1] := 'n' : \quad A[1,2] := 'evalf\left(-\frac{1}{n^3}\right)' : \quad A[1,3] := 'evalf\left(f\left(-\frac{1}{n^3}\right)\right)' :$$

$$A[1,4] := 'evalf\left(\frac{1}{n^3}\right)' : \quad A[1,5] := 'evalf\left(f\left(\frac{1}{n^3}\right)\right)' :$$

for n **from** 2 **to** 21 **do** $A[n,1] := n-1$: $\quad A[n,2] := evalf\left(-\dfrac{1}{A[n,1]^3}\right)$:

$A[n,3] := evalf\left(f\left(A[n,2]\right)\right)$: $\quad A[n,4] := evalf\left(\dfrac{1}{A[n,1]^3}\right)$: $\quad A[n,5] := evalf\left(f\left(A[n,4]\right)\right)$

end do:

$print(A)$

n	$evalf\left(-\dfrac{1}{n^3}\right)$	$evalf\left(f\left(-\dfrac{1}{n^3}\right)\right)$	$evalf\left(\dfrac{1}{n^3}\right)$	$evalf\left(f\left(\dfrac{1}{n^3}\right)\right)$
1	−1.	0.8414709848	1.	0.8414709848
2	−0.1250000000	0.9973978672	0.1250000000	0.9973978672
3	−0.03703703704	0.9997713921	0.03703703704	0.9997713921
4	−0.01562500000	0.9999593101	0.01562500000	0.9999593101
5	−0.008000000000	0.9999893334	0.008000000000	0.9999893334
6	−0.004629629630	0.9999964278	0.004629629630	0.9999964278
7	−0.002915451895	0.9999985834	0.002915451895	0.9999985834
8	−0.001953125000	0.9999993641	0.001953125000	0.9999993641
9	−0.001371742112	0.9999996865	0.001371742112	0.9999996865
10	−0.001000000000	0.9999998333	0.001000000000	0.9999998333
11	−0.0007513148009	0.9999999059	0.0007513148009	0.9999999059
12	−0.0005787037037	0.9999999442	0.0005787037037	0.9999999442
13	−0.0004551661356	0.9999999655	0.0004551661356	0.9999999655
14	−0.0003644314869	0.9999999778	0.0003644314869	0.9999999778
15	−0.0002962962963	0.9999999855	0.0002962962963	0.9999999855
16	−0.0002441406250	0.9999999902	0.0002441406250	0.9999999902
17	−0.0002035416243	0.9999999931	0.0002035416243	0.9999999931
18	−0.0001714677641	0.9999999953	0.0001714677641	0.9999999953
19	−0.0001457938475	0.9999999966	0.0001457938475	0.9999999966
20	−0.0001250000000	0.9999999976	0.0001250000000	0.9999999976

$$(2.64)$$

From column 3 of the table above and the graph of f, a possible estimate of $\lim\limits_{x \to 0-} f(x)$ is 1 and from column 5, a possible estimate of $\lim\limits_{x \to 0+} f(x)$ is also 1. Hence we predict that $\lim\limits_{x \to 0} f(x)$ exists and an estimated value of this limit is 1. We will now use *Maple* to verify our answers.

$$\lim_{x \to 0-} f(x)$$

$$1 \qquad\qquad (2.65)$$

$$\lim_{x \to 0+} f(x)$$

$$1 \qquad\qquad (2.66)$$

$$\lim_{x \to 0} f(x)$$

$$1 \qquad\qquad (2.67)$$

Therefore, $\lim_{x \to 0} f(x) = 0$. Furthermore, since 0 is not in the domain of f, $x = 0$ is a removable discontinuity of f. We will now graph f to verify this.

$$plot\Big(f(x), x = -10 \cdot \pi .. 10 \cdot \pi, discont = \big[showremovable = \big[color = blue,$$

$$symbol = solidcircle \big] \big] \Big)$$

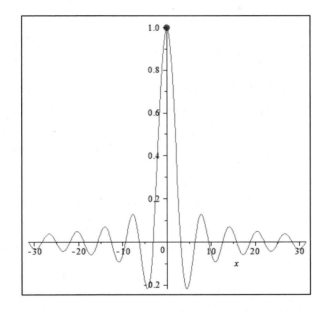

The Precise Definition of a Limit

In the following two examples, the limit obtained using a step by step procedure in Example 5 will now be verified using the precise definition of a limit.

Example 11: From the graph of $f(x) = 3x - x^2$ shown below, find the value of δ such that if $|x - 1| < \delta$ then $|f(x) - 2| = |(3x - x^2) - 2| < 0.1$.

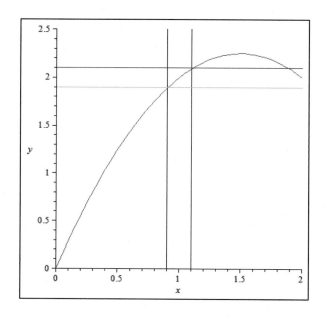

Solution: The inequality $\left|(3x-x^2)-2\right| < 0.1$ can be rewritten as $-0.1 < (3x-x^2)-2 < 0.1$ so that $1.9 < 3x-x^2 < 2.1$ or $1.9 < f(x) < 2.1$. From the graph above we see that if x is any number in the interval $0.9 < x < 1.1$ then $1.9 < f(x) < 2.1$. There may be other intervals, but we are only interested in the interval that is closest to $x=1$. This does not answer our question since, we need to find the value of δ that satisfies the condition that if $-\delta < x-1 < \delta$, then $1.9 < f(x) < 2.1$. Geometrically speaking, this is equivalent to finding a symmetric interval centered about $x=1$, that guarantees that $1.9 < f(x) < 2.1$. Now the distance from $x=1$ to $x=0.9$ is 0.1, the distance from $x=1$ to $x=1.1$ is 0.1, clearly any value of $\delta \leq 0.1$ will guarantee that $1.9 < f(x) < 2.1$. If we compare the above problem with the precise definition of limits, we see that this geometric analysis involves finding the value of δ when $\varepsilon = 0.1$ for the limit problem: $\lim\limits_{x \to 1}(3x-x^2) = 2$. Notice that $c = 1$ and $L = 2$ in this problem.

Remark: The example shown above gives a value for δ using the graph. We will now outline the steps necessary with *Maple* commands to find values of δ for different values of ε. It should be noted that since we want the limit at $x=1$, our analysis is restricted to some subinterval of $[0.5, 1.5]$.

Example 12: Let $f(x) = 3x - x^2$. Find the values of δ for each value of ε using the epsilon-delta definition to find $\lim\limits_{x \to 1} f(x)$: (i) $\varepsilon = 0.1$ (ii) $\varepsilon = 0.01$ (iii) $\varepsilon = 0.001$ (iv) any $\varepsilon > 0$.

Solutions: Since f is a continuous function, $\lim_{x \to 1} f(x) = f(1) = 2$. We first enter f as a *Maple* function, evaluate it when $x = 1$ and plot it. Note that the graph of this function is that of a parabola with vertex $(1.5, 2.25)$.

$f := x \to 3 \cdot x - x^2$

$$x \to 3x - x^2 \qquad\qquad\qquad \textbf{(2.68)}$$

$f(1)$

$$2 \qquad\qquad\qquad \textbf{(2.69)}$$

$plot(f(x), x = -1..4)$

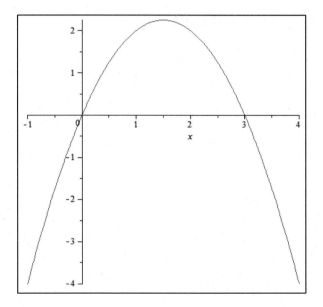

(i) $\varepsilon = 0.1$. Then $2 + \varepsilon = 2.1$ and $2 - \varepsilon = 1.9$ To obtain the value of δ for the value of $\varepsilon = 0.1$ we first solve for $f(x) = 2.1$ and $f(x) = 1.9$ for any x in the interval 0.5 to 1.5 (since we are interested in the limit $\lim_{x \to 1} f(x)$). Note that as $f(1.5) = 2.25$ and $f(0.5) = 1.25$ (as seen below),

$f(1.5)$

$$2.25 \qquad\qquad\qquad \textbf{(2.70)}$$

$f(0.5)$

$$1.25 \qquad\qquad\qquad \textbf{(2.71)}$$

Hence by the Intermediate Value Theorem that there are values c_1 and c_2 in the interval (0.5, 1.5) such that $f(c_1) = 2.1$ and $f(c_2) = 1.9$. We will do this by using the *Roots* command. Do not forget to activate the $with(Student[Calculus1])$ package.

$with(Student[Calculus1]):$
$Roots(f(x)=2.1, x=0.5..1.5)$

$$[1.112701665] \qquad\qquad (2.72)$$

$Roots(f(x)=1.9, x=0.5..1.5)$

$$[0.9083920217] \qquad\qquad (2.73)$$

We will now plot the function for the values of x between 0.5 and 1.5 along with the horizontal lines $y=2.1$ and $y=1.9$ and the vertical lines $x=1.112701665$ and $x=0.9083920217$. In order to plot the vertical lines we will use the *implicitplot* command. But first we must activate the *with(plots)* command.

$with(plots):$
$implicitplot\left(\left[y=3 \cdot x - x^2, y=2.1, y=1.9, x=1.112701665, x=0.9083920217\right]\right.$

$\left. x=0.5..1.5, y=1.25..2.25, color=[red, blue, green, brown, black]\right)$

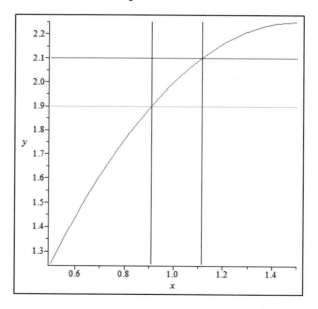

Since,

$|1 - 1.112701665|$

$$0.112701665 \qquad\qquad (2.74)$$

$|1 - 0.9083920217|$

$$0.0916079783 \qquad\qquad (2.75)$$

we choose $\delta \le 0.0916079783$ (as explained in example 11 above)and note that for any value of x such that $|x-1| < \delta$, we see that $|f(x) - f(1)| < 0.1$. A zoomed in graph of this can be made with the vertical line $x = 1$ and a horizontal line $y = 2$ to illustrate the choice of $\delta \le 0.0916079783$.

One can show by algebraic means that $\delta \le \left(\dfrac{\sqrt{35} - 5}{10} \right) \approx 0.0916079783$.

(ii) $\varepsilon = 0.01$. Then $2 + \varepsilon = 2.01$ and $2 - \varepsilon = 1.99$ To obtain the value of δ for the value of $\varepsilon = 0.01$ we first solve for $f(x) = 2.01$ and $f(x) = 1.99$ for any x in the interval 0.5 to 1.5 (since we are interested in the limit $\lim\limits_{x \to 1} f(x)$). Note that as $f(1.5) = 2.25$ and $f(0.5) = 1.25$ (as seen in part(i)), we see by the Intermediate Value Theorem that there are values c_1 and c_2 in the interval (0.5, 1.5) such that $f(c_1) = 2.01$ and $f(c_2) = 1.99$. We will again solve this by using the *Roots* command. We do not need to activate the $with(Student[Calculus1])$ and $with(plots)$ packages as these have been activated in part (i) of this problem.

$$Roots\left(f(x) = 2.01, x = 0.5 .. 1.5\right)$$
$$[1.010102051] \tag{2.76}$$
$$Roots\left(f(x) = 1.99, x = 0.5 .. 1.5\right)$$
$$[0.9900980486] \tag{2.77}$$

We will now plot the function for the values of x between 0.98 and 1.02 along with the horizontal lines $y = 2.01$ and $y = 1.99$ and the vertical lines $x = 1.010102051$ and $x = 0.9900980486$. In order to plot the vertical lines we will use the *implicitplot* command. But first we must activate the $with(plots)$ command.

$$implicitplot\left(\left[y = 3 \cdot x - x^2, y = 2.01, y = 1.99, x = 1.010102051, x = 0.9900980486 \right]\right.$$
$$\left. x = 0.98 .. 1.02, y = 1.98 .. 2.02, color = \left[red, blue, green, brown, black\right]\right)$$

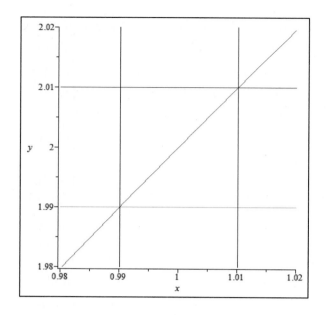

Since,

$$|1 - 1.010102051|$$

$$0.010102051 \qquad\qquad\qquad\textbf{(2.78)}$$

$$|1 - 0.9900980486|$$

$$0.0099019514 \qquad\qquad\qquad\textbf{(2.79)}$$

the distance from 1 to 1.1010102051 and 0.9900980486 are 0.0101012051 and 0.0099019514 respectively, we choose $\delta \le 0.0099019514$. Note that for any value of x such that $|x-1| < \delta$, we see that $|f(x) - f(1)| < 0.01$.

(iii) $\varepsilon = 0.001$. Then $2 + \varepsilon = 2.001$ and $2 - \varepsilon = 1.999$ To obtain the value of δ for the value of $\varepsilon = 0.001$ we first solve for $f(x) = 2.001$ and $f(x) = 1.999$ for any x in the interval 0.5 to 1.5 (since we are interested in the limit $\lim\limits_{x \to 1} f(x)$). Note that as $f(1.5) = 2.25$ and $f(0.5) = 1.25$ (as seen in part(i)), we see by the Intermediate Value Theorem that there are values c_1 and c_2 in the interval (0.5, 1.5) such that $f(c_1) = 2.001$ and $f(c_2) = 1.999$. We will again solve this by using the *Roots* command. We do not need to activate the $with(Student[Calculus1])$ and $with(plots)$ packages as these have been activated in part (i) of this problem.

$$Roots(f(x) = 2.001, x = 0.5..1.5)$$

$$[1.001001002] \qquad\qquad\qquad\textbf{(2.80)}$$

$$Roots(f(x) = 1.999, x = 0.5..1.5)$$

$$[0.9990009980] \qquad\qquad \textbf{(2.81)}$$

We will now plot the function for the values of x between 0.98 and 1.02 along with the horizontal lines $y = 2.01$ and $y = 1.99$ and the vertical lines $x = 1.001001002$ and $x = 0.9990009980$. In order to plot the vertical lines we will use the *implicitplot* command. But first we must activate the *with(plots)* command.

$$implicitplot\left(\left[\, y = 3 \cdot x - x^2, y = 2.001, y = 1.999, x = 1.001001002, x = 0.9990009980\,\right]\right.$$
$$\left. x = 0.998\,..\,1.002, y = 1.9985\,..\,2.0015, color = \left[\, red, blue, green, brown, black\,\right]\right)$$

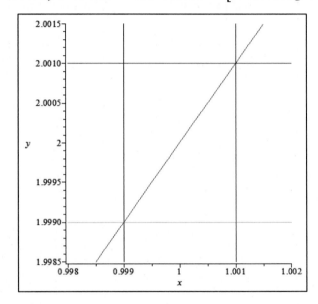

Since,

$$\left|1 - 1.001001002\right|$$

$$0.001001002 \qquad\qquad \textbf{(2.82)}$$

$$\left|1 - 0.9990009980\right|$$

$$0.0009990020 \qquad\qquad \textbf{(2.83)}$$

the distance from 1 to 1.001001002 and 0.9990009980 is 0.001001002 and 0.0009990020 respectively we can choose any $\delta \le 0.0009990020$. Note that for any value of x such that $\left|x-1\right| < \delta$, we see that $\left|f(x) - f(1)\right| < 0.001$.

(iv)   Any $\varepsilon > 0$. To obtain the value of δ for the value of $\varepsilon > 0$ we first solve for $f(x) = 2 + \varepsilon$ and $f(x) = 2 - \varepsilon$. We will restrict our analysis to the interval $[0.5, 1.5]$. We see by the Intermediate Value Theorem that there are values c_1 and

c_2 in the interval (0.5, 1.5) such that $f(c_1)=2+\varepsilon$ and $f(c_2)=2-\varepsilon$. We will do this by using the *solve* command.

$solve(f(x)=2+\varepsilon,\{x\})$

$$\left\{x=\frac{3}{2}+\frac{1}{2}\sqrt{1-4\varepsilon}\right\},\left\{x=\frac{3}{2}-\frac{1}{2}\sqrt{1-4\varepsilon}\right\} \qquad (2.84)$$

$solve(f(x)=2-\varepsilon,\{x\})$

$$\left\{x=\frac{3}{2}+\frac{1}{2}\sqrt{1+4\varepsilon}\right\},\left\{x=\frac{3}{2}-\frac{1}{2}\sqrt{1+4\varepsilon}\right\} \qquad (2.85)$$

Since we are interested in the values of x near the point 1, $c_1=\frac{3}{2}-\frac{1}{2}\sqrt{1-4\varepsilon}$ and

$c_2=\frac{3}{2}-\frac{1}{2}\sqrt{1+4\varepsilon}$. (Note: $0<\varepsilon<0.25$) Furthermore, as

$$\left|1-\left(\frac{3}{2}-\frac{1}{2}\sqrt{1-4\varepsilon}\right)\right|$$

$$\left|-\frac{1}{2}+\frac{1}{2}\sqrt{1-4\varepsilon}\right| \qquad (2.86)$$

$$\left|1-\left(\frac{3}{2}-\frac{1}{2}\sqrt{1+4\varepsilon}\right)\right|$$

$$\left|-\frac{1}{2}+\frac{1}{2}\sqrt{1+4\varepsilon}\right| \qquad (2.87)$$

we see that the distance from 1 to c_1 and c_2 are $d_1=\left|-\frac{1}{2}+\frac{\sqrt{1-4\varepsilon}}{2}\right|$ and

$d_2=\left|-\frac{1}{2}+\frac{\sqrt{1+4\varepsilon}}{2}\right|$, respectively. Choose any $\delta\leq\min\{d_1,d_2\}=$

$\left|-\frac{1}{2}+\frac{\sqrt{1+4\varepsilon}}{2}\right|$. Thus, for any value of x such that $|x-1|<\delta$, we see that

$|f(x)-f(1)|<\varepsilon.$

Example 13: Let $f(x)=3x-x^2$. Write a procedure to obtain a table for the values of $\varepsilon=0.2$, $\varepsilon=\frac{1}{2\pi}$, $\varepsilon=0.1$, $\varepsilon=0.01$, $\varepsilon=0.001$, and $\varepsilon=0.0001$ and their corresponding δ's.

Solutions: From Example 12 (iv), we know that for each $\varepsilon > 0$ the value of $\delta \le \left| -\dfrac{1}{2} + \dfrac{\sqrt{1+4\varepsilon}}{2} \right|$. We will use this value of δ in the procedure below.

$f := x \to 3 \cdot x - x^2$

$$x \to 3\,x - x^2 \tag{2.88}$$

$A := array(1..7, 1..2)$

$A[1,1] := '\varepsilon':$

$A[1,2] := '\delta':$

$A[2,1] := 0.2: \quad A[3,1] := \dfrac{1}{2\pi}: \quad A[4,1] := 0.1: \quad A[5,1] := 0.01: \quad A[6,1] := 0.001:$

$A[7,1] := 0.0001:$

for n **from** 2 **to** 7 **do** $A[n,2] := evalf\left(\left| -\dfrac{1}{2} + \dfrac{\sqrt{1+4 \cdot A[n,1]}}{2} \right|\right)$ **end do**:

$print(A)$

$$\begin{bmatrix} \varepsilon & \delta \\ 0.2 & 0.1708203930 \\ \dfrac{1}{2\pi} & 0.1396522045 \\ 0.1 & 0.0916079785 \\ 0.01 & 0.0099019515 \\ 0.001 & 0.0009990020 \\ 0.0001 & 0.0000999900 \end{bmatrix} \tag{2.89}$$

Remark: one can make a cruder estimate for δ using the following argument:
Since $|f(x) - f(1)| = |3x - x^2 - 2| = |x-1||x-2| < 2$, we can choose $\delta \le 1$. Then $|x-1| < 1 \implies -1 < x-1 < 1 \implies -2 < x-2 < 0$, it then follows that $|x-2| < 2$, and $|x-1||x-2| < 2\delta < \varepsilon$, so that $\delta = \min\left\{1, \dfrac{\varepsilon}{2}\right\}$ will suffice.

Limits at Infinity

Definition of a limit at ∞: Let f be a function. Let L be a real number. We say that the *limit of $f(x)$ as x approaches* ∞ is L, denoted by $\lim\limits_{x \to \infty} f(x) = L$, if for every $\varepsilon > 0$ there is a corresponding $M > 0$ such that if $x > M$ then $|f(x) - L| < \varepsilon$.

Definition of a limit at $-\infty$: Let f be a function. Let L be a real number. We say that the *limit of* $f(x)$ *as* x *approaches* $-\infty$ is L, denoted by $\lim\limits_{x \to -\infty} f(x) = L$, if for every $\varepsilon > 0$ there is a corresponding $M < 0$ such that if $x < M$ then $|f(x) - L| < \varepsilon$.

Example 14: Use the step-by-step procedure in *Maple* to evaluate the limit $\lim\limits_{x \to \infty} \dfrac{5x^2 + 3x + 2}{x^2 - 1}$.

Solution: We will begin by executing the *restart* command. Activate the $with\big(Student[Calculus1]\big)$ package and set $infolevel\big[Student[Calculus1]\big] := 1$. We will also use the inert form of the limit which is the *Limit* command. This command will return the limit unevaluated.

$restart$
$with\big(Student[Calculus1]\big):$
$infolevel\big[Student[Calculus1]\big] := 1:$
$Rule[\ \ ]\left(Limit\left(\dfrac{5 \cdot x^2 + 3 \cdot x + 2}{x^2 - 1}, x = \infty\right)\right)$

```
Creating problem #1
```

$$\lim_{x \to \infty} \frac{5\,x^2 + 3\,x + 2}{x^2 - 1} = \lim_{x \to \infty} \frac{5\,x^2 + 3\,x + 2}{x^2 - 1} \qquad \textbf{(2.90)}$$

We will divide the numerator and the denominator by x^2 and use the *rewrite* command to find the limit of the difference.

$$Rule\left[rewrite, \frac{5 \cdot x^2 + 3 \cdot x + 2}{x^2 - 1} = \frac{5 + \dfrac{3}{x} + \dfrac{2}{x^2}}{1 - \dfrac{1}{x^2}}\right]\big((\textbf{2.90})\big)$$

$$\lim_{x \to \infty} \frac{5\,x^2 + 3\,x + 2}{x^2 - 1} = \lim_{x \to \infty} \frac{5 + \dfrac{3}{x} + \dfrac{2}{x^2}}{1 - \dfrac{1}{x^2}} \qquad \textbf{(2.91)}$$

Use the *quotient* command to find the limit of the quotient.

$Rule[quotient]\big((\textbf{2.91})\big)$

$$\lim_{x\to\infty}\frac{5\,x^2+3\,x+2}{x^2-1}=\frac{\lim\limits_{x\to\infty}\left(5+\dfrac{3}{x}+\dfrac{2}{x^2}\right)}{\lim\limits_{x\to\infty}\left(1-\dfrac{1}{x^2}\right)} \tag{2.92}$$

Use the *sum* command to begin finding the limit of the numerator.

$Rule[sum]((\mathbf{2.92}))$

$$\lim_{x\to\infty}\frac{5\,x^2+3\,x+2}{x^2-1}=\frac{\lim\limits_{x\to\infty}5+\lim\limits_{x\to\infty}\dfrac{3}{x}+\lim\limits_{x\to\infty}\dfrac{2}{x^2}}{\lim\limits_{x\to\infty}\left(1-\dfrac{1}{x^2}\right)} \tag{2.93}$$

The *constant* rule will be used to find $\lim\limits_{x\to\infty}5$.

$Rule[constant]((\mathbf{2.93}))$

$$\lim_{x\to\infty}\frac{5\,x^2+3\,x+2}{x^2-1}=\frac{5+\lim\limits_{x\to\infty}\dfrac{3}{x}+\lim\limits_{x\to\infty}\dfrac{2}{x^2}}{\lim\limits_{x\to\infty}\left(1-\dfrac{1}{x^2}\right)} \tag{2.94}$$

The *constantmultiple* rule will be used to find $\lim\limits_{x\to\infty}\dfrac{3}{x}$.

$Rule[constantmultiple]((\mathbf{2.94}))$

$$\lim_{x\to\infty}\frac{5\,x^2+3\,x+2}{x^2-1}=\frac{5+3\lim\limits_{x\to\infty}\left(\dfrac{1}{x}\right)+\lim\limits_{x\to\infty}\dfrac{2}{x^2}}{\lim\limits_{x\to\infty}\left(1-\dfrac{1}{x^2}\right)} \tag{2.95}$$

The *power* rule is now used to find $\lim\limits_{x\to\infty}\dfrac{1}{x}$.

$Rule[power]((\mathbf{2.95}))$

$$\lim_{x\to\infty}\frac{5\,x^2+3\,x+2}{x^2-1}=\frac{5+\dfrac{3}{\lim\limits_{x\to\infty}x}+\lim\limits_{x\to\infty}\dfrac{2}{x^2}}{\lim\limits_{x\to\infty}\left(1-\dfrac{1}{x^2}\right)} \tag{2.96}$$

The *identity* rule is now used to find $\lim\limits_{x\to\infty} x$.

$Rule\big[identity\big]\big((\mathbf{2.96})\big)$

$$\lim_{x\to\infty}\frac{5\,x^2+3\,x+2}{x^2-1}=\frac{5+\lim\limits_{x\to\infty}\dfrac{2}{x^2}}{\lim\limits_{x\to\infty}\left(1-\dfrac{1}{x^2}\right)} \tag{2.97}$$

We will repeat the same steps for finding $\lim\limits_{x\to\infty}\dfrac{2}{x^2}$ as we did for finding $\lim\limits_{x\to\infty}\dfrac{3}{x}$.

$Rule\big[constantmultiple\big]\big((\mathbf{2.97})\big)$

$$\lim_{x\to\infty}\frac{5\,x^2+3\,x+2}{x^2-1}=\frac{5+2\left(\lim\limits_{x\to\infty}\dfrac{1}{x^2}\right)}{\lim\limits_{x\to\infty}\left(1-\dfrac{1}{x^2}\right)} \tag{2.98}$$

$Rule\big[power\big]\big((\mathbf{2.98})\big)$

$$\lim_{x\to\infty}\frac{5\,x^2+3\,x+2}{x^2-1}=\frac{5+\dfrac{2}{\left(\lim\limits_{x\to\infty} x\right)^2}}{\lim\limits_{x\to\infty}\left(1-\dfrac{1}{x^2}\right)} \tag{2.99}$$

$Rule\big[identity\big]\big((\mathbf{2.99})\big)$

$$\lim_{x\to\infty}\frac{5\,x^2+3\,x+2}{x^2-1}=\frac{5}{\lim\limits_{x\to\infty}\left(1-\dfrac{1}{x^2}\right)} \tag{2.100}$$

The limit in the denominator will be calculated by first using the *difference* command.

$Rule\big[difference\big]\big((\mathbf{2.100})\big)$

$$\lim_{x\to\infty}\frac{5\,x^2+3\,x+2}{x^2-1}=\frac{5}{\lim\limits_{x\to\infty}1-\left(\lim\limits_{x\to\infty}\dfrac{1}{x^2}\right)} \tag{2.101}$$

The constant command will be used to find $\lim\limits_{x\to\infty}1$.

$Rule\big[constant\big]\big((\mathbf{2.101})\big)$

$$\lim_{x \to \infty} \frac{5\,x^2 + 3\,x + 2}{x^2 - 1} = \frac{5}{1 - \left(\lim\limits_{x \to \infty} \dfrac{1}{x^2}\right)} \tag{2.102}$$

Use the power and identity commands successively to find $\lim\limits_{x \to \infty} \dfrac{1}{x^2}$.

$Rule\big[power\big]\big((\mathbf{2.102})\big)$

$$\lim_{x \to \infty} \frac{5\,x^2 + 3\,x + 2}{x^2 - 1} = \frac{5}{1 - \dfrac{1}{\left(\lim\limits_{x \to \infty} x\right)^2}} \tag{2.103}$$

$Rule\big[identity\big]\big((\mathbf{2.103})\big)$

$$\lim_{x \to \infty} \frac{5\,x^2 + 3\,x + 2}{x^2 - 1} = 5 \tag{2.104}$$

Hence $\lim\limits_{x \to \infty} \dfrac{5x^2 + 3x + 2}{x^2 - 1} = 5$.

Example 15: Use the step-by-step procedure in *Maple* to evaluate the limit $\lim\limits_{x \to -\infty} \dfrac{1 - e^{2x}}{8 + e^{x}}$.

Solution: We will begin by executing the *restart* command. Activate the $with\big(Student[Calculus1]\big)$ package and set $infolevel\big[Student[Calculus1]\big] := 1$. We will also use the inert form of the limit which is the *Limit* command. This command will return the limit unevaluated.

$restart$
$with\big(Student[Calculus1]\big):$
$infolevel\big[Student[Calculus1]\big] := 1:$
$Rule\big[\ \ \big]\left(Limit\left(\dfrac{1 - e^{2 \cdot x}}{8 + e^{x}}, x = -\infty\right)\right)$
```
Creating problem #1
```

$$\lim_{x \to -\infty} \frac{1 - e^{2\,x}}{8 + e^{x}} = \lim_{x \to -\infty} \frac{1 - e^{2\,x}}{8 + e^{x}} \tag{2.105}$$

Use the *quotient* command to find the limit of the quotient.

Rule[*quotient*]((**2.105**))

$$\lim_{x \to -\infty} \frac{1 - e^{2x}}{8 + e^{x}} = \frac{\displaystyle\lim_{x \to -\infty}\left(1 - e^{2x}\right)}{\displaystyle\lim_{x \to -\infty}\left(8 + e^{x}\right)} \tag{2.106}$$

Use the *difference* command to begin finding the limit of the numerator.

Rule[*difference*]((**2.106**))

$$\lim_{x \to -\infty} \frac{1 - e^{2x}}{8 + e^{x}} = \frac{\displaystyle\lim_{x \to -\infty} 1 - \left(\lim_{x \to -\infty} e^{2x}\right)}{\displaystyle\lim_{x \to -\infty}\left(8 + e^{x}\right)} \tag{2.107}$$

The *constant* rule will be used to find $\lim\limits_{x \to -\infty} 1$.

Rule[*constant*]((**2.107**))

$$\lim_{x \to -\infty} \frac{1 - e^{2x}}{8 + e^{x}} = \frac{1 - \left(\displaystyle\lim_{x \to -\infty} e^{2x}\right)}{\displaystyle\lim_{x \to -\infty}\left(8 + e^{x}\right)} \tag{2.108}$$

The *exp* rule will be used to find $\lim\limits_{x \to -\infty} e^{2x}$.

Rule[*exp*]((**2.108**))

$$\lim_{x \to -\infty} \frac{1 - e^{2x}}{8 + e^{x}} = \frac{1 - \left(e^{\lim\limits_{x \to -\infty} 2x}\right)}{\displaystyle\lim_{x \to -\infty}\left(8 + e^{x}\right)} \tag{2.109}$$

The *constantmultiple* rule will be used to find $\lim\limits_{x \to -\infty}\left(2x\right)$.

Rule[*constantmultiple*]((**2.109**))

$$\lim_{x \to -\infty} \frac{1 - e^{2x}}{8 + e^{x}} = \frac{1 - \left(e^{2\lim\limits_{x \to -\infty} x}\right)}{\displaystyle\lim_{x \to -\infty}\left(8 + e^{x}\right)} \tag{2.110}$$

The *identity* rule is now used to find $\lim\limits_{x \to -\infty} x$.

Rule[*identity*]((**2.110**))

$$\lim_{x \to -\infty} \frac{1-e^{2x}}{8+e^x} = \frac{1}{\lim\limits_{x \to -\infty}\left(8+e^x\right)} \tag{2.111}$$

Use the sum rule for finding the limit of the denominator.

Rule[*sum*]((**2.111**))

$$\lim_{x \to -\infty} \frac{1-e^{2x}}{8+e^x} = \frac{1}{\lim\limits_{x \to -\infty} 8 + \left(\lim\limits_{x \to -\infty} e^x\right)} \tag{2.112}$$

The *constant* rule will be used to find $\lim\limits_{x \to -\infty} 8$.

Rule[*constant*]((**2.112**))

$$\lim_{x \to -\infty} \frac{1-e^{2x}}{8+e^x} = \frac{1}{8 + \left(\lim\limits_{x \to -\infty} e^x\right)} \tag{2.113}$$

The *exp* rule will be used to find $\lim\limits_{x \to -\infty} e^x$.

Rule[*exp*]((**2.113**))

$$\lim_{x \to -\infty} \frac{1-e^{2x}}{8+e^x} = \frac{1}{8} \tag{2.114}$$

Hence $\lim\limits_{x \to -\infty} \dfrac{1-e^{2x}}{8+e^x} = \dfrac{1}{8}$.

<u>Asymptotes</u>

Definition of an infinite limit: Let f be a function. Let c be a number so that f is defined on an open interval containing c (except possibly for c).

(a)   We say that the *limit of $f(x)$ as x approaches c is* ∞, denoted by $\lim\limits_{x \to c} f(x) = \infty$, if for every $M > 0$ there is a corresponding $\delta > 0$ such that if $|x-c| < \delta$ then $f(x) > M$.

(b)   We say that the *limit of $f(x)$ as x approaches c is* $-\infty$, denoted by $\lim\limits_{x \to c} f(x) = -\infty$, if for every $M < 0$ there is a corresponding $\delta > 0$ such that if $|x-c| < \delta$ then $f(x) < M$.

(c)   To obtain the *infinite limit from the right* ($\lim\limits_{x \to c^+} f(x) = \infty$ or $\lim\limits_{x \to c^+} f(x) = -\infty$) we replace $|x - c| < \delta$ by $c < x < c + \delta$. To obtain the *infinite limit from the left* ($\lim\limits_{x \to c^-} f(x) = \infty$ or $\lim\limits_{x \to c^-} f(x) = -\infty$), we replace $|x - c| < \delta$ by $c < x < c + \delta$.

Definition of an asymptote: Let f be a function. An *asymptote* is a curve whose distance to the graph of f approaches 0 as x gets closer to a given number or gets arbitrarily large. An asymptote may or may not intersect the graph of f.

Definition of a horizontal asymptote: Let f be a function. Let L be a real number. We say that the line $y = L$ is a *horizontal asymptote* of the graph of f if $\lim\limits_{x \to \infty} f(x) = L$ or $\lim\limits_{x \to -\infty} f(x) = L$.

Definition of a vertical asymptote: Let f be a function. Let c be a real number. We say that the line $x = c$ is a *vertical asymptote* of the graph of f if $f(x)$ approaches ∞ (or $-\infty$) when x approaches c from the left or the right.

Example 16: Let $f(x) = \dfrac{5x^2 + 3x + 2}{x^2 - 1}$. What is the domain of f ? Use *Maple* commands to find the zeros, horizontal and vertical asymptotes of f, if they exist.

Solution: We will begin by entering f as a *Maple* function.

$$f := x \to \frac{5 \cdot x^2 + 3 \cdot x + 2}{x^2 - 1}$$

$$x \to \frac{5 \cdot x^2 + 3 \cdot x + 2}{x^2 - 1} \tag{2.115}$$

We will first find the domain of f by setting its denominator equal to zero.

$$solve\left(denom\left(f(x)\right) = 0, \{x\}\right)$$

$$\{x = 1\}, \{x = -1\} \tag{2.116}$$

Hence the domain of f is the set of all real numbers except 1 and -1. To find the zeros of f set its numerator equal to zero.

$$solve\left(numer\left(f(x)\right) = 0, \{x\}\right)$$

$$\left\{x = -\frac{3}{10} + \frac{1}{10} I \sqrt{31}\right\}, \left\{x = -\frac{3}{10} - \frac{1}{10} I \sqrt{31}\right\} \tag{2.117}$$

Therefore, f has no real zeros. To find the horizontal asymptotes we will find $\lim_{x \to \infty} f(x)$ and $\lim_{x \to -\infty} f(x)$ using the *Expression* and *Common* Symbols palettes.

$$\lim_{x \to \infty} f(x)$$

$$5 \qquad\qquad\qquad \textbf{(2.118)}$$

$$\lim_{x \to -\infty} f(x)$$

$$5 \qquad\qquad\qquad \textbf{(2.119)}$$

The horizontal asymptote is $y = 5$. The vertical asymptotes will be found by finding $\lim_{x \to 1+} f(x)$, $\lim_{x \to 1-} f(x)$, $\lim_{x \to -1+} f(x)$, and $\lim_{x \to -1-} f(x)$.

$$\lim_{x \to 1+} f(x)$$

$$\infty \qquad\qquad\qquad \textbf{(2.120)}$$

$$\lim_{x \to 1-} f(x)$$

$$-\infty \qquad\qquad\qquad \textbf{(2.121)}$$

$$\lim_{x \to -1+} f(x)$$

$$-\infty \qquad\qquad\qquad \textbf{(2.122)}$$

$$\lim_{x \to -1-} f(x)$$

$$\infty \qquad\qquad\qquad \textbf{(2.123)}$$

The vertical asymptotes are $x = 1$ and $x = -1$. Note that $\lim_{x \to 1+} f(x)$ and $\lim_{x \to -1+} f(x)$ do not exist (why?). We will plot the function f along with its asymptotes. The vertical asymptotes are automatically plotted with the function. However we need to include the horizontal asymptote in the plot command.

$$plot\left(\left[f(x), 5 \right], x = -4..8, y = -10..15, color = \left[red, black \right]\right)$$

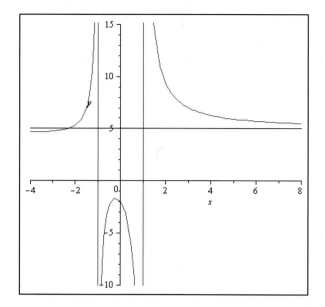

Notice that the horizontal asymptote intersects the curve at a point. How would you find this point?

Exercises

1. Graph the function $f(x) = \dfrac{3x^2 - 2x - 8}{x + 2}$. Find the domain of f. Find left-hand limit $\lim\limits_{x \to 2-} f(x)$ and the right hand limit $\lim\limits_{x \to 2+} f(x)$, if they exist. Use these numbers to determine whether $\lim\limits_{x \to 2} f(x)$ exists, if it exists. Is f continuous at $x = 2$? Give reasons for your answer.

2. Graph the function $f(x) = \dfrac{5x^2 - 2x + 3}{x - 4}$. Find the domain of f. Find left-hand limit $\lim\limits_{x \to -4-} f(x)$ and the right hand limit $\lim\limits_{x \to -4+} f(x)$, if they exist. Use these numbers to determine whether $\lim\limits_{x \to -4} f(x)$ exists, if it exists. Is f continuous at $x = -4$? Give reasons for your answer.

3. Graph the function $f(x) = \dfrac{6x^2 - x - 12}{3x - 4}$. Find the domain of f. Find left-hand limit $\lim\limits_{x \to \frac{4}{3}-} f(x)$ and the right hand limit $\lim\limits_{x \to \frac{4}{3}+} f(x)$, if they exist. Use these numbers to determine whether $\lim\limits_{x \to \frac{4}{3}} f(x)$ exists, if it exists. Is f continuous at $x = \dfrac{4}{3}$? Give reasons for your answer.

4. Graph the function $f(x) = \begin{cases} 4x - 6 & \text{if} \quad -6 \le x < 0 \\ 2x^2 - 6 & \text{if} \quad 0 \le x \le 4 \end{cases}$. Find the domain of f. Find left-hand limit $\lim\limits_{x \to 0-} f(x)$ and the right hand limit $\lim\limits_{x \to 0+} f(x)$, if they exist. Use these numbers to determine whether $\lim\limits_{x \to 0} f(x)$ exists, if it exists. Is f continuous at $x = 0$? Give reasons for your answer.

5. Graph the function $f(x) = \begin{cases} -12x - 112 & \text{if} \quad -10 \le x \le -6 \\ -x^2 - 4 & \text{if} \quad -6 < x \le 4 \end{cases}$. Find the domain of f. Find left-hand limit $\lim\limits_{x \to -6-} f(x)$ and the right hand limit $\lim\limits_{x \to -6+} f(x)$, if they exist. Use these numbers to determine whether $\lim\limits_{x \to -6} f(x)$ exists, if it exists. Is f continuous at $x = -6$? Give reasons for your answer.

6. Graph the function $f(x) = \begin{cases} 3x^2 + 4 & \text{if} \quad -2 \le x < 4 \\ -6x - 17 & \text{if} \quad 4 \le x \le 8 \end{cases}$. Find the domain of f.

   Find left-hand limit $\lim\limits_{x \to 4-} f(x)$ and the right hand limit $\lim\limits_{x \to 4+} f(x)$, if they exist. Use these numbers to determine whether $\lim\limits_{x \to 4} f(x)$ exists, if it exists. Is f continuous at $x = 4$? Give reasons for your answer.

7. Graph the function $f(x) = \begin{cases} -2x^2 - 8x + 28 & \text{if} \quad -2 \le x < 3 \\ -3x - 5 & \text{if} \quad 3 \le x \le 6 \end{cases}$. Find the domain of f. Find left-hand limit $\lim\limits_{x \to 3-} f(x)$ and the right hand limit $\lim\limits_{x \to 3+} f(x)$, if they exist. Use these numbers to determine whether $\lim\limits_{x \to 3} f(x)$ exists, if it exists. Is f continuous at $x = 3$? Give reasons for your answer.

8. Graph the function $f(x) = \begin{cases} 3x + 20 & \text{if} \quad -10 \le x \le 2 \\ 30 & \text{if} \quad 2 < x \le 12 \end{cases}$. Find the domain of f.

   Find left-hand limit $\lim\limits_{x \to 2-} f(x)$ and the right hand limit $\lim\limits_{x \to 2+} f(x)$, if they exist. Use these numbers to determine whether $\lim\limits_{x \to 2} f(x)$ exists, if it exists. Is f continuous at $x = 2$? Give reasons for your answer.

9. Graph the function $f(x) = \begin{cases} -2x + 2 & \text{if} \quad -4 \le x \le 0 \\ -4x - 2 & \text{if} \quad 0 < x \le 3 \end{cases}$. Find the domain of f.

   Find left-hand limit $\lim\limits_{x \to 0-} f(x)$ and the right hand limit $\lim\limits_{x \to 0+} f(x)$, if they exist. Use these numbers to determine whether $\lim\limits_{x \to 0} f(x)$ exists, if it exists. Is f continuous at $x = 0$? Give reasons for your answer.

10. Graph the function $f(x) = (-1)^{\text{floor}(x)}$. Find the domain and range of f. Find the points at which the function is discontinuous. Give reasons for your answer.

11. Graph the function $f(x) = \dfrac{3x^2 - 2x - 8}{x + 2}$. Find the domain of f. Find left-hand limit $\lim\limits_{x \to 2-} f(x)$ and the right hand limit $\lim\limits_{x \to 2+} f(x)$, if they exist. Use these numbers to determine whether $\lim\limits_{x \to 2} f(x)$ exists, if it exists. Is f continuous at $x = 2$? Give reasons for your answer.

12. Graph the function $f(x) = \dfrac{3x^2 - 2x - 8}{x - 2}$. Find the domain of f. Find left-hand limit $\lim\limits_{x \to -4-} f(x)$ and the right hand limit $\lim\limits_{x \to -4+} f(x)$, if they exist. Use these numbers to determine whether $\lim\limits_{x \to -4} f(x)$ exists, if it exists. Is f continuous at $x = -4$? Give reasons for your answer.

13. Graph the function $f(x) = \dfrac{6x^3 + x^2 - 31x + 24}{2x - 3}$. Find the domain of f. Find left-hand limit $\lim\limits_{x \to \frac{3}{2}-} f(x)$ and the right hand limit $\lim\limits_{x \to \frac{3}{2}+} f(x)$, if they exist. Use these numbers to determine whether $\lim\limits_{x \to \frac{3}{2}} f(x)$ exists, if it exists. Is f continuous at $x = \dfrac{3}{2}$? Give reasons for your answer.

14. Construct an array for $\left\{ 1 + \dfrac{1}{n} \right\}$ using the values of n from 1 till 20 to indicate the direction of convergence of the sequence $\left\{ 1 + \dfrac{1}{n} \right\}$. Can you use this to estimate the value of $\lim\limits_{n \to \infty} \left(1 + \dfrac{1}{n} \right)$? Give reasons for your answer. Use *Maple* commands to verify your answer.

15. Graph the function $f(x) = \dfrac{\cos(x) - 1}{x}$. Find the domain of f. Construct an array of 16 rows and 5 columns in such a way that the first row consists of the labels of the columns, the first column is the value of n (for the values of n from 1 through 15), the second column is the floating point approximation of $-\dfrac{1}{3^n}$, the third column is the floating point approximation of $f\left(-\dfrac{1}{3^n} \right)$, the fourth column is the floating point approximation of $\dfrac{1}{3^n}$, and the fifth column is the floating point approximation of $f\left(\dfrac{1}{3^n} \right)$. Using the values in the array and the graph of f, can you estimate the values of the left-hand limit $\lim\limits_{x \to 0-} f(x)$ and the right-hand limit $\lim\limits_{x \to 0+} f(x)$, if they exist? Can you use these numbers to determine an estimate of

$\lim\limits_{x\to 0} f(x)$, if it exists? Give reasons for your answer. Verify your answers using *Maple* commands.

16. Graph the function $f(x) = \dfrac{\tan(x)}{x}$. Find the domain of f. Construct an array of 21 rows and 5 columns in such a way that the first row consists of the labels of the columns, the first column is the value of n (for the values of n from 1 through 20), the second column is the floating point approximation of $-\dfrac{1}{7n}$, the third column is the floating point approximation of $f\left(-\dfrac{1}{7n}\right)$, the fourth column is the floating point approximation of $\dfrac{1}{7n}$, and the fifth column is the floating point approximation of $f\left(\dfrac{1}{7n}\right)$. Using the values in the array and the graph of f, can you estimate the values of the left-hand limit $\lim\limits_{x\to 0-} f(x)$ and the right-hand limit $\lim\limits_{x\to 0+} f(x)$, if they exist? Can you use these numbers to determine an estimate of $\lim\limits_{x\to 0} f(x)$, if it exists? Give reasons for your answer. Verify your answers using *Maple* commands.

17. Graph the function $f(x) = \dfrac{x^2 + 4x - 12}{x - 2}$. Find the domain of f. Construct an array of 12 rows and 5 columns in such a way that the first row consists of the labels of the columns, the first column is the value of n (for the values of n from 1 through 11), the second column is the floating point approximation of $2 - \dfrac{1}{2^n}$, the third column is the floating point approximation of $f\left(2 - \dfrac{1}{2^n}\right)$, the fourth column is the floating point approximation of $2 + \dfrac{1}{2^n}$, and the fifth column is the floating point approximation of $f\left(2 + \dfrac{1}{2^n}\right)$. Using the values in the array and the graph of f, can you estimate the values of the left-hand limit $\lim\limits_{x\to 2-} f(x)$ and the right-hand limit $\lim\limits_{x\to 2+} f(x)$, if they exist? Can you use these numbers to determine an estimate of $\lim\limits_{x\to 2} f(x)$, if it exists? Give reasons for your answer. Verify your answers using *Maple* commands.

18. Graph the function $f(x) = \dfrac{x^2 + 9x + 8}{x + 1}$. Find the domain of f. Construct an array of 21 rows and 5 columns in such a way that the first row consists of the labels of the columns, the first column is the value of n (for the values of n from 1 through 20), the second column is the floating point approximation of $-1 - \dfrac{1}{10^n}$, the third column is the floating point approximation of $f\left(-1 - \dfrac{1}{10^n}\right)$, the fourth column is the floating point approximation of $-1 + \dfrac{1}{10^n}$, and the fifth column is the floating point approximation of $f\left(-1 + \dfrac{1}{10^n}\right)$. Using the values in the array and the graph of f, can you estimate the values of the left-hand limit $\lim\limits_{x \to -1-} f(x)$ and the right-hand limit $\lim\limits_{x \to -1+} f(x)$, if they exist? Can you use these numbers to determine an estimate of $\lim\limits_{x \to -1} f(x)$, if it exists. Give reasons for your answer. Verify your answers using *Maple* commands.

19. Graph the function $f(x) = \dfrac{x^2 - 2x - 3}{x - 3}$. Find the domain of f. Construct an array of 11 rows and 5 columns in such a way that the first row consists of the labels of the columns, the first column is the value of n (for the values of n from 1 through 10), the second column is the floating point approximation of $3 - \dfrac{1}{25^n}$, the third column is the floating point approximation of $f\left(3 - \dfrac{1}{25^n}\right)$, the fourth column is the floating point approximation of $3 + \dfrac{1}{25^n}$, and the fifth column is the floating point approximation of $f\left(3 + \dfrac{1}{25^n}\right)$. Using the values in the array and the graph of f, can you estimate the values of the left-hand limit $\lim\limits_{x \to 3-} f(x)$ and the right-hand limit $\lim\limits_{x \to 3+} f(x)$, if they exist? Can you use these numbers to determine an estimate of $\lim\limits_{x \to 3} f(x)$, if it exists? Give reasons for your answer. Verify your answers using *Maple* commands.

20. Graph the function $f(x) = \dfrac{|x|}{x}$. Find the domain of f. Construct an array of 18 rows and 5 columns in such a way that the first row consists of the labels of the columns, the first column is the value of n (for the values of n from 1 through

17), the second column is the floating point approximation of $-\dfrac{1}{n}$, the third column is the floating point approximation of $f\left(-\dfrac{1}{n}\right)$, the fourth column is the floating point approximation of $\dfrac{1}{n}$, and the fifth column is the floating point approximation of $f\left(\dfrac{1}{n}\right)$. Using the values in the array and the graph of f, can you estimate the values of the left-hand limit $\lim\limits_{x\to 0-} f(x)$ and the right-hand limit $\lim\limits_{x\to 0} f(x)$, if they exist? Can you use these numbers to determine an estimate of $\lim\limits_{x\to 0} f(x)$, if it exists? Give reasons for your answer. Verify your answers using *Maple* commands.

21. Let $f(x)=2x+5$. Use the Intermediate Value Theorem to find a number c in the open interval $(-5,-3)$ such that $f(c)=-3$.

22. Let $f(x)=\sqrt{x}$. Use the Intermediate Value Theorem to find a number c in the open interval $(7,10)$ such that $f(c)=3$.

23. Let $f(x)=\dfrac{1}{x^2}$. Use the Intermediate Value Theorem to find a number c in the open interval $(-2,0.1)$ such that $f(c)=1$.

24. Let $f(x)=\sin\left(\dfrac{1}{x}\right)$. Use the Intermediate Value Theorem to find a number c in the open interval $(0.25,0.41)$ such that $f(c)=0$.

25. Show by algebraic means that $\delta \le \dfrac{\sqrt{35}-5}{10}\approx 0.0916$ in example 12 (i).

26. Let $f(x)=\begin{cases} 3x-x^2 & \text{if } x\ne 1 \\ 2009 & \text{if } x=1 \end{cases}$. Find the values of δ for each value of ε using the epsilon-delta definition to find $\lim\limits_{x\to 1} f(x)$:

    (a) $\varepsilon=0.3$
    (b) $\varepsilon=0.02$
    (c) $\varepsilon=0.001$
    (d)   any $\varepsilon>0$.

(Hint: Use the results from example 12 carefully and explain why you can do this.)

27. Let $f(x) = 2x + 5$. Find the values of δ for each value of ε using the epsilon-delta definition to find $\lim\limits_{x \to -4} f(x)$:

    (a) $\varepsilon = 0.3$
    (b) $\varepsilon = 0.02$
    (c) $\varepsilon = 0.001$
    (d)   any $\varepsilon > 0$.

28. Let $f(x) = \sqrt{x}$. Find the values of δ for each value of ε using the epsilon-delta definition to find $\lim\limits_{x \to 9} f(x)$:

    (a) $\varepsilon = 0.1$
    (b) $\varepsilon = 0.05$
    (c) $\varepsilon = 0.0002$
    (d)   any $\varepsilon > 0$.

29. Let $f(x) = \dfrac{1}{x^2}$. Find the values of δ for each value of ε using the epsilon-delta definition to find $\lim\limits_{x \to -1} f(x)$:

    (a) $\varepsilon = 0.4$
    (b) $\varepsilon = 0.03$
    (c) $\varepsilon = 0.001$
    (d)   any $\varepsilon > 0$.

30. Let $f(x) = \sin\left(\dfrac{1}{x}\right)$. Find the values of δ for each value of ε using the epsilon-delta definition to find $\lim\limits_{x \to \frac{1}{\pi}} f(x)$:

    (a) $\varepsilon = 0.1$
    (b) $\varepsilon = 0.01$
    (c) $\varepsilon = 0.001$
    (d)   any $\varepsilon > 0$.

31. Let $f(x) = \dfrac{\sin(x)}{x}$. Find the values of δ for each value of ε using the epsilon-delta definition to find $\lim\limits_{x \to 1} f(x)$:

    (a) $\varepsilon = 0.1$
    (b) $\varepsilon = 0.01$
    (c) $\varepsilon = 0.001$
    (d)   any $\varepsilon > 0$.

Chapter 2 Limits

32. Find the domain of the function $f(x) = \sqrt{5-x}$ and graph it. Use this graph (and the domain) to evaluate the left-hand limit $\lim_{x \to 5-} \sqrt{5-x}$ and the right-hand limit $\lim_{x \to 5+} \sqrt{5-x}$. Does $\lim_{x \to 5} \sqrt{5-x}$ exist? Give reasons for your answer. Use *Maple* commands to find $\lim_{x \to 5} \sqrt{5-x}$. Explain why *Maple* commands give an incorrect answer.

33. (a) Let $\lambda \le p \le \lambda$, where λ and p are real numbers. Give the value of p in terms of λ. State the theorem you are using.

    (b) Evaluate $\lim_{x \to 0}\left(x^4 \sin\left(x^2 - \dfrac{2010}{x^2} \right) \right)$. Outline every step of your solutions.

    (c) Illustrate your solutions graphically with the help of *Maple* commands.

34 – 40 Evaluate the following limits using a step-by-step approach.

34. $\lim_{x \to 0}\left(\dfrac{\sin(x)}{x^5 - 9x} \right)$

35. $\lim_{x \to 0} \dfrac{e^x - 1}{2x}$

36. $\lim_{x \to 1} \dfrac{\sqrt{x} - 1}{x - 1}$

37. $\lim_{x \to 0} \dfrac{\cos(9x)}{2x}$

38. $\lim_{x \to \infty} \dfrac{\cos(9x)}{2x}$

39. $\lim_{x \to 0}\left(\dfrac{\sqrt{x^2 + 4} - 2}{5x^2} \right)$

40. $\lim_{x \to 0} \dfrac{\sin(7x)}{\sin(3x)}$

41. Find the limit $\lim_{x \to 1} \dfrac{2^x - 1}{x - 1}$, if it exists.

42. Find the limit $\lim_{x \to 2}\left(\sqrt{2-x}\right)$, if it exists.

43. Find the limit $\lim_{x \to 0}\left(\sqrt{|2-x|}\right)$, if it exists.

44. Find the limit $\lim_{x \to \infty}\left(1-\dfrac{3}{2n}\right)^{n}$, if it exists.

45. Graph the function f that satisfies the following three conditions $f(0)=2$, $\lim_{x \to 0^{-}} f(x)=3$, and $\lim_{x \to 0^{+}} f(x)=-1$. Does $\lim_{x \to 0} f(x)$ exist? Give reasons for your answer.

46. Find the limit $\lim_{m \to \infty} \lim_{n \to \infty}\left|\cos(m!\pi x)^{n}\right|$, if it exists. (Hint: Consider separately, when x is a rational number and when x is an irrational number.)

47. Let $f(x) = \sin\left(\dfrac{1}{x}\right)$. Find the domain of f.

    (a) Construct an array of 21 rows and 5 columns in such a way that the first row consists of the labels of the columns, the first column is the value of n (for the values of n from 1 through 20), the second column is the floating point approximation of $-\dfrac{1}{n\pi}$, the third column is the floating point approximation of $f\left(-\dfrac{1}{n\pi}\right)$, the fourth column is the floating point approximation of $\dfrac{1}{n\pi}$, and the fifth column is the floating point approximation of $f\left(\dfrac{1}{n\pi}\right)$. Using the values in the array and the graph of f, can you estimate the values of the left-hand limit $\lim_{x \to 0-} f(x)$ and the right-hand limit $\lim_{x \to 0+} f(x)$, if they exist? Can you use these numbers to determine an estimate of $\lim_{x \to 0} f(x)$, if it exists? Give reasons for your answer.

    (b) Construct an array of 21 rows and 5 columns in such a way that the first row consists of the labels of the columns, the first column is the value of n (for the values of n from 1 through 20), the second column is the floating point approximation of $-\dfrac{2}{(4n+1)\pi}$, the third column is the floating point approximation of $f\left(-\dfrac{2}{(4n+1)\pi}\right)$, the fourth column is the

floating point approximation of $\dfrac{2}{(4n+1)\pi}$, and the fifth column is the

floating point approximation of $f\left(\dfrac{2}{(4n+1)\pi}\right)$. Using the values in the

array and the graph of f, can you estimate the values of the left-hand limit $\lim\limits_{x\to 0-} f(x)$ and the right-hand limit $\lim\limits_{x\to 0+} f(x)$, if they exist? Can you use these numbers to determine an estimate of $\lim\limits_{x\to 0} f(x)$, if it exists? Give reasons for your answer.

(c) Based on the results in parts (i) and (ii) what can you conclude about $\lim\limits_{x\to 0} f(x)$? Give reasons for your answer.

(d) Find a sequence $\{a_n\}$ such that $\lim\limits_{n\to\infty} a_n = 0$ and $\lim\limits_{n\to\infty} f(a_n) = \dfrac{1}{2}$.

48. Below is a list of some commands used in this chapter. Describe the significance of each command. Can you find examples where each command is used?

(a) $discont(f(x), x)$

(b) $implicitplot(implicit\ equation,\ x = a\,..\,b,\ y = c\,..\,d)$

(c) $infolevel\big[\,Student[Calculus1]\,\big] := n$

(d) $limit(f(x), x = a)$

(e) $piecewise(function)$

(f) $plot(f(x), x = a\,..\,b, y = c\,..\,d, discont = [showremovable])$

(g) $Rule[\ ](Limit(f(x), x = a))$

(h) $Rule[name\ of\ rule]((label\ number))$

(i) $ShowSolution(Limit(f(x), x = a))$

(j) $with(Student[Calculus1])$

49. Can you list new *Maple* commands that were used in this chapter but were not listed in exercise 48?

Chapter 3   Derivatives

The object of this chapter is to find the derivatives both explicitly and implicitly and to find the equations of tangent lines. Functions will be differentiated using a step-by-step procedure. Plotting curves that are implicit functions will also be discussed.

Definition of the Derivative

Definition: Let f be a function and x be a real number. If the limit

$$\lim_{h \to 0}\left(\frac{f(x+h)-f(x)}{h} \right)$$

exists then f is said to be differentiable at x. In this case the limit is said to be the derivative of the function f at x and is denoted by $f'(x)$. If f is differentiable at all points in its domain then f is said to be a differentiable function.

Example 1: Let $f(x) = \cos(x)$. Use the definition of the derivative to show that $f'(x) = -\sin(x)$.

Solution: We will enter f as a *Maple* function, evaluate $\dfrac{f(x+h)-f(x)}{h}$, right click on the answer and then click on *Expand*. Do not forget to put x in parenthesis when entering $\cos x$ as a *Maple* function.

$f := x \to \cos(x)$

$$x \to \cos(x) \tag{3.1}$$

$\dfrac{f(x+h)-f(x)}{h}$

$$\frac{\cos(x+h)-\cos(x)}{h} \tag{3.2}$$

expand
=

$$\frac{\cos(x)\cos(h)}{h} - \frac{\sin(x)\sin(h)}{h} - \frac{\cos(x)}{h} \tag{3.3}$$

This expression can be rewritten as $\dfrac{\cos(x)(\cos(h)-1)}{h} - \dfrac{\sin(x)\sin(h)}{h}$. Since $\lim_{h \to 0} \dfrac{\sin(h)}{h} = 1$ (see Example 7 in Chapter 2), $\lim_{h \to 0} \dfrac{\cos(h)-1}{h} = 0$ (see Exercise 15 in Chapter 2), and the properties of limits, we have

$$\lim_{h \to 0} \left(\frac{\cos(x)(\cos(h)-1)}{h} - \frac{\sin(x)\sin(h)}{h} \right) = \cos(x)\lim_{h \to 0} \frac{\cos(h)-1}{h} - \sin(x)\lim_{h \to 0} \frac{\sin(h)}{h}$$

$$= -\sin(x)$$

This will be verified using *Maple* commands.

$$\lim_{h \to 0} \left(\frac{f(x+h)-f(x)}{h} \right)$$

$$-\sin(x) \tag{3.4}$$

Remark: Let $f(x) = |x|$. Then f is a continuous function of x. However, since

$$\lim_{h \to 0} \left(\frac{f(h)-f(0)}{h} \right) = \lim_{h \to 0} \left(\frac{|h|-|0|}{h} \right) = \lim_{h \to 0} \left(\frac{|h|}{h} \right)$$

is undefined (Exercise 20 in Chapter 2), $f(x) = |x|$ is not differentiable at the point 0. In Exercises 28 and 29, we will consider continuous functions that are not differentiable at some points. Examples like this one can eventually give rise to a continuous function that is nowhere differentiable.

Equation of a Tangent Line

We will now use *Maple* commands to find the derivative of various functions. This will be used to find the equation of a tangent line at a particular point.

Remark: The equation of the tangent line to the curve $y = f(x)$ at the point (x_1, y_1) is given by $y = y_1 + f'(x_1) \cdot (x - x_1)$.

Example 2: Let $f(x) = 2x^4 + 3x^3 - 20x^2 - 24x + 32$. Find the equation of the tangent line to the graph of f when $x = -1$. Graph f along with its tangent line.

Solution: We will enter f as a *Maple* function and find the derivative of this polynomial using the $D(f)$ command.

$$f := x \to 2 \cdot x^4 + 3 \cdot x^3 - 20 \cdot x^2 - 24 \cdot x + 32$$

$$x \to 2x^4 + 3x^3 - 20x^2 - 24x + 32 \tag{3.5}$$

$$f1 := D(f)$$

$$x \to 8x^3 + 9x^2 - 40x - 24 \tag{3.6}$$

Note that the derivative is a *Maple* function. In order to find the slope of the tangent line we will evaluate $f1(-1)$.

$f1(-1)$

$$17 \tag{3.7}$$

We will now find the equation of the tangent line.

$L := x \rightarrow f(-1) + f1(-1) \cdot (x+1)$

$$x \rightarrow f(-1) + f1(-1) \cdot (x+1) \tag{3.8}$$

$expand(L(x))$

$$52 + 17x \tag{3.9}$$

Therefore, the equation of the tangent line is $y = 17x + 52$. Now plot the function along with its tangent line at $x = -1$.

$plot\left(\left[f(x), L(x)\right], x = -4 .. 4, y = -40 .. 125, color = \left[red, black\right]\right)$

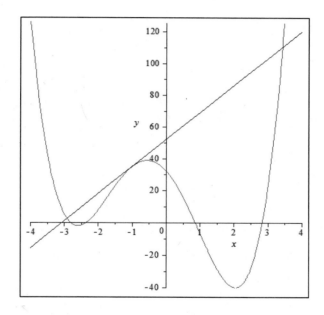

Remark: We can find the derivative of the polynomial using other techniques.

1.   Enter f as a *Maple* function, right click on the expression and click on *Differentiate* to find the derivative. Then right click on the derivative, click on *Assign to a Name* (we will call the derivative $f1$), enter $f1$ in the box next to *To name* and click on OK. Click on *Yes* when you see the following prompt:
        "*The object $f1$ already has an assigned name. Overwrite it?*"
Notice that this name appears as the output.

$f := x \rightarrow 2 \cdot x^4 + 3 \cdot x^3 - 20 \cdot x^2 - 24 \cdot x - 32$

$$x \to 2\,x^4 + 3\,x^3 - 20\,x^2 - 24\,x - 32 \qquad \textbf{(3.10)}$$

$$\xrightarrow{\text{differentiate}}$$

$$x \to 8\,x^3 + 9\,x^2 - 40\,x - 24 \qquad \textbf{(3.11)}$$

$$\xrightarrow{\text{assign to a name}}$$

$$f1 \qquad \textbf{(3.12)}$$

In order to find the slope of the tangent line we evaluate $f1(-1)$.

$$f1(-1)$$

$$17 \qquad \textbf{(3.13)}$$

2. Alternately we can use $f1 := \dfrac{\mathrm{d}}{\mathrm{d}x} f$ from the Expression palette to evaluate the derivative. Note that the answer obtained is a *Maple* expression and not a *Maple* function.

$$f1 := \frac{\mathrm{d}}{\mathrm{d}x} f(x)$$

$$8\,x^3 + 9\,x^2 - 40\,x - 24 \qquad \textbf{(3.14)}$$

In order to find the slope of the tangent line when $x = -1$, right click on the derivative, click on *Evaluate at a Point*, enter the number -1 and then click on OK.

$$\xrightarrow{\text{evaluate at a point}}$$

$$17 \qquad \textbf{(3.15)}$$

One can change $f1$ obtained above to a *Maple* function by using the *unapply* command.

$$unapply(f1, x)$$

$$x \to 8\,x^3 + 9\,x^2 - 40\,x - 24 \qquad \textbf{(3.16)}$$

3. We can use the *diff* command to find the derivative of any function. Note that the output is now a *Maple* expression. We can get the value of the derivative of a particular point by using any of the methods mentioned above.

$$f := x \to 2 \cdot x^4 + 3 \cdot x^3 - 20 \cdot x^2 - 24 \cdot x - 32$$

$$x \to 2\,x^4 + 3\,x^3 - 20\,x^2 - 24\,x - 32 \qquad \textbf{(3.17)}$$

$$diff(f(x), x)$$

$$8\,x^3 + 9\,x^2 - 40\,x - 24 \qquad \textbf{(3.18)}$$

4. The *Tangent* command can be used to find the tangent line to any curve at a particular point. In order to use this command we will need to activate the *with* (*Student*[*Calculus1*]) command.

$$f := x \rightarrow 2 \cdot x^4 + 3 \cdot x^3 - 20 \cdot x^2 - 24 \cdot x - 32$$
$$x \rightarrow 2\,x^4 + 3\,x^3 - 20\,x^2 - 24\,x - 32 \qquad \textbf{(3.19)}$$

with (*Student*[*Calculus1*]) :

Tangent $(f(x), x = -1)$

$$17\,x - 12 \qquad \textbf{(3.20)}$$

Note that if we want to find the slope of the tangent line or to plot the function along with its tangent we need to add options *output = slope* and *output = plot* .

Tangent $(f(x), x = -1, output = slope)$

$$17 \qquad \textbf{(3.21)}$$

Tangent $(f(x), x = -1, output = plot)$

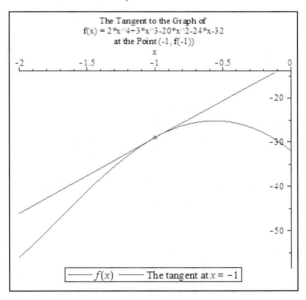

Step-by-step Differentiation

We will use the different rules for differentiation to find the derivatives of various functions using a step-by-step procedure. Just as we did in chapter 2, we will set the *infolevel* at 1. The inert form of the derivative is *Diff* . The rules of differentiation used are below. The *Maple* commands needed for step-by-step differentiation are given below.

Chapter 3 Derivatives

Rules of differentiation:

1. Let f and g be differentiable at x and c be any constant. Then

Rule for derivative	*Maple* rule for step-by-step differentiation
$(cf)'(x) = cf'(x)$	*constantmultiple*
$(f+g)'(x) = f'(x) + g'(x)$	*sum*
$(f-g)'(x) = f'(x) - g'(x)$	*difference*
$(fg)'(x) = g(x)f'(x) + f(x)g'(x)$	*product*
$\left(\dfrac{f}{g}\right)'(x) = \dfrac{g(x)f'(x) - f(x)g'(x)}{\left[g(x)\right]^2}$	*quotient*

2. (Chain rule) Let g be differentiable at x and f be differentiable at $g(x)$. Then the composite function $f \circ g$ is differentiable at x and

Rule for derivative	*Maple* rule for step-by-step differentiation
$(f \circ g)'(x) = f'(g(x)) \cdot g'(x)$	*chain*

3. (Derivatives of certain functions) Some of the functions listed below do not have specific *Maple* rules for step-by-step differentiation. These can be rewritten as a composition of two functions for which there is a *Maple* command for step-by-step differentiation. Note that there are no rules for $\left(a^x\right)' = a^x \ln(a)$ and $\left(\log_a(x)\right)' = \dfrac{1}{x \ln(a)}$.

Rule for derivative	*Maple* rule for step-by-step differentiation
$c' = 0$, where c is a constant	*constant*
$(x)' = 1$	*identity*
$\left(x^n\right)' = nx^{n-1}$, where n is a real number	*power*
$(\sin(x))' = \cos(x)$	*sin*

Rule for derivative	*Maple* rule for step-by-step differentiation
$\left(\cos(x)\right)' = -\sin(x)$	cos
$\left(\tan(x)\right)' = \sec^2(x)$	tan
$\left(\cot(x)\right)' = -\csc^2(x)$	cot
$\left(\sec(x)\right)' = \sec(x)\tan(x)$	sec
$\left(\csc(x)\right)' = \csc(x)\cot(x)$	csc
$\left(e^x\right)' = e^x$	exp
$\left(a^x\right)' = a^x \ln(a)$, where a is a positive real number and $a \neq 1$.	
$\left(\ln(x)\right)' = \dfrac{1}{x}$	ln
$\left(\log_a(x)\right)' = \dfrac{1}{x\ln(a)}$, where a is a positive real number and $a \neq 1$.	
$\left(\sin^{-1}(x)\right)' = \dfrac{1}{\sqrt{1-x^2}}$	arcsin
$\left(\cos^{-1}(x)\right)' = -\dfrac{1}{\sqrt{1-x^2}}$	arccos
$\left(\tan^{-1}(x)\right)' = \dfrac{1}{1+x^2}$	arctan
$\left(\cot^{-1}(x)\right)' = -\dfrac{1}{1+x^2}$	arccot
$\left(\sec^{-1}(x)\right)' = \dfrac{1}{x\sqrt{x^2-1}}$	arcsec
$\left(\csc^{-1}(x)\right)' = -\dfrac{1}{x\sqrt{x^2-1}}$	arccsc

Example 3: Use the step-by-step procedure in *Maple* to differentiate $f(x) = x^3 e^x$.

Solution: We will begin by executing the *restart* command. Activate the $with\left(Student\left[Calculus1\right]\right)$ package and set $infolevel\left[Student\left[Calculus1\right]\right] := 1$. We will also use the inert form of the derivative which is the *Diff* command. This command will return the derivative unevaluated.

restart

$with\left(Student\left[Calculus1\right]\right):$

$infolevel\left[\,Student\left[Calculus1\right]\right]:=1:$

$Rule\left[\;\right]\left(Diff\left(x^{3}\cdot e^{x},x\right)\right)$

```
Creating problem #1
```

$$\frac{d}{dx}\left(x^{3}\,e^{x}\right)=\frac{d}{dx}\left(x^{3}\,e^{x}\right) \tag{3.22}$$

Do not use the letter e on the keyboard but use e^{a} from the Expression palette. Since $f(x)=x^{3}e^{x}$ is the product of two functions, use the product rule to differentiate this function. We use the *product* command here.

$Rule\left[\,product\right]\left(\left(\mathbf{3.22}\right)\right)$

$$\frac{d}{dx}\left(x^{3}\,e^{x}\right)=\left(\frac{d}{dx}\left(x^{3}\right)\right)e^{x}+x^{3}\left(\frac{d}{dx}\,e^{x}\right) \tag{3.23}$$

We use the *power* rule to differentiate x^{3}.

$Rule\left[\,power\right]\left(\left(\mathbf{3.23}\right)\right)$

$$\frac{d}{dx}\left(x^{3}\,e^{x}\right)=3\,x^{2}\,e^{x}+x^{3}\left(\frac{d}{dx}\,e^{x}\right) \tag{3.24}$$

In order to complete the problem use the rule *exp* to differentiate e^{x}.

$Rule\left[\exp\right]\left(\left(\mathbf{3.24}\right)\right)$

$$\frac{d}{dx}\left(x^{3}\,e^{x}\right)=3\,x^{2}\,e^{x}+x^{3}\,e^{x} \tag{3.25}$$

Therefore, the derivative of $f(x)=x^{3}e^{x}$ is $3x^{2}e^{x}+x^{3}e^{x}$. Finally we can show the step-by-step computation of evaluating this derivative using the *ShowSteps* command.

$ShowSteps(\;)$

$$\frac{d}{dx}\left(x^{3}\,e^{x}\right)=\frac{d}{dx}\left(x^{3}\right)e^{x}+x^{3}\,\frac{d}{dx}\,e^{x} \tag{3.26}$$
$$=3\,x^{2}\,e^{x}+x^{3}\,\frac{d}{dx}\left(e^{x}\right)$$
$$=3\,x^{2}\,e^{x}+x^{3}\,e^{x}$$

Example 4: Use the step-by-step procedure in *Maple* to differentiate $f(x) = x \arcsin(x) + \sqrt{1-x^2}$.

Solution: As was the case in Example 3, we will begin by executing the *restart* command, activating the $with(Student[Calculus1])$ package and setting the $infolevel[Student[Calculus1]] := 1$.

restart

$with(Student[Calculus1]):$

$infolevel[Student[Calculus1]] := 1:$

$Rule[\ ]\left(Diff\left(x \cdot \arcsin(x) + \sqrt{1-x^2}, x\right)\right)$

```
Creating problem #1
```

$$\frac{d}{dx}\left(x \arcsin(x) + \sqrt{1-x^2}\right) = \frac{d}{dx}\left(x \arcsin(x) + \sqrt{1-x^2}\right) \qquad (3.27)$$

Since $f(x) = x \arcsin x + \sqrt{1-x^2}$ is the sum of two functions we use the *sum* command first.

$Rule[sum]((3.27))$

$$\frac{d}{dx}\left(x \arcsin(x) + \sqrt{1-x^2}\right) = \frac{d}{dx}\left(x \arcsin(x)\right) + \frac{d}{dx}\left(\sqrt{1-x^2}\right) \qquad (3.28)$$

As the first derivative is a product, we use the *product* command.

$Rule[product]((3.28))$

$$\frac{d}{dx}\left(x \arcsin(x) + \sqrt{1-x^2}\right) = \left(\frac{d}{dx}x\right)\arcsin(x) + x\left(\frac{d}{dx}\arcsin(x)\right) + \frac{d}{dx}\left(\sqrt{1-x^2}\right)$$
$$(3.29)$$

We use the *identity* command to differentiate x.

$Rule[identity]((3.29))$

$$\frac{d}{dx}\left(x \arcsin(x) + \sqrt{1-x^2}\right) = \arcsin(x) + x\left(\frac{d}{dx}\arcsin(x)\right) + \frac{d}{dx}\left(\sqrt{1-x^2}\right) \qquad (3.30)$$

We use the *arcsin* command to the second term.

$Rule[arcsin]((3.30))$

Chapter 3 Derivatives

$$\frac{d}{dx}\left(x\arcsin(x)+\sqrt{1-x^2}\right)=\arcsin(x)+\frac{x}{\sqrt{1-x^2}}+\frac{d}{dx}\left(\sqrt{1-x^2}\right) \qquad (3.31)$$

Since the last term is the composition of two functions (what are these functions?) we use the *chain* command to differentiate it.

$Rule\left[chain\right]\left(\left(\mathbf{3.31}\right)\right)$

$$\frac{d}{dx}\left(x\arcsin(x)+\sqrt{1-x^2}\right)=\arcsin(x)+\frac{x}{\sqrt{1-x^2}}+\left(\frac{d}{d\_X}\left(\sqrt{\_X}\right)\Big|_{\_X=1-x^2}\right)\left(\frac{d}{dx}\left(1-x^2\right)\right) \qquad (3.32)$$

Notice how the chain rule is used here as the derivative of the outside (the square root function) times the derivative of the inside $(1-x^2)$. We first differentiate $1-x^2$ using the *difference* (since it is a difference of 1 and x^2), *constant* (to differentiate 1), and *power* (to differentiate x^2) commands in that order.

$Rule\left[difference\right]\left(\left(\mathbf{3.32}\right)\right)$

$$\frac{d}{dx}\left(x\arcsin(x)+\sqrt{1-x^2}\right)=\arcsin(x)+\frac{x}{\sqrt{1-x^2}}+\left(\frac{d}{d\_X}\left(\sqrt{\_X}\right)\Big|_{\_X=1-x^2}\right)\left(\frac{d}{dx}1-\frac{d}{dx}\left(x^2\right)\right) \qquad (3.33)$$

$Rule\left[constant\right]\left(\left(\mathbf{3.33}\right)\right)$

$$\frac{d}{dx}\left(x\arcsin(x)+\sqrt{1-x^2}\right)=\arcsin(x)+\frac{x}{\sqrt{1-x^2}}-\left(\frac{d}{d\_X}\left(\sqrt{\_X}\right)\Big|_{\_X=1-x^2}\right)\left(\frac{d}{dx}\left(x^2\right)\right) \qquad (3.34)$$

$Rule\left[power\right]\left(\left(\mathbf{3.34}\right)\right)$

$$\frac{d}{dx}\left(x\arcsin(x)+\sqrt{1-x^2}\right)=\arcsin(x)+\frac{x}{\sqrt{1-x^2}}-2\left(\frac{d}{d\_X}\left(\sqrt{\_X}\right)\Big|_{\_X=1-x^2}\right)x \qquad (3.35)$$

Notice that the function x^2 is differentiated before $\sqrt{\_X}$. This is because we first used the *difference* rule to differentiate $1-x^2$. Now in order to complete the problem we use the *power* command to differentiate the square root function.

$Rule\left[power\right]\left(\left(\mathbf{3.35}\right)\right)$

$$\frac{d}{dx}\left(x\arcsin(x)+\sqrt{1-x^2}\right)=\arcsin(x) \qquad (3.36)$$

Therefore, the derivative of $f(x)=x\arcsin(x)+\sqrt{1-x^2}$ is $\arcsin(x)$.

Example 5: Use the step-by-step procedure in *Maple* to differentiate $y = x^{\frac{1}{x}}$.

Solution: As was the case in Example 3, we will begin by executing the *restart* command, activating the $with\left(Student\left[Calculus1\right]\right)$ package and setting the $infolevel\left[Student\left[Calculus1\right]\right]:=1$.

$restart$

$with\left(Student\left[Calculus1\right]\right):$

$infolevel\left[Student\left[Calculus1\right]\right]:=1:$

$Rule[\ ]\left(Diff\left(x^{\frac{1}{x}},x\right)\right)$

Creating problem #1

$$\frac{d}{dx}\left(x^{\frac{1}{x}}\right)=\frac{d}{dx}\left(x^{\frac{1}{x}}\right) \tag{3.37}$$

Since the exponent of $y = x^{\frac{1}{x}}$ is a function of x, we cannot use the *power* rule. Take logarithms of both sides and use the properties of logarithms to obtain

$$\ln(y)=\frac{1}{x}\ln(x)=\frac{\ln(x)}{x}$$

so that

$$y=e^{\frac{\ln(x)}{x}}.$$

Use the *rewrite* command first and put the function in this form in order to find its derivative.

$$Rule\left[rewrite,x^{\frac{1}{x}}=e^{\frac{\ln(x)}{x}}\right]\left(\left(3.37\right)\right)$$

$$\frac{d}{dx}x^{\frac{1}{x}}=\frac{d}{dx}e^{\frac{\ln(x)}{x}} \tag{3.38}$$

Utilize the *chain* command first.

$$Rule\left[chain\right]\left(\left(3.38\right)\right)$$

Chapter 3 Derivatives

$$\frac{d}{dx}x^{\frac{1}{x}} = \left(\frac{d}{d\_X}\left(e^{-X}\right)\Big|_{\_X=\frac{\ln(x)}{x}}\right)\left(\frac{d}{dx}\left(\frac{\ln(x)}{x}\right)\right) \tag{3.39}$$

We use the *quotient* command to differentiate $\dfrac{\ln(x)}{x}$.

Rule[*quotient*]((**3.39**))

$$\frac{d}{dx}x^{\frac{1}{x}} = \frac{\left(\frac{d}{d\_X}\left(e^{-X}\right)\Big|_{\_X=\frac{\ln(x)}{x}}\right)\left(\left(\frac{d}{dx}\ln(x)\right)x - \ln(x)\left(\frac{d}{dx}x\right)\right)}{x^2} \tag{3.40}$$

Now use the *ln* command to differentiate $\ln(x)$.

Rule[*ln*]((**3.40**))

$$\frac{d}{dx}x^{\frac{1}{x}} = \frac{\left(\frac{d}{d\_X}\left(e^{-X}\right)\Big|_{\_X=\frac{\ln(x)}{x}}\right)\left(1 - \ln(x)\left(\frac{d}{dx}x\right)\right)}{x^2} \tag{3.41}$$

We use the *identity* command to differentiate x.

Rule[*identity*]((**3.41**))

$$\frac{d}{dx}x^{\frac{1}{x}} = \frac{\left(\frac{d}{d\_X}\left(e^{-X}\right)\Big|_{\_X=\frac{\ln(x)}{x}}\right)\left(1 - \ln(x)\right)}{x^2} \tag{3.42}$$

Finally in order to complete the problem we use the *exp* command to differentiate the function e^{-X}.

Rule[*exp*]((**3.42**))

$$\frac{d}{dx}x^{\frac{1}{x}} = \frac{e^{\frac{\ln(x)}{x}}\left(1 - \ln(x)\right)}{x^2} \tag{3.43}$$

Therefore, the derivative of $y = x^{\frac{1}{x}}$ is $\dfrac{e^{\frac{\ln(x)}{x}}\left(1-\ln(x)\right)}{x^2} = \dfrac{x^{\frac{1}{x}}\left(1-\ln(x)\right)}{x^2} = x^{\frac{1}{x}-2}\left(1-\ln(x)\right)$

(since $y = x^{\frac{1}{x}} = e^{\frac{\ln(x)}{x}}$).

Applications of the Derivative

The following example is an application of the derivative.

Example 6: An object suspended to a string is pulled and released. Assume that the spring is perfect and we neglect friction, then the object has an oscillatory motion. If $x(t)$ is the position of the object at time t seconds, then $x(t) = 5\sin\left(10\sqrt{5}t + \dfrac{\pi}{2}\right)$. Find

(a) the velocity and acceleration of the object,
(b) the position and velocity at time $t = 0$ seconds.

Solution: Note that $x(t) = 5\sin\left(10\sqrt{5}t + \dfrac{\pi}{2}\right) = 5\cos\left(10\sqrt{5}t\right)$. We enter this formula of $x(t)$ as a *Maple* function.

$x := t \rightarrow 5 \cdot \cos\left(10 \cdot \sqrt{5} \cdot t\right)$

$$t \rightarrow 5\cos\left(10\sqrt{5}\,t\right) \tag{3.44}$$

(a) Let the velocity of the object be $v(t)$. Then $v(t)$ is the derivative of the position of the object $x(t)$. We will use *Maple* to find the derivative of $x(t)$.

$v := D(x)$

$$t \rightarrow -50\sin\left(10\sqrt{5}\,t\right)\sqrt{5} \tag{3.45}$$

If the acceleration of the object is denoted by $a(t)$, then $a(t) = v'(t) = x''(t)$. We will use *Maple* to find $a(t)$.

$a := D(v)$

$$t \rightarrow -500\cos\left(10\sqrt{5}\,t\right)\left(\sqrt{5}\right)^2 \tag{3.46}$$

Now simplify this expression by right clicking on it, selecting *Evaluate Procedure*, and clicking on *Enter*.

evaluate procedure →

$$-2500\cos\left(10\sqrt{5}\,t\right) \tag{3.47}$$

(b)   To find the position and velocity at time $t = 0$ we will evaluate $x(0)$ and $v(0)$.

$x(0)$

$$5 \tag{3.48}$$

$v(0)$

$$0 \tag{3.49}$$

Hence the position of the object at $t = 0$ seconds is 5 units and the velocity is 0 units per seconds.

Implicit Differentiation

In the following examples, we will find the derivative of implicit functions.

Example 9: Find the derivative $\dfrac{dy}{dx}$ of $4x^2 + 5y^3 = 20$ both implicitly and explicitly.

Solution: We will enter $4x^2 + 5y^3 = 20$ as a *Maple* expression and assign a name g to this expression. We will first find the derivative implicitly using the *implicitdiff* command to find the derivative as shown below. We will call the derivative $g1x$.

$g := 4x^2 + 5y^3 = 20$

$$4x^2 + 5y^3 = 20 \tag{3.50}$$

$g1x := implicitdiff\,(g, y, x)$

$$-\frac{8}{15}\frac{x}{y^2} \tag{3.51}$$

By using implicit differentiation, we see that the derivative $\dfrac{dy}{dx} = -\dfrac{8}{15}\dfrac{x}{y^2}$. In order to find the derivative explicitly, we will use $4x^2 + 5y^3 = 20$ and solve for y in terms of x using the *solve* command. Note that there will be three solutions as the exponent of y is 3.

$A := \left[\,solve\,(g, y)\right]$

$$\left[\frac{1}{5}\left(-100x^2 + 500\right)^{1/3}, -\frac{1}{10}\left(-100x^2 + 500\right)^{1/3} - \frac{1}{10}I\sqrt{3}\left(-100x^2 + 500\right)^{1/3}, -\frac{1}{10}\left(-100x^2 + 500\right)^{1/3}\right.$$
$$\left.+\frac{1}{10}I\sqrt{3}\left(-100x^2 + 500\right)^{1/3}\right] \tag{3.52}$$

We only want to consider the real solutions and delete the solutions that are complex. This can be achieved by using the *remove* command.

$B := remove(has, A, I)$

$$\left[\frac{1}{5} \left(-100x^2 + 500 \right)^{1/3} \right] \tag{3.53}$$

The derivative is now calculated for this solution using the Expression palette.

$\dfrac{d}{dx} B$

$$\left[-\frac{40}{3} \frac{x}{\left(-100x^2 + 500 \right)^{2/3}} \right] \tag{3.54}$$

Therefore, $\dfrac{dB}{dx} = -\dfrac{40}{3} \dfrac{x}{\left(-100x^2 + 500 \right)^{2/3}}$ is calculated explicitly. To check if the derivatives found explicitly and implicitly are the same, substitute B for y into the equation for $g1x$ and compare the results.

$subs(y = B, g1x)$

$$-\frac{8}{15} \frac{x}{\left[\frac{1}{5} \left(-100x^2 + 500 \right)^{1/3} \right]^2} \tag{3.55}$$

Since $-\dfrac{8}{15} \times \dfrac{1}{\left(\dfrac{1}{5} \right)^2} = -\dfrac{8}{15} \times \dfrac{25}{1} = -\dfrac{40}{3}$, both methods give the same answer.

Remark: Alternately, we can enter $4x^2 + 5y^3 = 20$ as a *Maple* expression and assign a name g to this expression. We can differentiate implicitly by right clicking on the expression $4x^2 + 5y^3 = 20$ and then selecting *Differentiate Implicitly*. Make sure that the independent variable is x and the dependent variable is y. The rest of the boxes (isolate for the derivative, include only right-hand side, and strip dependencies in the output) must be checked.

$g := 4x^2 + 5y^3 = 20$

$$4x^2 + 5y^3 = 20 \tag{3.56}$$

$\xrightarrow{\text{implicit differentiation}}$

$$-\frac{8}{15}\frac{x}{y^2} \qquad (3.57)$$

Just as we did in the remark after Example 2, we could assign the name $g1x$ to it.

$$\xrightarrow{\text{assign to a name}}$$

$$g1x \qquad (3.58)$$

Recall: The equation of the tangent line to the curve g at the point (x_1, y_1) is given by

$$y = y_1 + g1(x_1, y_1) \cdot (x - x_1)$$

where $g1(x_1, y_1)$ is the derivative of g with respect to x evaluated at the point (x_1, y_1).

Example 10: Use *Maple* commands to find the equation of the tangent line to the graph of $x^2 + y^2 = 100$ at the point (6, 8). Graph the curve along with this tangent line.

Solution: We will enter $x^2 + y^2 = 100$ as a *Maple* expression and assign a name to this expression.

$$g := x^2 + y^2 = 100$$

$$x^2 + y^2 = 100 \qquad (3.59)$$

Note that $x^2 + y^2 = 100$ is the equation of a circle with center (0, 0) and radius 10. The circle is drawn using the *implicitplot* command. However, before using this command, activate the *with(plots)* library.

$with(plots):$

$implicitplot(g, x = -10 .. 10, y = -10 .. 10)$

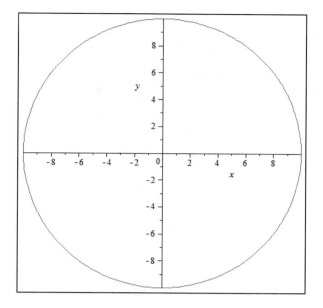

The slope of the tangent line is obtained by first finding the derivative $\dfrac{dy}{dx}$ using the *implicitdiff* command (this will be assigned the name *g1x*).

$g1x := implicitdiff\left(g, y, x\right)$

$$-\frac{x}{y} \tag{3.60}$$

Therefore, the derivative $\dfrac{dy}{dx} = -\dfrac{x}{y}$. The slope *m* of the tangent line at the point (6, 8) is found by substituting $x = 6$ and $y = 8$ in *g1x*.

$m := subs\left(x = 6, y = 8, g1x\right)$

$$-\frac{3}{4} \tag{3.61}$$

Hence the slope of the tangent line at the point (6, 8) is $-\dfrac{3}{4}$. The equation of the tangent line is found using the formula given before this example and then using the isolate command.

$L := y = 8 + m \cdot \left(x - 6\right)$

$$y = -\frac{3}{4}x + \frac{25}{2} \tag{3.62}$$

Therefore, the equation of the tangent line at the point (6, 8) is $y = -\dfrac{3}{4}x + \dfrac{25}{2}$. The graphs of both the circle and the tangent line are plotted below.

$implicitplot\left([g,L], x = -10..20, y = -10..18, color = [red, black]\right)$

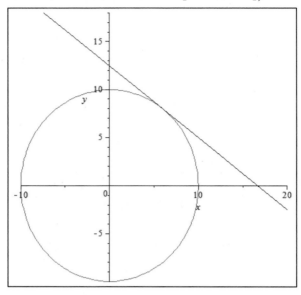

Notice that the circle in the graph above looks more like an ellipse. The reason for this distortion is that the scales are set by *Maple* to fit the window by default. This graph can be "corrected" by using the *scaling = constrained* in the *implicitplot* command as shown below.

$implicitplot\left([g,L], x = -10..20, y = -10..18, color = [red, black], scaling = constrained\right)$

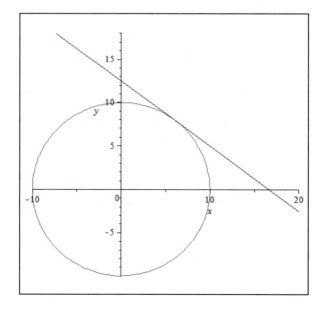

The example below gives a method of finding the equations of the horizontal and vertical tangent lines to a curve (if they exist).

Recall: The slope of a horizontal line is zero so that the tangent line is horizontal when the derivative $\dfrac{dy}{dx}$ is zero. Furthermore, the slope of a vertical line is undefined so that the tangent line is vertical when the derivative $\dfrac{dy}{dx}$ is undefined. The equation of a horizontal line is of the form $y = k$ and that of a vertical line is $x = h$.

Example 11: Use *Maple* commands to find the equations of the horizontal and vertical tangent lines to the graph of $x^2 + 4x + 4y^2 - 8y = 1$. Graph the curve along with these tangent lines.

Solution: We will enter $x^2 + 4x + 4y^2 - 8y = 1$ as a *Maple* expression and assign a name to this expression.

$g := x^2 + 4 \cdot x + 4 \cdot y^2 - 8 \cdot y = 1$

$$x^2 + 4\,x + 4\,y^2 - 8\,y = 1 \tag{3.63}$$

Note that $x^2 + 4x + 4y^2 - 8y = 1$ is the equation of an ellipse. This ellipse is sketched using the *implicitplot* command. Do not forget to activate the *with*(*plots*) library.

with(*plots*):

implicitplot$\left(g, x = -5..1, y = -0.5..2.5, scaling = constrained\right)$

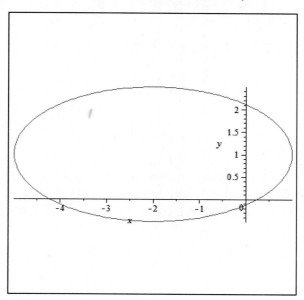

In order to find the equation of the horizontal and vertical tangent lines we first need to find the derivative $\dfrac{dy}{dx}$ using the *implicitdiff* command.

$g1x := implicitdiff\left(g, y, x\right)$

$$-\frac{1}{4}\frac{x+2}{y-1} \tag{3.64}$$

Thus, $\dfrac{dy}{dx} = -\dfrac{1}{4}\dfrac{x+2}{y-1}$. To find the horizontal tangent line we set the derivative $\dfrac{dy}{dx}$ equal to zero, solve for x, substitute this value of x in g and then solve for y.

$solve\left(g1x = 0, x\right)$

$$-2 \tag{3.65}$$

$subs\left(x = -2, g\right)$

$$-4 + 4y^2 - 8y = 1 \tag{3.66}$$

$\xrightarrow{\text{solve for } y}$

$$\left[\left[y = \frac{5}{2}\right], \left[y = -\frac{1}{2}\right]\right] \tag{3.67}$$

Hence the equations of the horizontal tangent lines are $y = \dfrac{5}{2}$ and $y = -\dfrac{1}{2}$. In order to find the vertical tangent lines, we first set the denominator of the derivative equal to zero, solve for y, substitute this value of y in g and then solve for x.

$solve\left(denom\left(g1x\right) = 0, y\right)$

$$1 \tag{3.68}$$

$subs\left(y = 1, g\right)$

$$x^2 + 4x - 4 = 1 \tag{3.69}$$

$\xrightarrow{\text{solve for } x}$

$$\left[\left[x = 1\right], \left[x = -5\right]\right] \tag{3.70}$$

Hence the equations of the vertical tangent lines are $x = 1$ and $x = -5$. The graph of the ellipse along with its horizontal and vertical tangent lines is shown below. We will increase the number of points that are used to plot a function by using the *grid* command.

$$implicitplot\left(\left[g, y = \frac{5}{2}, y = -\frac{1}{2}, x = 1, x = -5\right], x = -5.5..1.5, y = -1..3, scaling = constrained,\right.$$

$$\left. color = [red, green, blue, black, brown],\quad grid = [300, 300], thickness = 2\right)$$

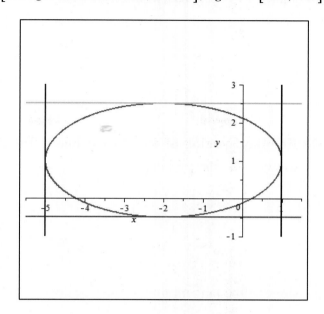

Linear Approximation

Let $y = f(x)$ be differentiable at x_1. Then the equation of the tangent line to the curve $y = f(x)$ at the point (x_1, y_1) is given by $y = y_1 + f'(x_1) \cdot (x - x_1)$. If the value of x is sufficiently close to x_1, then the graph of the function $f(x)$ looks like the graph of the tangent line. Hence the value of $f(x)$ is approximately $f(x) \approx f(x_1) + f'(x_1) \cdot (x - x_1)$. This process is called the linear approximation of the function f at x_1. It is used to approximate nonlinear functions with linear functions. The next example will illustrate this process.

Example 12: Consider the function $f(x) = x^{-\frac{1}{3}}$.

(a)   Find the equation of the tangent line to the curve $f(x) = x^{-\frac{1}{3}}$ at the point $x = 1$.

(b)   Graph $f(x) = x^{-\frac{1}{3}}$ along with this tangent line.

(c)   Find a linear approximation to the function at $x = 1$.

(d)   Use this linear approximation to approximate $(1.02)^{-\frac{1}{3}}$.

(e)   Use *Maple* commands to find the value of $(1.02)^{-\frac{1}{3}}$.

(f)   Compare your answers in parts (d) and (e). Is the answer in part (d) a good estimate of your answer in part (e)? Give reasons for your answer.

Solution: We will enter $f(x) = x^{-\frac{1}{3}}$ as a *Maple* function.

$$f := x \rightarrow x^{-\frac{1}{3}}$$

$$x \rightarrow \frac{1}{x^{1/3}} \qquad \textbf{(3.71)}$$

(a)   In order to find the equation of the tangent line to the curve $f(x) = x^{-\frac{1}{3}}$ at the point $x = 1$ we will find the first derivative

$$f1 := D(f)$$

$$x \rightarrow -\frac{1}{3\,x^{4/3}} \qquad \textbf{(3.72)}$$

To find the slope of the tangent line we will evaluate $f1(1)$.

$$f1(1)$$

$$-\frac{1}{3} \qquad \textbf{(3.73)}$$

Hence the slope of the tangent line when $x = 1$ is $-\frac{1}{3}$. Next we find the equation of the tangent line.

$$L := x \rightarrow f(1) + f1(1) \cdot (x-1)$$

$$x \rightarrow f(1) + f1(1)(x-1) \qquad \textbf{(3.74)}$$

$$expand(L(x))$$

$$\frac{4}{3} - \frac{1}{3}x \qquad \textbf{(3.75)}$$

Therefore, the equation of the tangent line is $y = \frac{4}{3} - \frac{1}{3}x$. Plot the function along with its tangent line at $x = 1$.

$$plot\left(\left[f(x), L(x)\right], x = 0..1.7, y = 0.8..1.5, color = \left[red, black\right]\right)$$

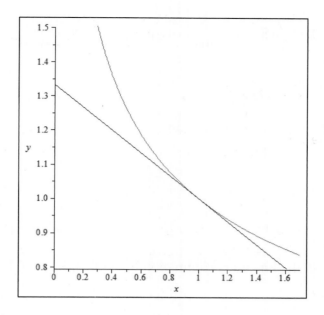

We will consider a very small interval around $x = 1$, and plot the graph of f and its tangent line. Notice that, in the plot below, the graphs of the function $f(x) = x^{-\frac{1}{3}}$ and its tangent line $y = \dfrac{4}{3} - \dfrac{1}{3}x$ when $x = 1$ are almost identical.

$$plot\left(\left[f(x), L(x)\right], x = 0.95\,..\,1.05, y = 0.985\,..\,1.015, color = \left[red, black\right]\right)$$

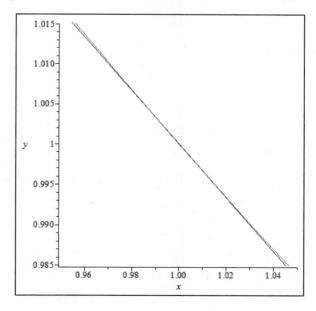

(c)   Hence the linear approximation to the function $f(x) = x^{\frac{1}{3}}$ at $x = 1$ is

$$x^{-\frac{1}{3}} \approx \frac{4}{3} - \frac{1}{3}x.$$

(d)   Using the approximation in part (c) we can find an approximation for $(1.02)^{-\frac{1}{3}}$ by evaluating $L(1.02)$.

$L(1.02)$

$$0.9933333333 \tag{3.76}$$

Therefore, the linear approximation for $(1.02)^{-\frac{1}{3}}$ is $(1.02)^{-\frac{1}{3}} \approx 0.9933333333$.

(e)   We will now use *Maple* commands to evaluate $(1.02)^{-\frac{1}{3}}$.

$f(1.02)$

$$0.9934208622 \tag{3.77}$$

Hence, $(1.02)^{-\frac{1}{3}} \approx 0.9934208622$, using *Maple* commands.

(f)   Since

$$\left| f(1.02) - L(1.02) \right|$$

$$0.0000875289 \tag{3.78}$$

the error in the two calculations is 0.0000875289 which is a relatively small number. Hence we can conclude that the answer given in part (d) is a good approximation to the answer in part (e). In other words, this linear approximation is a good approximation for the value of $x^{-\frac{1}{3}}$ as long as x is near 1.

Differentials

Let f be a differentiable function of x. Define the differential dx as a real valued independent variable. The differential dy is then defined as $dy = f'(x)dx$. Let P be a point $(x, y) = (x, f(x))$ on the graph of f. Let Q be the point $(x + \Delta x, f(x + \Delta x))$ on the graph of f that is sufficiently close to P so that the change in x, namely Δx, is small. Then the change in y, denoted by Δy, as we move from the point P to the point Q along the graph of f is $\Delta y = f(x + \Delta x) - f(x)$. Hence $(x + \Delta x, y + \Delta y) = (x + \Delta x, f(x) + \Delta y) =$

$(x+\Delta x, f(x+\Delta x))$ are the coordinates of the point Q. Now assume that $dx = \Delta x$ so that the differential dx is also small. Let R be the point $(x+dx, y+dy)$. Then R is on the tangent line to the curve at the point P and is sufficiently close to P. As the differential dx is small, the point Q is close to the point R. Therefore, $\Delta y \approx dy$, which implies that $f(x+\Delta x) \approx f(x)+dy$. This is another way to write the linear approximation for f using differentials. This is illustrated in the plot below where the function $f(x) = \sin(x)$ is graphed along with its tangent line at $x = \dfrac{\pi}{3}$.

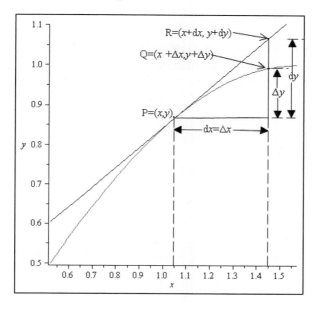

Example 13: Consider the function $f(x) = \sin(x)$.

(a)   Find the equation of the tangent line to the curve $f(x) = \sin(x)$ at the point $x = \dfrac{\pi}{3}$.

(b)   Graph $f(x) = \sin(x)$ along with this tangent line.

(c)   Find the differential dy.

(d)   Find the value of dy when the value of x changes from $x = \dfrac{\pi}{3}$ to $x = 0.34\pi$. Use this value to approximate $\sin(0.34\pi)$.

(e)   Use *Maple* commands to find the value of $\sin(0.34\pi)$.

(f)   Compare your answers in parts (d) and (e). Is the answer in part (d) a good estimate of your answer in part (e)? Give reasons for your answer.

Solution: We will enter $f(x) = \sin(x)$ as a *Maple* function.

$$f := x \rightarrow \sin(x)$$

$$x \rightarrow \sin(x) \qquad \textbf{(3.79)}$$

(a)   In order to find the equation of the tangent line to the curve $f(x) = \sin(x)$ at the point $x = \dfrac{\pi}{3}$ we will find the first derivative

$$f1 := D(f)$$

$$x \rightarrow \cos \qquad \textbf{(3.80)}$$

Note that the angle x is not included in the output for the derivative. This is a *Maple* notation. In order to find the slope of the tangent line we will evaluate $f1\left(\dfrac{\pi}{3}\right)$.

$$f1\left(\frac{\pi}{3}\right)$$

$$\frac{1}{2} \qquad \textbf{(3.81)}$$

Hence the slope of the tangent line when $x = \dfrac{\pi}{3}$ is $\dfrac{1}{2}$. We will now find the equation of the tangent line.

$$L := x \rightarrow f\left(\frac{\pi}{3}\right) + f1\left(\frac{\pi}{3}\right) \cdot \left(x - \frac{\pi}{3}\right)$$

$$x \rightarrow f\left(\frac{1}{3}\pi\right) + f1\left(\frac{1}{3}\pi\right)\left(x - \frac{1}{3}\pi\right) \qquad \textbf{(3.82)}$$

$expand(L(x))$

$$\frac{1}{2}\sqrt{3} + \frac{1}{2}x - \frac{1}{6}\pi \qquad \textbf{(3.83)}$$

Therefore, the equation of the tangent line is $y = \dfrac{1}{2}\sqrt{3} + \dfrac{1}{2}x - \dfrac{1}{6}\pi$. Plot the function along with its tangent line at $x = \dfrac{\pi}{3}$.

$$plot\left(\left[f(x),L(x)\right],x=\frac{\pi}{6}..\frac{\pi}{2},y=0.5..1,color=\left[red,black\right]\right)$$

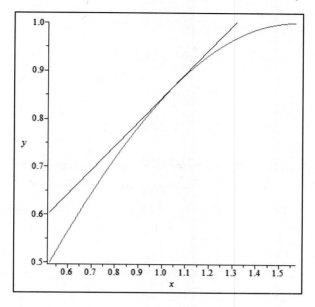

(c)   Since $dy=f'(x)dx$, we see that the differential dy is given by $dy=\cos(x)dx$.

(d)   To compare the values of dy and Δy when the value of x changes from $x=\frac{\pi}{3}$ to $x=0.34\pi$ we will first find Δx which is the same as dx, using *Maple*.

$$evalf\left(0.34\cdot\pi-\frac{\pi}{3}\right)$$

$$0.02094395113 \hspace{4cm} \textbf{(3.84)}$$

Therefore, when $x=\frac{\pi}{3}$ changes to $x=0.34\pi$ we see that the change in x is $\Delta x=dx=0.02094395113$.

$$evalf\left(f1\left(\frac{\pi}{3}\right)\cdot0.02094395113\right)$$

$$0.01047197556 \hspace{4cm} \textbf{(3.85)}$$

Hence, as $dy = \cos(x)dx = f1(x)dx$, $evalf\left(f1\left(\dfrac{\pi}{3}\right) \cdot 0.02094395113\right)$

$= 0.01047197556$ and as we see that $dy = 0.01047197556$. Since

$f(0.34\pi) \approx f\left(\dfrac{\pi}{3}\right) + dy$ and as

$evalf\left(f\left(\dfrac{\pi}{3}\right) + 0.01047197556\right)$

$$0.8764973796 \qquad\qquad\qquad\text{(3.86)}$$

we see that $\sin(0.34\pi) \approx 0.8764973796$.

(e) We will now use *Maple* commands to evaluate $\sin(0.34\pi)$.

$evalf\left(f\left(0.34 \cdot \pi\right)\right)$

$$0.8763066997 \qquad\qquad\qquad\text{(3.87)}$$

Hence $\sin(0.34\pi) \approx 0.8763066997$.

(f) Since

$\left|evalf\left(f\left(0.34 \cdot \pi\right)\right) - 0.8764973796\right|$

$$0.0001906997 \qquad\qquad\qquad\text{(3.88)}$$

the error in this case is 0.0001906997. Hence the answer given in part (d) is a good approximation to the answer in part (e).

Exercises

1. Let $f(x) = -3x + 7$. Use *Maple* commands to find the difference quotient $\dfrac{f(x+h)-f(x)}{h}$. How does your answer relate to the slope of the line $y = -3x + 7$? Give reasons for your answer. Use the definition of the derivative to find the derivative $f'(x)$.

2. Let $f(x) = 12x^3 - 8x^2 - 11x - 65$. Use *Maple* commands to find the difference quotient $\dfrac{f(x+h)-f(x)}{h}$. Use the definition of the derivative to find the derivative $f'(x)$.

3. Let $f(x) = \sqrt{x}$. Use *Maple* commands to find the difference quotient $\dfrac{f(x+h)-f(x)}{h}$. Can you use *Maple* commands to rationalize the denominator of this difference quotient? Give reasons for your answer. Use the definition of the derivative to find the derivative $f'(x)$.

4. Let $f(x) = \sqrt[5]{32x^5 + 1}$. Use *Maple* commands to find the difference quotient $\dfrac{f(x+h)-f(x)}{h}$. Can you use *Maple* commands to rationalize the denominator of this difference quotient? Give reasons for your answer. Use the definition of the derivative to find the derivative $f'(x)$.

5 – 7   Use *Maple* commands to find the difference quotient $\dfrac{f(x+h)-f(x)}{h}$. Use the definition of the derivative to find the derivative $f'(x)$.

5. $f(x) = \tan(x)$.

6. $f(x) = 11x - \sec(x)$.

7. $f(x) = \sin(\cot(x))$.

8. Let $f(x) = 4x - \cos(x)$. Find the equation of the tangent line to the graph of f when $x = -\dfrac{4\pi}{3}$. Graph f along with this tangent line.

9. Let $f(x) = \sin(5x)$. Find the equation of the tangent line to the graph of f when $x = \dfrac{7\pi}{6}$ and $x = \dfrac{\pi}{10}$. Graph f along with these tangent lines.

10. Let $f(x) = \dfrac{2x+7}{3x^2+1}$. Find the equation of the tangent line to the graph of f when $x = -2$ and $x = \dfrac{\sqrt{3}}{3}$. Graph f along with these tangent lines.

11. Use the step-by-step procedure in *Maple* to differentiate $f(x) = 11^x$.

12. Use the step-by-step procedure in *Maple* to differentiate $f(x) = \ln(x^3 e^x)$.

13. Use the step-by-step procedure in *Maple* to differentiate $f(x) = 2x \operatorname{arccot}(x) - \ln(x^2+1)$.

14. Use the step-by-step procedure in *Maple* to differentiate $f(x) = \sin(\cot(x))$. Is this answer obtained different from the answer obtained in exercise 7? Give reasons for your answer.

15. Use the step-by-step procedure in *Maple* to differentiate $f(x) = (\csc(x))^x$.

16. Find the derivative $\dfrac{dy}{dx}$ of $2x^2 + 2xy^3 = 5$ both implicitly and explicitly.

17. Use *Maple* commands to find the equation of the tangent line to the graph of $x^2 + xy + y^2 = 1$ at the point (-1, 1). Graph the curve along with this tangent line.

18. Use *Maple* commands to find the equation of the tangent line to the graph of $x^2 + y^2 = 2$ at the point (-1, 1). Graph the curve along with this tangent line.

19. Use *Maple* commands to find the equation of the tangent line to the graph of $9x^2 + 4y^2 = 52$ at the point (2, -2). Graph the curve along with this tangent line.

20. Use *Maple* commands to find the equation of the tangent line to the graph of $9x^2 + 4y^2 + 36x - 8y = 60$ at the point (-4, 5). Graph the curve along with this tangent line.

21. Use *Maple* commands to find the equation of the tangent line to the graph of $x^5 + x^2 y + xy^3 + y^4 = 4$ at the point (1, 1). Graph the curve along with this tangent line.

22.   Let $f(x) = \sqrt[5]{x}$.

(a)   Find the equation of the tangent line to the curve $f(x) = \sqrt[5]{x}$ at the point $x = 32$.

(b)   Graph $f(x) = \sqrt[5]{x}$ along with this tangent line.

(c)   Find a linear approximation to the function at $x = 32$.

(d)   Use this to approximate $\sqrt[5]{32.1661}$.

(e)   Use the square root command in *Maple* to find the value of $\sqrt[5]{32.1661}$.

(f)   Compare your answers in parts (d) and (e). Are the answers in part (d) a good estimate of your answers in part (e)? Give reasons for your answer.

(g)   Use differentials to get the estimate for $\sqrt[5]{32.1661}$. Compare your answers with part (d). Are the methods used different?

23.   Let $f(x) = \sqrt{x}$.

(a)   Find the equation of the tangent line to the curve $f(x) = \sqrt{x}$ at the point $x = 16$.

(b)   Graph $f(x) = \sqrt{x}$ along with this tangent line.

(c)   Find a linear approximation to the function at $x = 16$.

(d)   Use this to approximate $\sqrt{16.42}$ and $\sqrt{15.31}$.

(e)   Use the square root command in *Maple* to find the value of $\sqrt{16.42}$ and $\sqrt{15.31}$.

(f)   Compare your answers in parts (d) and (e). Are the answers in part (d) a good estimate of your answers in part (e)? Give reasons for your answer.

(g)   Use differentials to get the estimates for $\sqrt{16.42}$ and $\sqrt{15.31}$. Compare your answers with part (d). Are the methods used different?

24.   Use differentials to estimate:

(a) $\sqrt{25.0134}$.

(b) $\sqrt[5]{243.07}$.

25.   Use differentials to estimate the increase in area of a soap bubble when its radius increases from 2 centimeters to 2.014 centimeters. What assumptions are you making?

26.   Estimate (using differentials) the volume of a spherical shell of inner radius 2 centimeters and outer radius 2.014 centimeters.

27.   The side of a cubical box is measured as 9.8 centimeters with a possible error of ± 0.01. Show that an estimate of the error of the volume of the box is 941 ± 3 cubic centimeters.

28.*   Find a function that is continuous on all real numbers, but is not differentiable at $x = -1$ and $x = 1$. Make a plot of your graph over the interval $[-2, 2]$.

29.* Let $f(x) = \begin{cases} 2x & 0 \le x \le 1 \\ 4 - 2x & 1 < x \le 2 \end{cases}$.

    (a) Graph the function f. What are the domain and range of f?

    (b) Evaluate $\dfrac{f(1+h) - f(1)}{h}$ when $h > 0$.

    (c) Evaluate $\dfrac{f(1+h) - f(1)}{h}$ when $h < 0$.

    (d) Use parts (b) and (c) and the definition of the derivative to show that $f'(x)$ does not exist.

    (e) Consider the function $g(x) = f\left(x - 2\text{floor}\left(\dfrac{x}{2}\right)\right)$. Plot the graph of $g(x)$.

    Is $g(x)$ a continuous function? Give reasons for your answer.

30.* Let $x^2 y^2 - 2x = 4 - 4y$.

    (a) Show that $y' = \dfrac{2 - 2xy^2}{2x^2 y + 4}$.

    (b) Find the values of y when $x = 1.8$. Find the equation of the tangent lines at these points. Graph the curve along with its tangent lines at these points using *Maple* commands. (Hint: Use the solve command with $x = \dfrac{9}{5}$. Then check your answer with $x = 1.8$)

    (c) Show that we need to solve $4y^3 - 4y^2 - 1 = 0$ to find the points where the tangent line to $x^2 y^2 - 2x = 4 - 4y$ is horizontal. (**Hint**: Start by setting $\dfrac{dy}{dx} = 0$, that is $\dfrac{2 - 2xy^2}{2x^2 y + 4} = 0$, solve for x and substitute your answer into $x^2 y^2 - 2x = 4 - 4y$.)

    (d) Solve the equation $4y^3 - 4y^2 - 1 = 0$ with the help of *Maple* commands. (Observe that two of the roots are complex with $I = \sqrt{-1}$.)

    (e) From part (d) show that the equation of the horizontal tangent line is given by $y = \left(\dfrac{\left(35 + 3\sqrt{129}\right)^{1/3}}{6} + \dfrac{2}{3\left(35 + 3\sqrt{129}\right)^{1/3}} + \dfrac{1}{3} \right)$.

(f)   Plot the graph of $x^2 y^2 - 2x = 4 - 4y$ and its tangent line at the point

$(2, -2)$ and at $y = \left(\dfrac{\left(35 + 3\sqrt{129} \right)^{1/3}}{6} + \dfrac{2}{3\left(35 + 3\sqrt{129} \right)^{1/3}} + \dfrac{1}{3} \right)$ on the same

set of axes using *Maple* commands. Use different colors for the three graphs.

31.*   Let $x^3 + 2x^2 + 2 = 0$.

(a)   Show that we need to solve $x^3 + 2x^2 + 2 = 0$ to find the points where the tangent line to $x^2 y^2 - 2x = 4 - 4y$ is vertical. (**Hint**: Start by setting the denominator of $\dfrac{dy}{dx} = 0$, that is $2x^2 y + 4 = 0$, solve for y and substitute your answer into $x^2 y^2 - 2x = 4 - 4y$.)

(b)   Solve the equation $x^3 + 2x^2 + 2 = 0$ with the help of *Maple* commands. (Observe that two of the roots are complex with $I = \sqrt{-1}$.)

32.*   Exploration problem:

(a)   Give an interval of values for which $\sin(x) \approx x$

(b)   Give an interval of values for which $\cos(x) \approx 1 - \dfrac{x^2}{2}$

(c)   Verify your answers to parts (a) and (b) by using appropriate tables and graphs.

33.   Below is a list of some commands used in this chapter. Describe the significance of each command. Can you find examples where each command is used?

(a) $D(f)$

(b) $infolevel\left[Student\left[Calculus1 \right] \right] := n$

(c) $implicitplot\left(f(x), x = a .. b, y = c .. d \right)$

(d) $restart$

(e) $ShowSteps(\ )$

(f) $with(plots)$

(g) $with\left(Student\left[Calculus1 \right] \right)$

34.   Can you list new *Maple* commands that were used in this chapter but were not listed in exercise 33?

Chapter 4   Graphs of Functions using Limits and Derivatives

Techniques used to find the limits and derivatives of functions in chapters 2 and 3 will be used to graph different functions.

Definitions

Definition: A real-valued function f has an *absolute maximum* if there is a number c in the domain of f such that $f(c) \geq f(x)$ for every real number x in the domain of f.

Definition: A real-valued function f has an *absolute minimum* if there is a number c in the domain of f such that $f(c) \leq f(x)$ for every real number x in the domain of f.

Definition: A real valued function f has a *local maximum* if there is a number c in the domain of f such that $f(c) \geq f(x)$ for every real number x near c.

Definition: A real valued function f has a *local minimum* if there is a number c in the domain of f such that $f(c) \leq f(x)$ for every real number x near c.

Definition: A number c in the domain of a function f is a *critical number* of f if either $f'(c) = 0$ or $f'(c)$ does not exist.

Theorem: Let f be a differentiable real-valued function.
(a) If $f'(x) > 0$ for every real number x in an interval I in the domain of f, then f is increasing on I.
(b) If $f'(x) < 0$ for every real number x in an interval I in the domain of f, then f is decreasing on I.

First Derivative Test: Let c be a critical number of a differentiable function f.
(a) If f' changes from positive to negative at c, then f has a local maximum at c.
(b) If f' changes from negative to positive at c, then f has a local minimum at c.
(c) If f' does not change signs at c, then f has no local maximum or local maximum at c.

Definition: A function f is *concave up* in an interval I if the graph of f lies above the tangent lines for all the values in the interval I.

Definition: A function f is *concave down* in an interval I if the graph of f lies below the tangent lines for all the values in the interval I.

Definition: A point $\left(s, f(s)\right)$ is a *point of inflection* of a function f if f is continuous at this point and changes its concavity from concave up to concave down or vice versa.

Second Derivative Test: Let f be a differentiable function such that f'' is continuous near the number c.
(a) If $f'(c)=0$ and $f''(c)>0$, then f has a local minimum at c.
(b) If $f'(c)=0$ and $f''(c)<0$, then f has a local maximum at c.

Theorem: Let f be a differentiable real-valued function.
(a) If $f''(x)>0$ for every real number x in an interval I in the domain of f, then f is concave up on I.
(b) If $f''(x)<0$ for every real number x in an interval I in the domain of f, then f is concave down on I.

Definition: A line $y=mx+b$ is said to be an *oblique asymptote* or a *slant asymptote* of a function f if either $\lim\limits_{x\to-\infty}\left[f(x)-(mx+b)\right]=0$ or $\lim\limits_{x\to\infty}\left[f(x)-(mx+b)\right]=0$.

Graphs of Functions

Example 1: Let $f(x)=x^3-4x^2-3x-36$.
(a)   Find the real zeros, first derivative, and the local extrema. Identify the intervals where the function is increasing and decreasing.
(b)   Find the second derivative and point(s) of inflection of this polynomial function. Identify the intervals where the function is concave up and down.
(c)   Graph the function f along with its horizontal tangent lines and its first derivative. Plot the function f with the quantitative information about its properties.

Solution: We will begin by entering f as a *Maple* function.

$$f := x \to x^3 - 4 \cdot x^2 - 3 \cdot x - 36$$
$$x \to x^3 - 4\,x^2 - 3\,x - 36 \qquad\qquad \textbf{(4.1)}$$

(a)   Since the degree of the polynomial function f it has a maximum of three real zeros. We will use the *fsolve* command to find these zeros.

$$fsolve\left(f(x)=0, x\right)$$
$$5.655832433 \qquad\qquad \textbf{(4.2)}$$

Hence the only real zero is 5.655832433. The other zeros of f are complex. If we use the *solve* command, the two complex zeros will be noted. We will find the first derivative $f'(x)$.

$f'(x)$

$$3x^2 - 8x - 3 \qquad\qquad\qquad (4.3)$$

Therefore, $f'(x) = 3x^2 - 8x - 3$. To find the critical number we will set $f'(x)$ equal to zero and solve for x.

$c := solve\left(f'(x) = 0, x\right)$

$$3, -\frac{1}{3} \qquad\qquad\qquad (4.4)$$

Hence the critical numbers are 3 and $-\dfrac{1}{3}$. We will evaluate f at these points to find the horizontal tangent lines.

$f\left(c[1]\right)$

$$-54 \qquad\qquad\qquad (4.5)$$

$f\left(c[2]\right)$

$$-\frac{958}{27} \qquad\qquad\qquad (4.6)$$

As a result the polynomial function f has horizontal tangent lines $y = -54$ and $y = -\dfrac{958}{27}$. We will now use the *solve* command to determine where the function f is increasing or decreasing.

$solve\left(f'(x) > 0, \{x\}\right)$

$$\left\{x < -\frac{1}{3}\right\}, \{3 < x\} \qquad\qquad\qquad (4.7)$$

$solve\left(f'(x) < 0, \{x\}\right)$

$$\left\{-\frac{1}{3} < x, x < 3\right\} \qquad\qquad\qquad (4.8)$$

From these results we can conclude that the function f is increasing in the interval $\left(-\infty, -\dfrac{1}{3}\right) \cup (3, \infty)$ and decreasing in the interval $\left(-\dfrac{1}{3}, 3\right)$. Furthermore, using the first derivative test we can conclude that the point $\left(-\dfrac{1}{3},-\dfrac{958}{27}\right)$ is a local maximum and the point $(3,-54)$ is a local minimum.

(b)   We will find the second derivative $f''(x)$.

$f''(x)$

$$6x-8 \tag{4.9}$$

Therefore, $f''(x)=6x-8$. In order to find the points of inflection, if they exist, we set $f''(x)$ equal to zero and solve for x.

$s := solve\left(f''(x)=0,x\right)$

$$\frac{4}{3} \tag{4.10}$$

$f(s)$

$$-\frac{1208}{27} \tag{4.11}$$

We will now find the intervals where the function f is concave up and concave down.

$solve\left(f''(x)>0,\{x\}\right)$

$$\left\{\frac{4}{3}<x\right\} \tag{4.12}$$

$solve\left(f''(x)<0,\{x\}\right)$

$$\left\{x<\frac{4}{3}\right\} \tag{4.13}$$

From these results we can conclude that the function f is concave up in the interval $\left(\dfrac{4}{3},\infty\right)$ and concave down in the interval $\left(-\infty,\dfrac{4}{3}\right)$. Hence the point $\left(\dfrac{4}{3},-\dfrac{1208}{27}\right)$ is a point of inflection of f.

(c)   We will now graph the function along with its derivative and horizontal tangent lines.

$$plot\left(\left[f(x),f'(x),-54,-\frac{958}{27}\right],x=-4..8,y=-80..120,color=\left[red,blue,brown,green\right]\right)$$

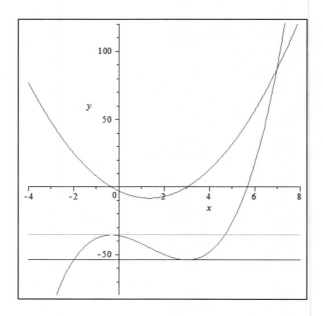

We can now plot this quantitative information using either the *FunctionChart* or the *FunctionPlot* command. In order to use this command we first need to activate the $with\left(Student\left[Calculus1\right]\right)$ package.

$with\left(Student\left[Calculus1\right]\right):$

$FunctionChart\left(f\left(x\right),x=-4..8\right)$

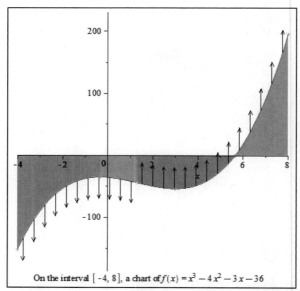

On the interval $[-4, 8]$, a chart of $f(x)=x^3-4x^2-3x-36$

The zeros are denoted by circles, extreme points (critical numbers and the "end points" on the graph) are denoted by diamonds and points of inflections are denoted by a cross and these are in green. The part of the curve that is increasing is in red and the part that is decreasing is in blue. The area where the function is concave up is in pink and is denoted by arrows pointing up. The area where the function is concave down is in brown and is denoted by arrows pointing down. These cannot be seen clearly on the plot above.

However, we can make this clearer by including some options. By setting the symbol size to 20 we can see the critical numbers and points of inflection on the chart above. The *thickness*(3, 3) command made the function plotted thicker. We can make changes in the color of the graph where the function is increasing or decreasing by using the option of *slope*. The color for the region where the function is concave up or down is done using the *concavity* option.

$$FunctionChart\left(f\left(x \right), x = -4..8, pointoptions = \left[symbolsize = 20, color = black \right], \right.$$

$$sign = \left[thickness\left(3,3 \right) \right], slope = color\left(magenta, blue \right),$$

$$\left. concavity = \left[filled\left(grey, aquamarine \right), arrow \right] \right)$$

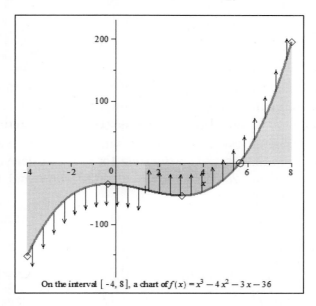

On the interval $[-4, 8]$, a chart of $f(x) = x^3 - 4x^2 - 3x - 36$

Remark:

1.   We can use the drawing techniques in chapter 1 to get the picture below. Note that p.o.i. denotes the point of inflection.

$$plot\left(\left[f\left(x \right), f'\left(x \right), -54, -\frac{958}{27} \right], x = -4..8, y = -80..120, color = \left[red, blue, brown, green \right] \right)$$

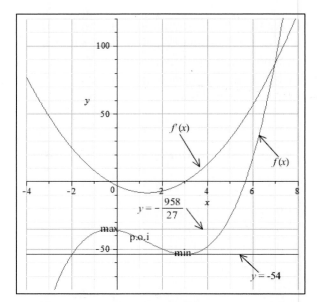

2. If one wants to plot the function along with its derivatives we can use the *DerivativePlot* command. The graph below shows the function along with its first two derivatives. The options for *color*, *linestyle* and *thickness* for the function f and its derivatives have also been altered. The original function is in brown, the first derivative is in blue, and the second derivative is in magenta. The *linestyle* for the first derivative is *dash* and for the second derivative is *dotdash*. Before executing the command we need to execute the $with\left(Student\left[Calculus1\right]\right)$ package.

$with\left(Student\left[Calculus1\right]\right):$

$DerivativePlot\left(f\left(x\right), x=-4..8, order=1..2, functionoptions=\left[color=brown\right],\right.$

$\quad derivativeoptions\left[1\right]=\left[color=blue, linestyle=dash, thickness=2\right],$

$\quad \left.derivativeoptions\left[2\right]=\left[color=magenta, linestyle=dotdash, thickness=3\right]\right)$

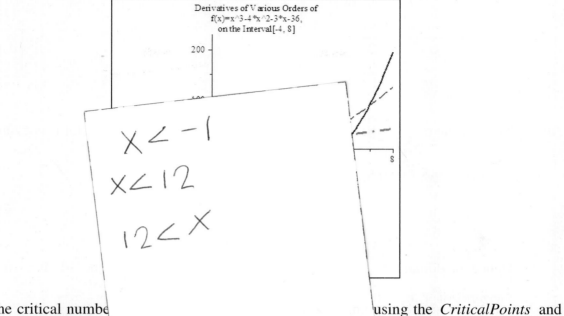

Derivatives of Various Orders of
f(x)=x^3-4*x^2-3*x-36,
on the Interval[-4, 8]

$X < -1$

$X < 12$

$12 < X$

3.   The critical numbe   using the *CriticalPoints* and
      InflectionPoints c   t be taken to activate the
      $with\left(Student\left[Calc\right.\right.$ we will use these commands in the next few
      examples.

$$with\left(Student\left[Calculus1\right]\right):$$
$$c := CriticalPoints\left(f\left(x\right)\right)$$

$$\left[-\frac{1}{3}, 3\right]$$ **(4.14)**

$$s := InflectionPoints\left(f\left(x\right)\right)$$

$$\left[\frac{4}{3}\right]$$ **(4.15)**

Example 2: Let $f\left(x\right) = \dfrac{6x^2 - 3x + 1}{4x^2 + x - 3}$.

(a)   Find the domain, real zeros, and the asymptotes of this function.
(b)   Find the first derivative and the local extrema. Identify the intervals where the function is
       increasing and decreasing.
(c)   Find the second derivative and point(s) of inflection of this rational function. Identify the
       intervals where the function is concave up and down.
(d)   Graph the function f along with its asymptotes. Plot the function f with the
       quantitative information about its properties.

Solution: We will begin by entering f as a *Maple* function.

$$f := x \rightarrow \frac{6 \cdot x^2 - 3 \cdot x + 1}{4 \cdot x^2 + x - 3}$$

$$x \rightarrow \frac{6 x^2 - 3 x + 1}{4 x^2 + x - 3} \tag{4.16}$$

(a)   The domain of f is the set of all real numbers where the denominator is not zero.

$$solve\left(denom\left(f\left(x\right)\right) = 0, x\right)$$

$$\frac{3}{4}, -1 \tag{4.17}$$

Hence the domain of f is the set of all real numbers except $\frac{3}{4}$ and -1. We will now find the zeros of f by setting the numerator of f equal to zero.

$$solve\left(numer\left(f\left(x\right)\right) = 0, x\right)$$

$$\frac{1}{4} + \frac{1}{12} I \sqrt{15}, \frac{1}{4} - \frac{1}{12} I \sqrt{15} \tag{4.18}$$

Hence f has no real zeros. We will now find the vertical asymptotes of f by finding $\lim\limits_{x \to \frac{3}{4}-} f\left(x\right)$, $\lim\limits_{x \to \frac{3}{4}+} f\left(x\right)$, $\lim\limits_{x \to -1-} f\left(x\right)$, and $\lim\limits_{x \to -1+} f\left(x\right)$.

$$\lim\limits_{x \to \frac{3}{4}-} f\left(x\right)$$

$$-\infty \tag{4.19}$$

$$\lim\limits_{x \to \frac{3}{4}+} f\left(x\right)$$

$$\infty \tag{4.20}$$

$$\lim\limits_{x \to -1-} f\left(x\right)$$

$$\infty \tag{4.21}$$

$$\lim\limits_{x \to -1+} f\left(x\right)$$

$$-\infty \tag{4.22}$$

Since $\lim\limits_{x \to \frac{3}{4}-} f\left(x\right) = -\infty$, $\lim\limits_{x \to \frac{3}{4}+} f\left(x\right) = \infty$, $\lim\limits_{x \to -1-} f\left(x\right) = \infty$, and $\lim\limits_{x \to -1+} f\left(x\right) = -\infty$, the lines $x = \frac{3}{4}$ and $x = -1$ are vertical asymptotes of f. We could come to the same conclusion by showing that either the limit from the left or the limit from the right is ∞ or $-\infty$. We will now find the horizontal asymptotes of f by finding $\lim\limits_{x \to -\infty} f(x)$ and $\lim\limits_{x \to \infty} f(x)$.

$\lim\limits_{x \to -\infty} f(x)$

$$\frac{3}{2} \tag{4.23}$$

$\lim\limits_{x \to \infty} f(x)$

$$\frac{3}{2} \tag{4.24}$$

Hence the line $y = \dfrac{3}{2}$ is a horizontal asymptote of f.

(b)   We will now find the first derivative of f.

$f'(x)$

$$\frac{12\,x-3}{4\,x^2+x-3} - \frac{\left(6\,x^2-3\,x+1\right)\left(8\,x+1\right)}{\left(4\,x^2+x-3\right)^2} \tag{4.25}$$

$simplify\left(f'(x)\right)$

$$\frac{2\left(-22\,x+4+9\,x^2\right)}{\left(4\,x^2+x-3\right)^2} \tag{4.26}$$

Hence the first derivative of f is $f'(x) = \dfrac{2\left(-22\,x+4+9\,x^2\right)}{\left(4\,x^2+x-3\right)^2}$. We will now find the critical numbers of f. Before finding the critical numbers do not forget to activate the $with\left(Student\left[Calculus1\right]\right)$ package.

$with\left(Student\left[Calculus1\right]\right):$
$c := CriticalPoints\left(f(x)\right)$

$$\left[-1, \frac{11}{9} - \frac{1}{9}\sqrt{85}, \frac{3}{4}, \frac{11}{9} + \frac{1}{9}\sqrt{85}\right] \tag{4.27}$$

$CriticalPoints\left(f(x), numeric\right)$

$$\left[-0.999999999999063, 0.1978283936, 0.750000000008129, 2.246616051\right] \tag{4.28}$$

$f\left(c[1]\right)$

```
Error, (in f) numeric exception; division by zero
```

$f\left(c[2]\right)$

$$\frac{6\left(\dfrac{11}{9}-\dfrac{1}{9}\sqrt{85}\right)^2-\dfrac{8}{3}+\dfrac{1}{3}\sqrt{85}}{4\left(\dfrac{11}{9}-\dfrac{1}{9}\sqrt{85}\right)^2-\dfrac{16}{9}-\dfrac{1}{9}\sqrt{85}}$$

(4.29)

$evalf\left(f\left(c[2]\right)\right)$

$$-0.2424117925$$

(4.30)

$f\left(c[3]\right)$

```
Error, (in f) numeric exception; division by zero
```

$f\left(c[4]\right)$

$$\frac{6\left(\dfrac{11}{9}+\dfrac{1}{9}\sqrt{85}\right)^2-\dfrac{8}{3}-\dfrac{1}{3}\sqrt{85}}{4\left(\dfrac{11}{9}+\dfrac{1}{9}\sqrt{85}\right)^2-\dfrac{16}{9}+\dfrac{1}{9}\sqrt{85}}$$

(4.31)

$evalf\left(f\left(c[4]\right)\right)$

$$1.262819956$$

(4.32)

From the calculations above, the critical numbers are -1, $\dfrac{11}{9}-\dfrac{1}{9}\sqrt{85}$, $\dfrac{3}{4}$, and $\dfrac{11}{9}+\dfrac{1}{9}\sqrt{85}$. Notice that when the numeric option was used, the number -1 was approximated by 0.999999999999063 and the number $\dfrac{3}{4}$ was approximated by 0.750000000008129. Note that these floating-point approximations may be slightly different when you work with your worksheet. *Maple* version 14 actually evaluates the function when $x = 0.999999999999063$ but earlier versions give an answer of $Float\left(\infty\right)$.

$solve\left(f'(x)>0.0,\{x\}\right)$

$$\{x<-1.\},\{-1.<x,x<0.1978283936\},\{2.246616051<x\}$$

(4.33)

$solve\left(f'(x)<0.0,\{x\}\right)$

$$\{0.1978283936<x,x<0.7500000000\},\{0.7500000000<x,x<2.246616051\}$$

(4.34)

We used the inequalities $f'(x)>0.0$ and $f'(x)<0.0$ rather than $f'(x)>0$ and $f'(x)<0$, respectively, so that our answers would be expressed as a decimal. Hence the function is increasing in the interval $(-\infty,-1)\cup(-1,0.1978283936)\cup(2.246616051,\infty)$

and decreasing in the interval $(0.1978283936, 0.75) \cup (0.75, 2.246616051)$. The local maximum is ($0.1978283936$, -0.242411792) and the local minimum is (2.246616051, 1.262819956).

(c)   We will now find the second derivative of f.

$f''(x)$

$$\frac{12}{4x^2+x-3} - \frac{2(12x-3)(8x+1)}{(4x^2+x-3)^2} + \frac{2(6x^2-3x+1)(8x+1)^2}{(4x^2+x-3)^3}$$
$$-\frac{8(6x^2-3x+1)}{(4x^2+x-3)^2}$$

(4.35)

$simplify\left(f''(x)\right)$

$$-\frac{4(36x^3-132x^2+48x-29)}{(4x^2+x-3)^3}$$

(4.36)

Therefore, the second derivative is $f''(x) = -\dfrac{4(36x^3-132x^2+48x-29)}{(4x^2+x-3)^3}$. We will

now find the points of inflection

$s := InflectionPoints\left(f(x)\right)$

$$\left[-1, \frac{3}{4}, \frac{1}{9} 1700^{1/3} + \frac{1}{180} 1700^{2/3} + \frac{11}{9}\right]$$

(4.37)

$InflectionPoints\left(f(x), numeric\right)$

$$[-0.999999999980536, 0.750000000007892, 3.339649174]$$

(4.38)

$f(s[1])$

Error, (in f) numeric exception; division by zero

$f(s[2])$

Error, (in f) numeric exception; division by zero

$f(s[3])$

$$-\frac{6\left(\frac{1}{9}1700^{1/3}+\frac{1}{180}1700^{2/3}+\frac{11}{9}\right)^2 - \frac{1}{3}1700^{1/3} - \frac{1}{60}1700^{2/3} - \frac{8}{3}}{4\left(\frac{1}{9}1700^{1/3}+\frac{1}{180}1700^{2/3}+\frac{11}{9}\right)^2 + \frac{1}{9}1700^{1/3} + \frac{1}{180}1700^{2/3} - \frac{16}{9}}$$

(4.39)

$evalf\left(f(s[3])\right)$

$$1.288034390 \qquad\qquad\qquad\qquad \textbf{(4.40)}$$

We will now find the intervals where the function is concave up and concave down.

$solve\left(f''(x)>0.0,\{x\}\right)$

$$\{x<-1.\},\{0.7500000000<x,x<3.339649174\} \qquad \textbf{(4.41)}$$

$solve\left(f''(x)<0.0,\{x\}\right)$

$$\{-1.<x,x<0.7500000000\},\{3.339649174<x\} \qquad \textbf{(4.42)}$$

Hence the function is concave up in the interval $(-\infty,-1)\cup(0.75,3.339649174)$ and concave down in the interval $(-1,0.75)\cup(3.339649174,\infty)$. The point of inflection of f is (3.339649174, 1.288034390).

(d)   The graph of f along with its asymptotes is sketched below.

$$plot\left(\left[f(x),\frac{3}{2}\right],x=-4..8,y=-3..4,color=\left[red,blue\right]\right)$$

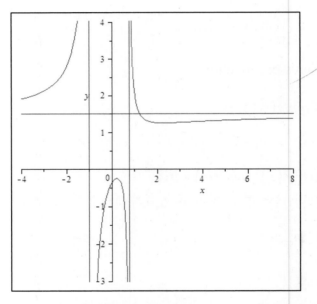

From this graph we can confirm that the local maximum is (0.1978283936, -0.242411792) and the local minimum is (2.246616051, 1.262819956). Observe that the color of the vertical asymptote is the same color as that of the original function. A quantitative chart for the function is sketched below.

$FunctionChart\left(f(x),x=-4..8,pointoptions=\left[symbolsize=20,color=black\right],\right.$

$\qquad sign=\left[thickness(3,3)\right],slope=color(magenta,blue),$

$\qquad\left. concavity=\left[filled(grey,aquamarine),arrow\right]\right)$

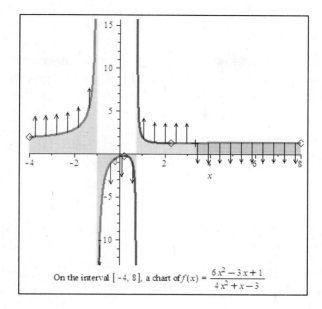

On the interval $[-4, 8]$, a chart of $f(x) = \dfrac{6x^2 - 3x + 1}{4x^2 + x - 3}$

Remark: One can use the *Asymptotes* command to find all the vertical, horizontal, and oblique asymptotes of the function f. Do not forget to activate the $with\big(Student[Calculus1]\big)$ command. If the values of $x = a .. b$ are specified, the command gives all the vertical asymptotes the interval $[a, b]$ and all the horizontal and oblique asymptotes of f.

$with\big(Student[Calculus1]\big):$

$Asymptotes\big(f(x), x\big)$

$$\left[y = \frac{3}{2}, x = -1, x = \frac{3}{4} \right]$$
(4.43)

Note that y is the default dependent variable. Since the independent variable, x, can be uniquely determined from the formula for f, one does not need to include that in the command. The authors caution the readers against not mentioning the independent variable.

$Asymptotes\big(f(x)\big)$

$$\left[y = \frac{3}{2}, x = -1, x = \frac{3}{4} \right]$$
(4.44)

We will now find the vertical asymptotes of the cosecant function.

$Asymptotes\big(csc(x), x\big)$

```
Warning, the expression has an infinity of asymptotes, some
examples of which are given
```

$$[x = -\pi, x = 0, x = \pi] \qquad\qquad \textbf{(4.45)}$$

Maple does not give all the vertical asymptotes of the cosecant function, since it has an infinite number of vertical asymptotes. Instead, a warning message along with some examples of vertical asymptotes appear in the output. The command below gives all the vertical asymptotes in the interval $[0, 7]$.

$Asymptotes\left(\csc(x), x = 0 .. 7\right)$

$$[x = 0, x = \pi, x = 2\,\pi] \qquad\qquad \textbf{(4.46)}$$

The *numeric* option could be used to find the asymptotes using floating-point computations as shown below. If this option is used and the values of $x = a .. b$ are not specified, then this command gives all the vertical asymptotes the interval $[-10, 10]$ and all the horizontal and oblique asymptotes of f.

$Asymptotes\left(\csc(x), x, numeric\right)$

$$[x = -9.42477796075921, x = -6.28318530717222, x = -3.14159265357375 \qquad \textbf{(4.47)}$$
$$x = 0., x = 3.14159265361170, x = 6.28318530719198, x = 9.42477796077085]$$

$Asymptotes\left(\csc(x), x = 0 .. 7, numeric\right)$

$$[x = 0., x = 3.14159265359966, x = 6.28318530720325] \qquad \textbf{(4.48)}$$

There are examples where the *Asymptotes* command does not give all the vertical asymptotes. The authors would advise the readers to use limits and graphs to verify their answers.

Example 3: Let $f(x) = x\sqrt{x + 3}$.
(a)   Find the real zeros, first derivative, and the local extrema. Identify the intervals where the function is increasing and decreasing.
(b)   Find the second derivative and point(s) of inflection of this polynomial function. Identify the intervals where the function is concave up and down.
(c)   Graph the function f along with its first derivative. Plot the function f with the quantitative information about its properties.

Solution: We will begin by entering f as a *Maple* function.

$$f := x \rightarrow x \cdot \sqrt{x + 3}$$

$$x \rightarrow x\sqrt{x + 3} \qquad\qquad \textbf{(4.49)}$$

(a)   We will first find the domain of f.

$solve\left(f\left(x\right)\geq 0,\{x\}\right)$

$$\{-3\leq x\} \tag{4.50}$$

Hence the domain of f is $[-3,\infty)$. We will now find the zeros of f.

$solve\left(f\left(x\right)=0,x\right)$

$$0,-3 \tag{4.51}$$

Hence 0 and -3 are the zeros of f.

(b)   We will now find the first derivative of f and its critical numbers. Before finding the critical numbers do not forget to activate the $with\left(Student\left[Calculus1\right]\right)$ package.

$f'\left(x\right)$

$$\sqrt{x+3}+\frac{1}{2}\frac{x}{\sqrt{x+3}} \tag{4.52}$$

$simplify\left(f'\left(x\right)\right)$

$$\frac{3}{2}\frac{x+2}{\sqrt{x+3}} \tag{4.53}$$

Hence the first derivative of f is $f'\left(x\right)=\dfrac{3}{2}\dfrac{x+2}{\sqrt{x+3}}$. We will now find the critical points of f.

$with\left(Student\left[Calculus1\right]\right):$
$c:=CriticalPoints\left(f\left(x\right)\right)$

$$\left[-3,-2\right] \tag{4.54}$$

$f\left(c[1]\right)$

$$0 \tag{4.55}$$

$f\left(c[2]\right)$

$$-2 \tag{4.56}$$

The local extrema are (-3, 0) and (-2, -2). Please note that if the answer for (4.54) is reversed, i.e. the output is $\left[-2,-3\right]$, then the output for $f\left(c[1]\right)$ would be -2 and the output for $f\left(c[2]\right)$ would be 0.

$solve\left(f'\left(x\right)>0.0,\{x\}\right)$

Chapter 4 Graphs of Functions using Limits and Derivatives

$$\{-2 < x\} \tag{4.57}$$

$solve(f'(x) < 0.0, \{x\})$

$$\{-3 < x, x < -2\} \tag{4.58}$$

Therefore the function is increasing in the interval $(-2, \infty)$ and decreasing in the interval $(-3, -2)$. The local maximum is (-3, 0) and the local minimum is (-2, -2).

(c) We will now find the second derivative of f and the points of inflection.

$f''(x)$

$$\frac{1}{\sqrt{x+3}} - \frac{1}{4}\frac{x}{(x+3)^{3/2}} \tag{4.59}$$

$simplify(f''(x))$

$$\frac{3}{4}\frac{x+4}{(x+3)^{3/2}} \tag{4.60}$$

Therefore, the second derivative is $f''(x) = \frac{3}{4}\frac{x+4}{(x+3)^{3/2}}$. We will now find the points of inflection of f.

$s := InflectionPoints(f(x), numeric)$

$$[\ ] \tag{4.61}$$

The function f has no inflection points.

$solve(f''(x) > 0.0, \{x\})$

$$\{-3 < x\} \tag{4.62}$$

$solve(f''(x) < 0.0, \{x\})$

Hence the function is concave up in the interval $(-3, \infty)$. Notice that there was no output when we solved for $f''(x) < 0.0$.

(d) The graph of f along with its first derivative is sketched below.

$plot([f(x), f'(x)], x = -3..6, y = -10..18, color = [red, blue])$

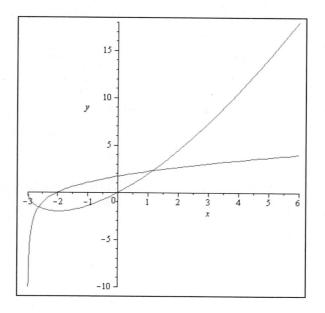

A quantitative chart for the function is sketched below.

$$FunctionChart\left(f\left(x\right), x=-3\,..\,6, pointoptions=\left[symbolsize=20, color=black\right],\right.$$
$$sign=\left[thickness\left(3,3\right)\right], slope=color\left(magenta, blue\right)$$
$$\left.concavity=\left[\,filled\left(grey, aquamarine\right), arrow\right]\right)\right)$$

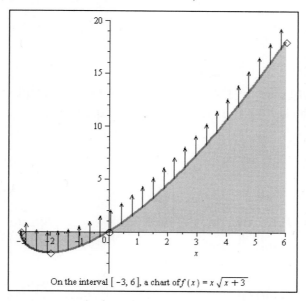

On the interval $\left[-3, 6\right]$, a chart of $f\left(x\right)=x\sqrt{x+3}$

Remark: Notice what happens if we try to evaluate $f\left(s[1]\right)$ or $f\left(s\right)$.

$f\left(s[1]\right)$

Error, invalid subscript selector

$f(s)$

$$[\ ]\sqrt{[\ ]}+3 \qquad\qquad\qquad\qquad \textbf{(4.63)}$$

Example 4: Let $f(x) = \dfrac{x^3+2x^2-23x-60}{x^2+8x+15}$.

(a)   Find the domain, real zeros, and the asymptotes of this function.

(b)   Find the first derivative and the local extrema. Identify the intervals where the function is increasing and decreasing.

(c)   Find the second derivative and point(s) of inflection of this rational function. Identify the intervals where the function is concave up and down.

(d)   Graph the function f along with its asymptotes. Plot the function f with the quantitative information about its properties.

Solution: We will begin by entering f as a *Maple* function.

$$f := x \rightarrow \frac{x^3+2\cdot x^2-23\cdot x-60}{x^2+8\cdot x+15}$$

$$x \rightarrow \frac{x^3+2\,x^2-23\,x-60}{x^2+8\,x+15} \qquad\qquad \textbf{(4.64)}$$

(a)   We will first find the domain and zeros of f.

$$solve\big(denom(f(x))=0,x\big)$$

$$-3,-5 \qquad\qquad\qquad\qquad \textbf{(4.65)}$$

Hence the domain of f is the set of all real numbers except -3 and -5. We will now find the zeros of f by setting the numerator of f equal to zero.

$$solve\big(numer(f(x))=0,x\big)$$

$$5,-4,-3 \qquad\qquad\qquad\qquad \textbf{(4.66)}$$

We will now find the asymptotes of f.   Do not forget to activate the $with\big(Student[Calculus1]\big)$ package.

$$with\big(Student[Calculus1]\big):$$
$$Asymptotes\big(f(x),x\big)$$

$$[y=x-6, x=-5] \qquad\qquad\qquad \textbf{(4.67)}$$

Hence the line $y = x - 6$ is an oblique asymptote of f and $x = -5$ is a vertical asymptote of f. Note that the function f has a removable discontinuity at $x = -3$. This is verified in the remark after this example.

(b)   We will now find the first derivative of f and its critical numbers. We do not need to activate the $with(Student[Calculus1])$ package as we have done so earlier.

$f'(x)$

$$\frac{3x^2 + 4x - 23}{x^2 + 8x + 15} - \frac{(x^3 + 2x^2 - 23x - 60)(2x + 8)}{(x^2 + 8x + 15)^2} \qquad (4.68)$$

$simplify(f'(x))$

$$\frac{x^2 + 10x + 15}{(x + 5)^2} \qquad (4.69)$$

Hence the first derivative of f is $f'(x) = \dfrac{x^2 + 10x + 15}{(x + 5)^2}$. We will now find the critical points of f.

$c := CriticalPoints(f(x))$

$$\left[-5 - \sqrt{10}, -5, -3, -5 + \sqrt{10} \right] \qquad (4.70)$$

Since -5, and -3 are not in the domain of f, the critical numbers of f are $-5 + \sqrt{10}$ and $-5 - \sqrt{10}$. Notice that -3 does not appear if the *numeric* option is added to the command.

$CriticalPoints(f(x), numeric)$

$$\left[-8.162277660, -5.000000000, -1.837722340 \right] \qquad (4.71)$$

$f(c[1])$

$$x \to \frac{\left(-5 - \sqrt{10} \right)^3 + 2\left(-5 - \sqrt{10} \right)^2 + 55 + 23\sqrt{10}}{\left(-5 - \sqrt{10} \right)^2 - 25 - 8\sqrt{10}} \qquad (4.72)$$

$evalf(f(c[1]))$

$$-17.32455532 \qquad (4.73)$$

$f(c[2])$

Error, (in f) numeric exception; division by zero

$f(c[3])$

```
Error, (in f) numeric exception; division by zero
```

$f(c[4])$

$$x \rightarrow \frac{\left(-5+\sqrt{10}\right)^3 + 2\left(-5+\sqrt{10}\right)^2 + 55 - 23\sqrt{10}}{\left(-5+\sqrt{10}\right)^2 - 25 + 8\sqrt{10}} \tag{4.74}$$

$evalf\left(f\left(c[4]\right)\right)$

$$-4.675444678 \tag{4.75}$$

Since -5 and -3 are not in the domain of f, we get the errors as answer for $f(-5)$ and $f(-3)$. The local extrema are (-8.162277660, -17.32455532) and (-1.837722340, -4.675444678).

$solve\left(f'(x) > 0.0, \{x\}\right)$

$$\{x < -8.162277660\}, \{-1.837722340 < x\} \tag{4.76}$$

$solve\left(f'(x) < 0.0, \{x\}\right)$

$$\{-8.162277660 < x, x < -5.\}, \{-5. < x, x < -3.\}, \{-3. < x, x < -1.837722340\} \tag{4.77}$$

Hence the function is increasing in the interval $(-\infty, -8.162277660) \cup (-1.837722340, \infty)$ and decreasing in the interval $(-8.162277660, -5) \cup (-5, -3) \cup (-3, -1.837722340)$. The local maximum is (-8.162277660, -17.32455532) and the local minimum is (-1.837722340, -4.675444678).

(c)   We will now find the second derivative of f and the points of inflection.

$f''(x)$

$$\frac{6x+4}{x^2+8x+15} - \frac{2\left(3x^2+4x-23\right)\left(2x+8\right)}{\left(x^2+8x+15\right)^2} + \frac{2\left(x^3+2x^2-23x-60\right)\left(2x+8\right)^2}{\left(x^2+8x+15\right)^2}$$

$$-\frac{2\left(x^3+2x^2-23x-60\right)\left(2x+8\right)^2}{\left(x^2+8x+15\right)^3} \tag{4.78}$$

$simplify\left(f''(x)\right)$

$$\frac{20}{\left(x+5\right)^3} \tag{4.79}$$

Therefore, the second derivative of f is $f''(x) = \dfrac{20}{(x+5)^3}$. We will now find the points of inflection of f.

$s := InflectionPoints\big(f(x)\big)$

$$[-5] \qquad\qquad (4.80)$$

$f\big(s[1]\big)$

Error, (in f) numeric exception; division by zero

Observe that the output for $f\big(s[1]\big)$ as -5 is not in the domain of f so the function does not have a point of inflection.

$solve\big(f''(x) > 0.0, \{x\}\big)$

$$\{-5. < x, x < -3.\}, \{-3. < x\} \qquad\qquad (4.81)$$

$solve\big(f''(x) < 0.0, \{x\}\big)$

$$\{x < -5.\} \qquad\qquad (4.82)$$

The function is concave up in the interval $(-5, -3) \cup (-3, \infty)$ and concave down in the interval $(-\infty, -5)$.

(d)   The graph of f along with its asymptotes is sketched below.

$plot\big(\big[f(x), x-6\big], x = -12..6, y = -30..10, color = [red, blue]\big)$

The plot below uses the option $discont = [showremovable]$ to show the removable discontinuity at $x = -3$. This option is available in version 14 and not in the earlier versions. Notice that in this case, we do not see the vertical asymptote.

$$plot([[f(x), x-6], x = -12..6, y = -30..10, color = [red, blue], discont = [showremovable]])$$

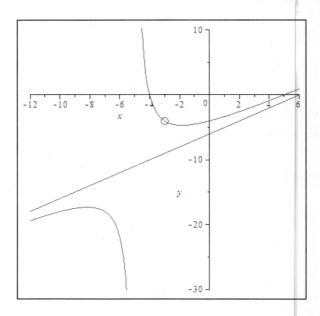

A quantitative chart for the function is sketched below.

$$FunctionChart(f(x), x = -12..6, pointoptions = [symbolsize = 20, color = black],$$
$$sign = [thickness(3,3)], slope = color(magenta, blue),$$
$$concavity = [filled(grey, aquamarine), arrow]$$

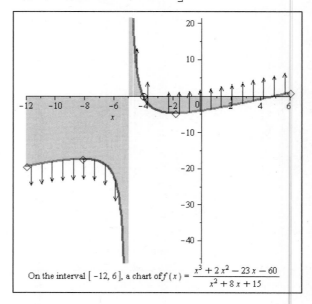

On the interval $[-12, 6]$, a chart of $f(x) = \dfrac{x^3 + 2x^2 - 23x - 60}{x^2 + 8x + 15}$

Chapter 4 Graphs of Functions using Limits and Derivatives

Remarks: We will first show that f has a removable discontinuity when $x = -3$.

$$\lim_{x \to -3-} f(x)$$
$$-4 \tag{4.83}$$

$$\lim_{x \to -3+} f(x)$$
$$-4 \tag{4.84}$$

From the calculations above we see that the limit $\lim_{x \to -3} f(x)$ exists and equals -4. Hence the line $x = -3$ is not a vertical asymptote. Since -3 is not in the domain of f (and hence f is not continuous when $x = -3$) but $\lim_{x \to -3} f(x) = 4$, f has a removable discontinuity when $x = -3$. The following steps verify the results for the asymptotes of f.

$$\lim_{x \to -5-} f(x)$$
$$-\infty \tag{4.85}$$

$$\lim_{x \to -5+} f(x)$$
$$\infty \tag{4.86}$$

Since $\lim_{x \to -5-} f(x) = -\infty$ and $\lim_{x \to -5-} f(x) = \infty$, the line $x = -5$ is a vertical asymptote. Since f is a rational function and the degree of the numerator of f is one more than the degree of the denominator of f, the rational function f has a slant asymptote. We will verify this by first dividing the numerator of f by the denominator of f and then calculating the limit.

$$quo\big(numer(f(x)), denom(f(x)), x\big)$$
$$x - 6 \tag{4.87}$$

$$\lim_{x \to -\infty}\big[f(x) - (x-6)\big]$$
$$0 \tag{4.88}$$

$$\lim_{x \to \infty}\big[f(x) - (x-6)\big]$$
$$0 \tag{4.89}$$

Hence the line $y = x - 6$ is an oblique asymptote of f.

Example 5: Let $f(x) = \dfrac{3 - e^x}{5 - 2e^{3x}}$.

(a) Find the domain, real zeros, and the asymptotes of this function.
(b) Find the first derivative and the local extrema. Identify the intervals where the function is increasing and decreasing.

(c)   Find the second derivative and point(s) of inflection of this rational function. Identify the intervals where the function is concave up and down.

(d)   Graph the function f along with its asymptotes. Plot the function f with the quantitative information about its properties.

Solution: We will begin by entering f as a *Maple* function.

$$f := x \rightarrow \frac{3-e^x}{5-2 \cdot e^{3 \cdot x}}$$

$$x \rightarrow \frac{3-e^x}{5-2\,e^{3\,x}} \qquad\qquad\qquad (4.90)$$

(a)   We will first find the domain and zeros of f.

$$solve\left(denom\left(f\left(x\right)\right)=0, x\right)$$

$$\frac{1}{3}\ln\left(\frac{5}{2}\right) \qquad\qquad\qquad (4.91)$$

Hence the domain of f is the set of all real numbers except $\frac{1}{3}\ln\left(\frac{5}{2}\right)$. We will now find the zeros of f by setting the numerator of f equal to zero.

$$solve\left(numer\left(f\left(x\right)\right)=0, x\right)$$

$$\ln\left(3\right) \qquad\qquad\qquad (4.92)$$

Hence $\ln\left(3\right)$ is a zero of f. We will now find the asymptotes of f. Do not forget to activate the $with\left(Student\left[Calculus1\right]\right)$ package.

$$with\left(Student\left[Calculus1\right]\right):$$
$$Asymptotes\left(f\left(x\right), x\right)$$

$$\left[y = 0, y = \frac{3}{5}, x = \frac{1}{3}\ln\left(5\right) - \frac{1}{3}\ln\left(2\right)\right] \qquad\qquad\qquad (4.93)$$

Hence the lines $y = \frac{3}{5}$ and $y = 0$ are horizontal asymptotes of f and the line $x = \frac{1}{3}\ln\left(\frac{5}{2}\right)$ (which is the same as $x = \frac{1}{3}\ln\left(5\right) - \frac{1}{3}\ln\left(2\right)$) is a vertical asymptote.

(b)   We will now find the first derivative of f and its critical numbers.

$f'(x)$

$$-\frac{e^x}{5-2\,e^{3x}}+\frac{6\left(3-e^x\right)e^{3x}}{\left(5-2\,e^{3x}\right)^2}$$

(4.94)

$simplify\left(f'(x)\right)$

$$-\frac{e^x\left(5+4\,e^{3x}-18\,e^{2x}\right)}{\left(-5+2\,e^{3x}\right)^2}$$

(4.95)

Hence the first derivative of f is $f'(x)=-\dfrac{e^x\left(5+4\,e^{3x}-18\,e^{2x}\right)}{\left(-5+2\,e^{3x}\right)^2}$. We will now find the

critical points of f.

$c:=CriticalPoints\left(f(x)\right)$

$$\left[-\ln(2)+\ln\left(5-\sqrt{15}\right),\frac{1}{3}\ln(5)-\frac{1}{3}\ln(2),-\ln(2)+\ln\left(5+\sqrt{15}\right)\right]$$

(4.96)

$f\left(c[1]\right)$

$$\frac{3-e^{-\ln(2)+\ln\left(5-\sqrt{15}\right)}}{5-2\,e^{-3\ln(2)+3\ln\left(5-\sqrt{15}\right)}}$$

(4.97)

$evalf\left(f\left(c[1]\right)\right)$

$$0.5248655564$$

(4.98)

$f\left(c[2]\right)$

$$\frac{3-e^{\frac{1}{3}\ln(5)-\frac{1}{3}\ln(2)}}{5-2\,e^{\ln(5)-\ln(2)}}$$

(4.99)

$simplify\left(f\left(c[2]\right)\right)$

Error, (in f) numeric exception; division by zero

$f\left(c[3]\right)$

$$\frac{3-e^{-\ln(2)+\ln\left(5+\sqrt{15}\right)}}{5-2\,e^{-3\ln(2)+3\ln\left(5+\sqrt{15}\right)}}$$

(4.100)

$evalf\left(f\left(c[3]\right)\right)$

$$0.008467776917$$

(4.101)

The local extrema could be $\left(f\left(c[1]\right),0.5248655564\right)$ and $\left(f\left(c[3]\right),0.008467776917\right)$.

Note that as $\dfrac{1}{3}\ln\left(\dfrac{5}{2}\right)$ is not in the domain of f, we use the simplify command to evaluate $f\left(c[2]\right)$. Again since $\dfrac{1}{3}\ln\left(\dfrac{5}{2}\right)=0.3054302440$ is not in the domain of f, we could get the answer for $f\left(0.3054302440\right)$ as $-Float\left(\infty\right)$ in *Maple* versions 13 and below. However, with version 14, we do not use the *evalf* command as that will yield an answer as shown below.

evalf $\left(f\left(c[2]\right)\right)$

$$8.213955960\,10^{8} \tag{4.102}$$

We will now find the intervals of increase and decrease for the function f.

solve $\left(f'(x)>0.0,\{x\}\right)$
$$\{-0.5735731685<x,x<0.3054302440\},\{0.3054302440<x,x<1.489863900\} \tag{4.103}$$
solve $\left(f'(x)<0.0,\{x\}\right)$
$$\{x<-0.5735731685\},\{1.489863900<x\} \tag{4.104}$$

Hence the function is increasing in the interval $(-0.5735731685,0.3054302440)\cup(0.3054302440,1.489863900)$ and decreasing in the interval $(-\infty,-0.5735731685)\cup(1.489863900,\infty)$. The local maximum is $\left(f\left(c[3]\right),0.008467776917\right)$ = (1.4898639000, 0.008467776917) and the local minimum is $\left(f\left(c[1]\right),0.5248655564\right)$ = (-0.57355731685, 0.5248655564). Check that the values of $c[1]=-0.57355731685$ and $c[3]=1.4898639000$ using the *CriticalPoints* $\left(f(x),numeric\right)$ command.

(c)   We will now find the second derivative of f and the points of inflection.

$f''(x)$

$$-\frac{e^{x}}{5-2\,e^{3x}}-\frac{12\,e^{x}\,e^{3x}}{\left(5-2\,e^{3x}\right)^{2}}+\frac{72\left(3-e^{x}\right)\left(e^{3x}\right)^{2}}{\left(5-2\,e^{3x}\right)^{3}}+\frac{18\left(3-e^{x}\right)e^{3x}}{\left(5-2\,e^{3x}\right)^{2}} \tag{4.105}$$

simplify $\left(f''(x)\right)$

$$\frac{e^x \left(25+130\, e^{3x} +16\, e^{6x} -108\, e^{5x} -270\, e^{2x}\right)}{\left(-5+2\, e^{3x}\right)^3} \tag{4.106}$$

Therefore, $f''(x) = \dfrac{e^x \left(25+130\, e^{3x} +16\, e^{6x} -108\, e^{5x} -270\, e^{2x}\right)}{\left(-5+2\, e^{3x}\right)^3}$. We will now find the

points of inflection of f.

$s := InflectionPoints\left(f(x), numeric\right)$

$$\left[-1.115517185, 0.305430243962148, 1.890506949\right] \tag{4.107}$$

$f\left(s[1]\right)$

$$0.5419115778 \tag{4.108}$$

$f\left(s[2]\right)$

$$-2.67343853392140\, 10^{10} \tag{4.109}$$

$f\left(s[3]\right)$

$$0.006289978358 \tag{4.110}$$

The points of inflection of f could be $(-1.111517185, 0.5419115778)$ and $(1.890506949, 0.006289978358)$. Note that $\left(0.3054302441, -2.67343853392140\, 10^{10}\right)$ is

not a point of inflection since $\dfrac{1}{3}\ln\left(\dfrac{5}{2}\right)$ is not in the domain of f and $\dfrac{1}{3}\ln\left(\dfrac{5}{2}\right)$ is

approximately 0.3054302441.

$solve\left(f''(x) > 0.0, \{x\}\right)$

$$\{-1.111517185 < x, x < 0.3054302440\}, \{1.890506949 < x\} \tag{4.111}$$

$solve\left(f''(x) < 0.0, \{x\}\right)$

$$\{x < -1.111517185\}, \{0.3054302440 < x, x < 1.890506949\} \tag{4.112}$$

Hence the function is concave up in the interval $(-1.111517185, 0.3054302441) \cup (1.890506949, \infty)$ and concave down in the interval $(-\infty, -1.111517185) \cup (0.3054302441, 1.890506949)$. Therefore, the points of inflection of f are $(-1.111517185, 0.5419115778)$ and $(1.890506949, 0.006289978358)$.

(d)   The graph of f along with its asymptotes is sketched below.

$$plot\left(\left[f(x),0,\frac{3}{5}\right], x=-1.8..2, y=-0.8..1.4, color=[red,blue,brown], thickness=3\right)$$

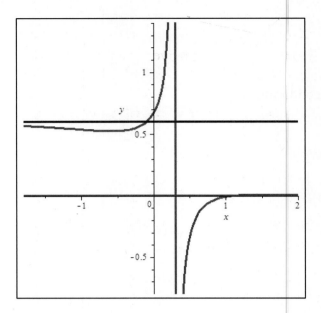

We will now sketch a quantitative chart for the function.

$$FunctionChart\left(f(x), x=-1.8..2, pointoptions=[symbolsize=20, color=black],\right.$$
$$sign=[thickness(3,3)], slope=color(magenta,blue),$$
$$concavity=\left[filled(grey,aquamarine),arrow\right]$$

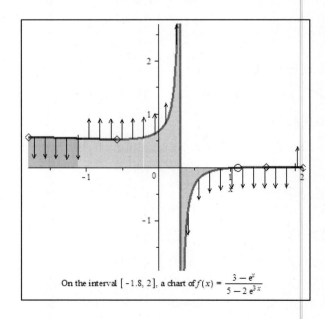

On the interval $[-1.8, 2]$, a chart of $f(x) = \dfrac{3-e^x}{5-2\,e^{3x}}$

Example 6: Let $f(x)=\sin(x)\sqrt{x}$ restricted to $[0,2\pi]$.

(a) Find the real zeros, first derivative, and the local extrema. Can you identify the intervals where the function is increasing and decreasing?

(b) Find the second derivative and point(s) of inflection of this polynomial function. Can you identify the intervals where the function is concave up and down?

(c) Graph the function f along with its horizontal tangent lines and its first derivative. Plot the function f with the quantitative information about its properties. Identify the intervals where the function is increasing and decreasing and is concave up and down.

Solution: We will begin by entering f as a *Maple* function.

$$f := x \rightarrow \sin(x) \cdot \sqrt{x}$$

$$x \rightarrow \sin(x) \sqrt{x} \tag{4.113}$$

(a) We will now find the zeros of f. We will use the *Roots* command as f is not a polynomial function. Do not forget to activate the $with\left(Student\left[Calculus1\right]\right)$ package.

$with\left(Student\left[Calculus1\right]\right)$:
$Roots\left(f\left(x\right) = 0, x = 0..2\pi\right)$

$$\left[0, \pi, 2\pi\right] \tag{4.114}$$

Hence 0, π, and 2π are the zeros of f. We will now find the first derivative of f and its critical numbers.

$f'\left(x\right)$

$$\cos\left(x\right)\sqrt{x} + \frac{1}{2}\frac{\sin\left(x\right)}{\sqrt{x}} \tag{4.115}$$

$simplify\left(f'\left(x\right)\right)$

$$\frac{1}{2}\frac{2\cos\left(x\right)x + \sin\left(x\right)}{\sqrt{x}} \tag{4.116}$$

Hence the first derivative of f is $f'\left(x\right) = \frac{1}{2} \cdot \frac{2\cos\left(x\right)\sqrt{x} + \sin\left(x\right)}{\sqrt{x}}$. We will now find the critical points of f.

$c := CriticalPoints\left(f\left(x\right), x = 0..2\cdot\pi\right)$

$$\left[0\right] \tag{4.117}$$

$c := CriticalPoints\left(f\left(x\right), x = 0..2\cdot\pi, numeric\right)$

$$\left[0., 1.836597203, 4.815842318\right] \tag{4.118}$$

$$f\left(c[1]\right)$$

$$0. \tag{4.119}$$

$$f\left(c[2]\right)$$

$$1.307619413 \tag{4.120}$$

$$f\left(c[3]\right)$$

$$-2.182769785 \tag{4.121}$$

The local extrema are $(0,0)$, $(1.836597203, 1.307619413)$, and $(4.815842318, -2.182769785)$. Furthermore, f has $x=0$, $x=1.307619413$, and $x=-2.182769785$ as its horizontal tangent lines (why?).

$$solve\left(f'(x) > 0.0, \{x\}\right)$$
Warning, solutions may have been lost
$$solve\left(f'(x) < 0.0, \{x\}\right)$$
Warning, solutions may have been lost

Hence we cannot find the intervals of increase and decrease for the function using the *solve* command.

(b)   We will now find the second derivative of f and the points of inflection.

$$f''(x)$$

$$-\sin(x)\sqrt{x} + \frac{\cos(x)}{\sqrt{x}} - \frac{1}{4}\frac{\sin(x)}{x^{3/2}} \tag{4.122}$$

$$simplify\left(f''(x)\right)$$

$$\frac{1}{4}\frac{-4\sin(x)x^2 + 4\cos(x)x - \sin(x)}{x^{3/2}} \tag{4.123}$$

Therefore, $f''(x) = \dfrac{1}{4}\dfrac{-4\sin(x)x^2 + 4\cos(x)x - \sin(x)}{x^{3/2}}$. We will now find the points of inflection of f.

$$s := InflectionPoints\left(f(x), numeric\right)$$

$$\left[0.7463497368, 3.420385489\right] \tag{4.124}$$

$$f\left(s[1]\right)$$

$$0.5865668774 \tag{4.125}$$

$$f\left(s[2]\right)$$

$$-0.5089540009 \tag{4.126}$$

The function f could have inflection points at $(0.7463497368, 0.5865668774)$ and $(3.420385489, -0.5089540009)$.

$solve(f''(x) > 0.0, \{x\})$

Warning, solutions may have been lost

$solve(f''(x) < 0.0, \{x\})$

Warning, solutions may have been lost

Hence we cannot find the intervals where the function is concave up or concave down using the *solve* command.

(c)   The graph of f along with its first derivative and horizontal tangent lines is sketched below.

$$plot\left(\left[f(x), f'(x), 0, 1.307619413, -2.182769785\right], x = 0..2 \cdot \pi,\right.$$
$$\left.color = \left[red, blue, brown, green, black\right], , thickness = 2\right)$$

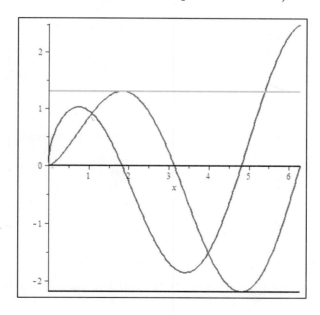

From the graph above the function is increasing in the interval $(0, 1.836597203) \cup (4.815842318, 2\pi)$ and is decreasing in the interval $(1.836597203, 4.815842318)$. The function f is concave up in the interval $(0, 0.7463497368) \cup (3.420385489, 2\pi)$ and is concave down in the interval $(0.7463497368, 3.420385489)$. This is verified from the quantitative chart for the function that is sketched below.

$$FunctionChart\left(f\left(x \right), x = 0\,..\,2 \cdot \pi,\, pointoptions = \left[symbolsize = 20, color = black \right],\right.$$
$$sign = \left[thickness\left(3,3 \right) \right], slope = color\left(magenta, blue \right)$$
$$\left. concavity = \left[filled\left(grey, aquamarine \right), arrow \right] \right)\right)$$

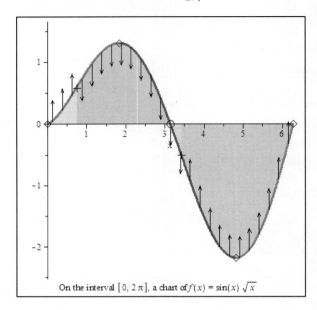

On the interval $\left[0,\, 2\,\pi \right]$, a chart of $f(x) = \sin(x)\,\sqrt{x}$

Example 7: Sketch a graph of a polynomial function whose first derivative is $f1\left(x \right) = x^2 - 3x - 18$.

Solution: We will begin by entering $f1$ as a *Maple* function and finding the critical numbers.

$f1 := x \rightarrow x^2 - 3x - 18$

$$x \rightarrow x^2 - 3x - 18 \qquad\qquad\qquad \textbf{(4.127)}$$

$solve\left(f1\left(x \right) = 0, x \right)$

$$6, -3 \qquad\qquad\qquad \textbf{(4.128)}$$

Hence the critical numbers are -3 and 6. Now find the intervals where f is increasing and decreasing.

$solve\left(f1\left(x \right) > 0, \left\{ x \right\} \right)$

$$\left\{ x < -3 \right\}, \left\{ 6 < x \right\} \qquad\qquad\qquad \textbf{(4.129)}$$

$solve\left(f1\left(x \right) < 0, \left\{ x \right\} \right)$

$$\left\{ -3 < x, x < 6 \right\} \qquad\qquad\qquad \textbf{(4.130)}$$

The polynomial function is increasing in the interval $(-\infty, -3) \cup (6, \infty)$ and decreasing in the interval (-3, 6). Therefore, f has a relative maximum when $x = -3$ and a relative minimum when $x = 6$. We now use the second derivative to find the points of inflection.

$f1'(x)$

$$2x - 3 \qquad\qquad\qquad\text{(4.131)}$$

$solve(f1'(x) = 0, x)$

$$x = \frac{3}{2} \qquad\qquad\qquad\text{(4.132)}$$

We now determine the intervals where f is concave up and concave down.

$solve(f1'(x) > 0, \{x\})$

$$\left\{ \frac{3}{2} < x \right\} \qquad\qquad\qquad\text{(4.133)}$$

$solve(f1'(x) < 0, \{x\})$

$$\left\{ x < \frac{3}{2} \right\} \qquad\qquad\qquad\text{(4.134)}$$

The polynomial function is concave up in $\left(\frac{3}{2}, \infty\right)$ and concave down in $\left(-\infty, \frac{3}{2}\right)$. The graph below gives a general shape of the polynomial f.

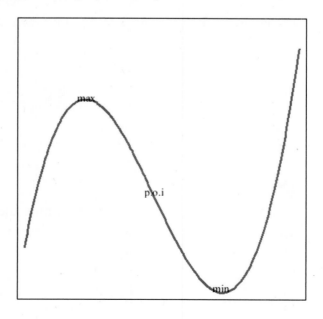

Example 8: Sketch a graph of a function whose first derivative is $f1(x)=1+\ln(x)$.

Solution: We will begin by entering $f1$ as a *Maple* function and finding the critical numbers.

$f1:=x\rightarrow1+\ln(x)$

$$x\rightarrow1+\ln(x) \tag{4.135}$$

$solve(f1(x)=0,x)$

$$\frac{1}{e} \tag{4.136}$$

Hence the critical number is $\dfrac{1}{e}$. Now find the intervals where f is increasing and decreasing.

$solve(f1(x)>0,\{x\})$

$$\left\{\frac{1}{e}<x\right\} \tag{4.137}$$

$solve(f1(x)<0,\{x\})$

$$\left\{0<x,x<\frac{1}{e}\right\} \tag{4.138}$$

The function f is increasing in the interval $\left(\dfrac{1}{e},\infty\right)$ and decreasing in the interval $\left(0,\dfrac{1}{e}\right)$.

Therefore, f has no relative maximum and a relative minimum when $x=\dfrac{1}{e}$. We now use the second derivative to find the points of inflection.

$f1'(x)$

$$\frac{1}{x} \tag{4.139}$$

$solve(f1'(x)=0,x)$

Note that there is no output in this case. Why? We now determine the intervals where f is concave up and concave down.

$solve(f1'(x)>0,\{x\})$

$$\{0<x\} \tag{4.140}$$

$$solve\big(f1'(x) < 0, \{x\}\big)$$

$$\{x < 0\} \tag{4.141}$$

The function f is concave up in $(0,\infty)$. Note that the function is not defined in the interval $(-\infty, 0)$ (why?). The graph below gives a general shape of the function f.

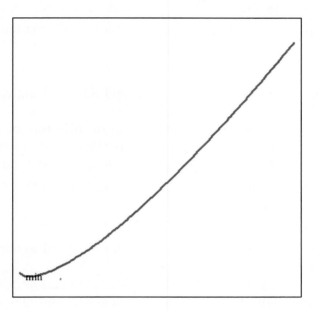

Exercises

1. Let $f(x) = -x^3 - 5x^2 + 48x - 81$.

   (a) Find the real zeros, first derivative, and the local extrema. Identify the intervals where the function is increasing and decreasing.

   (b) Find the second derivative and point(s) of inflection of this polynomial function. Identify the intervals where the function is concave up and down.

   (c) Graph the function f along with its horizontal tangent lines and its first derivative. Plot the function f with the quantitative information about its properties.

2. Let $f(x) = -3x^4 - 21x^3 + 162x^2 + 480x + 371$.

   (a) Find the real zeros, first derivative, and the local extrema. Identify the intervals where the function is increasing and decreasing.

   (b) Find the second derivative and point(s) of inflection of this polynomial function. Identify the intervals where the function is concave up and down.

   (c) Graph the function f along with its horizontal tangent lines and its first derivative. Plot the function f with the quantitative information about its properties.

3. Let $f(x) = 6x^4 + 128x^3 + 477x^2 + 2376x + 114$.

   (a) Find the real zeros, first derivative, and the local extrema. Identify the intervals where the function is increasing and decreasing.

   (b) Find the second derivative and point(s) of inflection of this polynomial function. Identify the intervals where the function is concave up and down.

   (c) Graph the function f along with its horizontal tangent lines and its first derivative. Plot the function f with the quantitative information about its properties.

4. Let $f(x) = x \ln(x)$.

   (a) Find the real zeros, first derivative, and the local extrema. Identify the intervals where the function is increasing and decreasing.

   (b) Find the second derivative and point(s) of inflection of this polynomial function. Identify the intervals where the function is concave up and down.

   (c) Graph the function f along with its horizontal tangent lines and its first derivative. Plot the function f with the quantitative information about its properties.

5. Let $f(x) = x \cos(x)$ restricted to the interval $[-\pi, \pi]$.

   (a) Find the real zeros, first derivative, and the local extrema. Identify the intervals where the function is increasing and decreasing.

   (b) Find the second derivative and point(s) of inflection of this polynomial function. Identify the intervals where the function is concave up and down.

   (c) Graph the function f along with its horizontal tangent lines and its first derivative. Plot the function f with the quantitative information about its properties.

6. Let $f(x) = \dfrac{2x^3 + 11x^2 - 31x - 180}{2x^2 - 3x - 54}$.

(a) Find the real zeros, first derivative, and the local extrema. Identify the intervals where the function is increasing and decreasing.

(b) Find the second derivative and point(s) of inflection of this polynomial function. Identify the intervals where the function is concave up and down.

(c) Graph the function f along with its horizontal tangent lines and its first derivative. Plot the function f with the quantitative information about its properties.

7. Let $f(x) = \dfrac{4x^3 + 43x^2 - 81x - 252}{4x^2 - 19x + 21}$.

(a) Find the domain, real zeros, and the asymptotes of this function.

(b) Find the first derivative and the local extrema. Identify the intervals where the function is increasing and decreasing.

(c) Find the second derivative and point(s) of inflection of this rational function. Identify the intervals where the function is concave up and down.

(d) Graph the function f along with its asymptotes. Plot the function f with the quantitative information about its properties.

8. Let $f(x) = \dfrac{5x^3 + 23x^2 - 17x - 35}{x^2 + 7x + 6}$.

(a) Find the domain, real zeros, and the asymptotes of this function.

(b) Find the first derivative and the local extrema. Identify the intervals where the function is increasing and decreasing.

(c) Find the second derivative and point(s) of inflection of this rational function. Identify the intervals where the function is concave up and down.

(d) Graph the function f along with its asymptotes. Plot the function f with the quantitative information about its properties.

9. Let $f(x) = \dfrac{6x^3 + 5x^2 - 68x + 32}{3x^3 + 17x^2 + 12x - 32}$.

(a) Find the domain, real zeros, and the asymptotes of this function.

(b) Find the first derivative and the local extrema. Identify the intervals where the function is increasing and decreasing.

(c) Find the second derivative and point(s) of inflection of this rational function. Identify the intervals where the function is concave up and down.

(d) Graph the function f along with its asymptotes. Plot the function f with the quantitative information about its properties.

10. Let $f(x) = \dfrac{4x^3 - 15x^2 + 6x + 11}{3x^2 + 4x - 54}$.

(a) Find the domain, real zeros, and the asymptotes of this function.

(b) Find the first derivative and the local extrema. Identify the intervals where the function is increasing and decreasing.

(c)   Find the second derivative and point(s) of inflection of this rational function. Identify the intervals where the function is concave up and down.

(d)   Graph the function f along with its asymptotes. Plot the function f with the quantitative information about its properties.

11.   Let $f(x) = \dfrac{5x}{4 - e^x}$.

(a)   Find the domain, real zeros, and the asymptotes of this function.

(b)   Find the first derivative and the local extrema. Identify the intervals where the function is increasing and decreasing.

(c)   Find the second derivative and point(s) of inflection of this rational function. Identify the intervals where the function is concave up and down.

(d)   Graph the function f along with its asymptotes. Plot the function f with the quantitative information about its properties.

12.   Let $f(x) = \dfrac{x}{\ln(x + 3)}$.

(a)   Find the domain, real zeros, and the asymptotes of this function.

(b)   Find the first derivative and the local extrema. Identify the intervals where the function is increasing and decreasing.

(c)   Find the second derivative and point(s) of inflection of this rational function. Identify the intervals where the function is concave up and down.

(d)   Graph the function f along with its asymptotes. Plot the function f with the quantitative information about its properties.

13.   Let $f(x) = \dfrac{x}{3 + \ln(x)}$.

(a)   Find the domain, real zeros, and the asymptotes of this function.

(b)   Find the first derivative and the local extrema. Identify the intervals where the function is increasing and decreasing.

(c)   Find the second derivative and point(s) of inflection of this rational function. Identify the intervals where the function is concave up and down.

(d)   Graph the function f along with its asymptotes. Plot the function f with the quantitative information about its properties.

14.   Sketch a graph of a polynomial function whose first derivative is $f1(x) = 6x^2 - 41x - 7$.

15.   Sketch a graph of a polynomial function whose first derivative is $f1(x) = -9x^2 + 109x + 560$.

16.   Sketch a graph of a polynomial function whose first derivative is $f1(x) = 2 - e^{-5x}$.

17. Sketch a graph of a polynomial function whose first derivative is $f1(x) = 6x^3 + 23x^2 + 11x - 12$.

18. Let $f(x) = x\cos(x)$ restricted to the interval $[0, \pi]$.
    (a) Find the domain and real zeros of this function by using the *Roots* command.
    (b) Find the vertical asymptotes using limits and your answers in part (a) of this problem (as we have in example 2).
    (c) Use the *Asymptotes* command to find the asymptotes of this function in the interval $[0, \pi]$. How many asymptotes are listed? What happens when you use the *numeric* option with this command?
    (d) Graph the function f in the interval $[0, \pi]$. Do you see the asymptote $x = 0$? What happens when you increase the viewing window for x to $x = -0.1..\pi$? What happens when you use the option *thickness* $= 3$ in this case?

19. Let $f(x) = \cot(2\sin(3x))$ restricted to the interval $[0, \pi]$.
    (a) Find the domain and real zeros of this function (restricted to the interval $[0, \pi]$) by using the *Roots* command.
    (b) Find the vertical asymptotes using limits and your answers in part (a) of this problem (as we have in example 2).
    (c) Use the *Asymptotes* command to find the asymptotes of this function in the interval $[0, \pi]$. How many asymptotes are listed? What happens when you use the *numeric* option with this command? Compare your answers with the ones from part (b).
    (d) Graph the function f in the interval $[0, \pi]$. Do you see the asymptote $x = 0$? What happens when you increase the viewing window for x to $x = -0.1..\pi$? What happens when you use the option *thickness* $= 3$ in this case?

20. Let $f(x) = \csc(x^2)$ restricted to the interval $[-2\pi, 2\pi]$.
    (a) Find the domain and real zeros of this function (restricted to the interval $[-2\pi, 2\pi]$) by using the *Roots* command.
    (b) Find the vertical asymptotes using limits and your answers in part (a) of this problem (as we have in example 2).
    (c) Use the *Asymptotes* command to find the asymptotes of this function in the interval $[-2\pi, 2\pi]$. How many asymptotes are listed? What happens when you use the *numeric* option with this command? Compare your answers with the ones from part (b).
    (d) Graph the function f in the interval $[-2\pi, 2\pi]$. Do you see all the vertical asymptotes? What happens when you use the option *thickness* $= 3$ in this case? What happens when you increase the viewing window for x to $x = -2.000000001\pi .. 2.000000001\pi$?

21.   Let $f(x) = \csc(x^2)$ restricted to the interval $[-\pi, 2\pi]$.

(a)   Find the domain and real zeros of this function (restricted to the interval $[-\pi, 2\pi]$) by using the *Roots* command.

(b)   Find the vertical asymptotes using limits and your answers in part (a) of this problem (as we have in example 2).

(c)   Use the *Asymptotes* command to find the asymptotes of this function in the interval $[-\pi, 2\pi]$. How many asymptotes are listed? What happens when you use the *numeric* option with this command?

(d)   Graph the function f in the interval $[-\pi, 2\pi]$. Do you see all the vertical asymptotes? What happens when you use the option *thickness* $= 3$ in this case? What happens when you increase the viewing window for x to $x = -1.0000000001\pi .. 2.0000000001\pi$?

22.   Below is a list of some commands used in this chapter. Describe the significance of each command. Can you find examples where each command is used?

(a)   *Asymptotes* $(f(x), x)$

(b)   *Asymptotes* $(f(x), x = a .. b, numeric)$

(c)   *CriticalPoints* $(f(x), numeric)$

(d)   *DerivativePlot* $(f(x), x = a .. b, order = 1 .. n)$

(e)   *DerivativePlot* $(f(x), x = a .. b, order = 1 .. 2, functionoptions = [color = color1],$

$derivativeoptions[1] = [color = color2, linestyle = linestyle1, thickness = n],$

$derivativeoptions[2] = [color = color3, linestyle = linestyle2, thickness = m])$

(f)   *FunctionChart* $(f(x), x = a .. b)$

(g)   *InflectionPoints* $(f(x))$

23.   Can you list new *Maple* commands that were used in this chapter but were not listed in exercise 22?

Chapter 5   Applications of Differentiation

The object of this chapter is to see applications of the derivative while calculating extrema and applying the Extreme Value Theorem, Rolle's Theorem, and the Mean Value Theorem. An application section will also be included on related rates, optimization, and finding roots via Newton's method.

Extreme Value Theorem

Extreme Value Theorem: Let f be a continuous function on the closed interval $[a,b]$. Then f has an absolute minimum value and absolute maximum value in the interval $[a,b]$.

Example 1: Find the domain and real zeros of the rational function $f(x) = \dfrac{x^2 + 2x + 1}{x^3 + 1}$. Determine if the Extreme Value Theorem can be applied to this function over the interval $[0,4]$. If so, find the maximum and minimum value of the function f in this interval.

Solution: We will enter f as a *Maple* function.

$$f := x \to \frac{x^2 + 2 \cdot x + 1}{x^3 + 1}$$

$$x \to \frac{x^2 + 2\,x + 1}{x^3 + 1} \tag{5.1}$$

To find the domain we set the denominator equal to zero and solve for x.

$$solve\left(denom\left(f(x)\right) = 0, x\right)$$

$$-1, \frac{1}{2} + \frac{1}{2}I\sqrt{3}, \frac{1}{2} - \frac{1}{2}I\sqrt{3} \tag{5.2}$$

Hence the domain of f is the set of all real numbers except -1. To find the zeros we set the numerator equal to zero and solve for x. Note that the zeros must be in the domain of f.

$$solve\left(numer\left(f(x)\right) = 0, x\right)$$

$$-1, -1 \tag{5.3}$$

The function f has no real zeros as -1 is not in the domain of f. The Extreme Value Theorem can be applied since f is continuous on $[0,4]$ (-1 is not in the interval $[0,4]$).

$plot\left(f\left(x\right),x=0..4\right)$

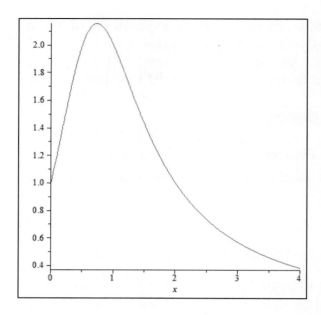

The graph shows that f has a maximum value when $0<x<4$. We will now use the *ExtremePoints* command found in the $with\left(Student\left[Calculus1\right]\right)$ package. Do not forget to activate this package. In order to get the floating-point computation use the *numeric* command.

$with\left(Student\left[Calculus1\right]\right):$

$c:=ExtremePoints\left(f\left(x\right),x=0..4\right)$

$$\left[0,-1+\sqrt{3},4\right] \tag{5.4}$$

$c:=ExtremePoints\left(f\left(x\right),x=0..4,numeric\right)$

$$\left[0.,0.7320508076,4.\right] \tag{5.5}$$

$f\left(c\left[1\right]\right);f\left(c\left[2\right]\right);f\left(c\left[3\right]\right);$

1.000000000
2.154700537
0.3846153846   (5.6)

Hence f has a minimum value, namely 0.3846153846, when $x=4$ and a maximum value, namely 2.154700537, when $x=0.7320508076$.

Rolle's Theorem

Rolle's Theorem: Let f be a continuous function on the closed interval $[a,b]$, a differentiable function on the open interval (a,b) such that $f(a) = f(b)$. Then there is a number c in the open interval (a,b) such that $f'(c) = 0$.

Example 2: Determine if the Rolle's Theorem can be applied to the function $f(x) = x^3 + x^2 - 17x + 15$ over the interval $[1,3]$. If so, find the value of c in the open interval $(1,3)$ such that $f'(c) = 0$.

Solution: Note that as f is a polynomial function it is continuous on the closed interval $[1,3]$ and differentiable on the open interval $(1,3)$. We will enter f as a *Maple* function and evaluate $f(1)$ and $f(3)$.

$f := x \rightarrow x^3 + x^2 - 17 \cdot x + 15$

$$x \rightarrow x^3 + x^2 - 17\,x + 15 \qquad \textbf{(5.7)}$$

$f(1); f(3);$

$$0$$
$$0 \qquad \textbf{(5.8)}$$

Therefore, Rolle's Theorem can be applied to f. Hence there is a number c in the open interval $(1,3)$ such that $f'(c) = 0$. We will now use the *RollesTheorem* command found in the $with(Student[Calculus1])$ package (by first activating this package) to find c.

$with(Student[Calculus1]):$

$RollesTheorem(f(x), x = 1..3)$

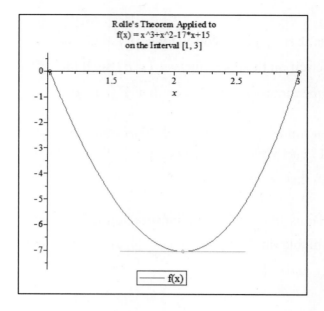

Notice that $RollesTheorem(f(x), x = 1..3)$ just plots the function in the interval $[1,3]$ and show a number c in the open interval $(1,3)$ on the graph where the tangent line is horizontal (so that $f'(c) = 0$). However, if we are interested in finding the value(s) of c we need to give this command more options. The *output = points* gives the value(s) of c and the *numeric* option added to this gives the floating-point computation of the answer.

$c := RollesTheorem(f(x), x = 1..3, output = points)$

$$-\frac{1}{3} + \frac{2}{3}\sqrt{13}$$ (5.9)

$c := RollesTheorem(f(x), x = 1..3, output = points, numeric)$

$$2.070367517$$ (5.10)

Remark: We could use the *fsolve* command to find the value of c as shown below.

$f'(x)$

$$x \rightarrow 3x^2 + 2x - 17$$ (5.11)

$c := fsolve(f'(x) = 0, x = 1..3)$

$$2.070367517$$ (5.12)

Mean Value Theorem

Mean Value Theorem: Let f be continuous on the closed interval $[a,b]$ and differentiable on the open interval (a,b). Then there is a number c in the open interval (a,b) such that $f'(c) = \dfrac{f(b)-f(a)}{b-a}$.

Example 3: Find point(s) c (if they exist) that satisfy the Mean Value Theorem for the function $f(x) = 12x^4 - 4x^3 - 40x^2$ over the interval $[-1,2]$. Find the equation(s) of the tangent line(s) at these points. Graph the function along with these tangent lines and the secant lines through the points $(-1, f(-1))$ and $(2, f(2))$.

Solution: Note that as f is a polynomial function it is continuous on the closed interval $[-1,2]$ and differentiable on the open interval $(-1,2)$. Hence the Mean Value Theorem can be applied to f. We will enter f as a *Maple* function.

$f := x \rightarrow 12 \cdot x^4 - 4 \cdot x^3 - 40 \cdot x^2$

$$x \rightarrow 12\,x^4 - 4\,x^3 - 40\,x^2 \qquad\qquad \textbf{(5.13)}$$

In order to give a geometric interpretation for the Mean Value Theorem, we need to find all points in the interval $(-1,2)$ such that the tangent lines at these points is parallel to the secant line through $(-1, f(-1))$ and $(2, f(2))$. We will use the *makeproc* command to find the equation of the secant line through $(-1, f(-1))$ and $(2, f(2))$. Before doing this we will activate the $with(student)$ package.

$with(student):$

$A := \left[-1, f(-1)\right]; B := \left[2, f(2)\right];$

$$[-1, -24]$$
$$[2, 0] \qquad\qquad \textbf{(5.14)}$$

$q := makeproc(A, B)$

$$x \rightarrow 8\,x - 16 \qquad\qquad \textbf{(5.15)}$$

Therefore, the equation of the secant line through the points $(-1, f(-1))$ and $(2, f(2))$ is $y = 8x - 16$. The graph of f and this secant line is given below.

$plot\left(\left[f(x), q(x)\right], x = -1..2, color = [red, blue]\right)$

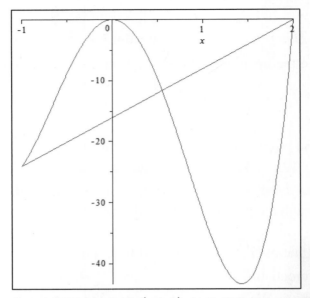

Notice that there are two points in the interval $(-1, 2)$ such that the tangent lines at these points are parallel to the secant line $y = 8x - 16$. To find these points we will find the first derivative f' of f and set $f'(x) = 8$ (as 8 is the slope of the secant line and hence also the slope of the tangent line) and solve for x.

$m := slope(A, B)$

$$8 \tag{5.16}$$

$f'(x)$

$$x \to 48\, x^3 - 12\, x^2 - 80\, x \tag{5.17}$$

$c := fsolve(f'(x) = m, x = -1..2)$

$$-0.1022075784, 1.465165823 \tag{5.18}$$

$f(c[1]); f(c[2]);$

$$-0.4122752381 \tag{5.19}$$
$$-43.14917358$$

Therefore the two values of c are -0.10220757841 and 1.465165823. We will now find the equation of the tangent line through these points using the *makeproc* command and then graphing it using the *showtangent* command.

$makeproc\left([c[1], f(c[1])], `slope` = m\right)$

$$x \to 8\, x + 0.4053853891 \tag{5.20}$$

Hence the equation of the tangent line at the point (-0.10220757841, -0.4122752381) is $y = 8x + 0.4053853891$. The plot below gives graph of the tangent line along with the function f.

showtangent $\left(f(x), x = c[1], x = -1..2 \right)$

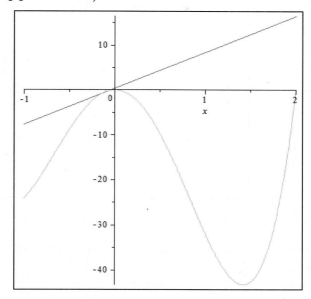

We will repeat the same procedure for the second point (1.465165823, -43.14917358).

makeproc $\left(\left[c[2], f(c[2]) \right], `slope` = m \right)$

$$x \rightarrow 8\,x - 54.87050016 \tag{5.21}$$

Hence the equation of the tangent line at the point (1.465165823, -43.14917358) is $y = 8x - 54.87050016$. The plot below gives graph of the graph tangent line along with the function f.

showtangent $\left(f(x), x = c[2], x = -1..2 \right)$

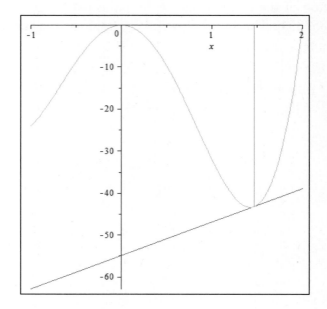

The three graphs can be now plotted together. This is shown below. Do not forget to activate the $with(plots)$ package.

$with(plots):$

$p1 := plot\left(\left[\,f(x),q(x)\right],x=-1..2\right)$

$p2 := showtangent\left(f(x),x=c[1],x=-1..2\right)$

$p3 := showtangent\left(f(x),x=c[2],x=-1..2\right)$

$display(\,p1,p2,p3)$

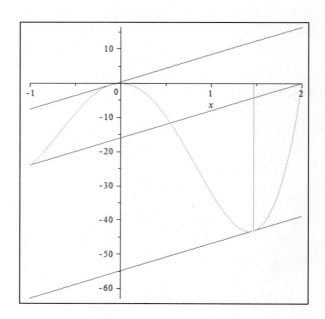

Remarks:

1. The equation of the tangent can be obtained by first activating the $with\left(Student\left[Calculus1\right]\right)$ package and using the *Tangent* command as shown below.

    $with\left(Student\left[Calculus1\right]\right):$

    $Tangent\left(f\left(x\right),x=c[1],-1..2,output=line\right)$

    $$7.999999997\,x+0.4053853888 \qquad\qquad \textbf{(5.22)}$$

    Notice that since $c[1]$ had a numerical answer, the slope and the *y*-intercept of the tangent line are slightly different from the answer given by the *makeproc* command.

2. We can use the *MeanValueTheorem* command in the $with\left(Student\left[Calculus1\right]\right)$ package to solve this problem.

    $$f := x \rightarrow 12\cdot x^4 - 4\cdot x^3 - 40\cdot x^2$$

    $$x \rightarrow 12\,x^4 - 4\,x^3 - 40\,x^2 \qquad\qquad \textbf{(5.23)}$$

    $with\left(Student\left[Calculus1\right]\right):$

    $MeanValueTheorem\left(f\left(x\right),x=-1..2\right)$

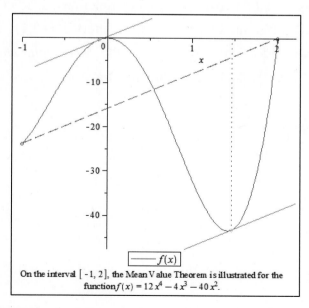

On the interval $[-1, 2]$, the Mean Value Theorem is illustrated for the function $f(x) = 12\,x^4 - 4\,x^3 - 40\,x^2$.

$$c := MeanValueTheorem\left(f\left(x\right),x=-1..2,output=points\right)$$

Warning, no points satisfying the Mean Value Theorem were found using symbolic techniques, try using the 'numeric = true option'

$$[\ ]$$ **(5.24)**

Note that that we do not see any output in this case. Let us use the *numeric* option.

$$c := MeanValueTheorem\left(f\left(x\right), x = -1\,..\,2, output = points, numeric\right)$$

$$\left[-0.1022075784, 1.465165823\right]$$ **(5.25)**

These answers are the same as those in (5.18).

Example 4: Find point(s) c (if they exist) that satisfy the Mean Value Theorem for the function $f\left(x\right) = x^2 \sin\left(x\right)$ over the interval $\left[-1,1\right]$. Find the equation(s) of the tangent line(s) at these point(s). Graph the function along with these tangent lines and the secant lines through the points $\left(-1, f\left(-1\right)\right)$ and $\left(1, f\left(1\right)\right)$.

Solution: Enter f as a *Maple* function.

$$f := x \rightarrow x^2 \cdot \sin\left(x\right)$$

$$x \rightarrow x^2 \sin\left(x\right)$$ **(5.26)**

Since f is the product of two differentiable functions, it is continuous on the closed interval $\left[-1,1\right]$ and differentiable on the open interval $\left(-1,1\right)$. Hence the Mean Value Theorem can be applied to f. We will use the *MeanValueTheorem* command in the $with\left(Student\left[Calculus1\right]\right)$ package as shown below.

$$with\left(Student\left[Calculus1\right]\right):$$
$$MeanValueTheorem\left(f\left(x\right), x = -1\,..\,1\right)$$

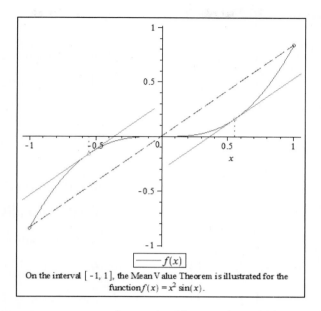

On the interval [-1, 1], the Mean Value Theorem is illustrated for the function $f(x) = x^2 \sin(x)$.

The graphs of the secant line connecting $\left(-1, f\left(-1\right)\right)$ and $\left(1, f\left(1\right)\right)$ and the tangent lines is shown in the plot above. By adding the option *output = points* and *numeric* we can calculate the values of c that satisfy the Mean Value Theorem.

$c := MeanValueTheorem\left(f\left(x\right), x = -1..1, output = points, numeric\right)$

$$[-0.5531168205, 0.5531168205]$$ **(5.27)**

$f\left(c[1]\right); f\left(c[2]\right);$

$$-0.1607221495$$
$$0.1607221495$$ **(5.28)**

We still need to find the equations of the tangent lines at the points $\left(-0.5531168205, -0.1607221495\right)$ and $\left(0.5531168205, 0.1607221495\right)$. The Tangent command will be used here. Note that we do not need to activate the $with\left(Student\left[Calculus1\right]\right)$ package as we have just done that.

$Tangent\left(f\left(x\right), x = c[1], -1..1, output = line\right)$

$$0.8414709849\, x + 0.3047096062$$ **(5.29)**

$Tangent\left(f\left(x\right), x = c[2], -1..1, output = line\right)$

$$0.8414709849\, x - 0.3047096062$$ **(5.30)**

Hence the equation of the tangent line at $(-0.5531168205, -0.1607221495)$ is $y = 0.8414709849\ x + 0.3047096062$ and at $(0.5531168205, 0.1607221495)$ is $y = 0.8414709849\ x - 0.3047096062$.

Remarks: We can find the equation of the secant line through the points $(-1, f(-1))$ and $(1, f(1))$ using the same method as in example 3. Do not forget to activate the $with(student)$ package.

Newton's Method

Newton's Method: Let f be continuous on the closed interval $[a, b]$ and differentiable on the open interval (a, b). *Newton's method* (or *Newton-Raphson's method*) gives an iterative method of finding an approximate solution to the equation $f(x) = 0$ where x is in the interval $[a, b]$. Let x_1 be any point in $[a, b]$. Define the points x_n iteratively as $x_{n+1} = x_n - \dfrac{f(x_n)}{f'(x_n)}$, $n = 1, 2, \ldots$ Then $\lim_{n \to \infty} x_n = c$, where c is a root of the equation $f(x) = 0$.

Example 5: Show that the function $f(x) = x^5 + x - 5$ has exactly one real root. Write a *Maple* procedure to find this root using fifteen iterations in Newton's method (approximate this root to nine decimal places).

Solution: Since f is a polynomial function it is continuous on the set of real numbers $\mathbb{R}$. We will enter f as a *Maple* function.

$f := x \to x^5 + x - 5$

$$x \to x^5 + x - 5 \tag{5.31}$$

We will now find $f'(x)$ to find the intervals of increase and decrease.

$f'(x)$

$$x \to 5\,x^4 + 1 \tag{5.32}$$

Since $f'(x) = 5\,x^4 + 1 > 0$ for all real numbers x, f is strictly increasing on the set of real numbers $\mathbb{R}$. Note that $f(x) \le 0$ when $x \le 0$. Since

$f(1)$

$$-3 \tag{5.33}$$

$f(2)$

$$29 \qquad \textbf{(5.34)}$$

We see that $f(1) = -3 < 0$ and $f(2) = 29 > 0$. Hence by the Intermediate Value Theorem f had at least one real root in $[1,2]$. Let us select an initial value of 1.1 and write a *Maple* procedure to apply Newton's method.

$$g := x \rightarrow x - \frac{f(x)}{D(f)(x)}$$

$$x \rightarrow x - \frac{f(x)}{D(f)(x)} \qquad \textbf{(5.35)}$$

$x[1] := 1.1$

$$1.1 \qquad \textbf{(5.36)}$$

for n **from** 1 **to** 15 **do** $x[n+1] := g(x[n])$ **end do**:

$seq(x[n], n = 1..15)$

> 1.1, 1.375162550, 1.306682555, 1.299233535, 1.299152802, 1.299152792, **(5.37)**
> 1.299152792, 1.299152792, 1.299152792, 1.299152792, 1.299152792,
> 1.299152792, 1.299152792, 1.299152792, 1.299152792

Hence 1.299152792 is a real root (approximated to nine decimal places) of the function $f(x) = x^5 + x - 5$. We will now plot the function to see the x-intercept and check our answer.

$plot(f(x), x = -2..3)$

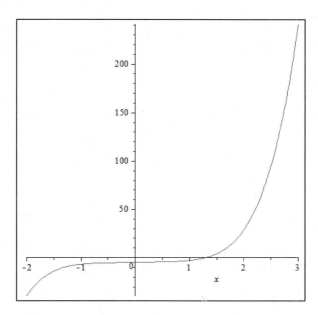

$$fsolve\left(f\left(x\right)=0,x\right)$$

$$1.299152792 \tag{5.38}$$

Remarks:

1. We used the operator $D(f)$ for the derivative when we defined g. This was done so that g could be defined as a function. The error, if we used f' instead, is shown below.

$$g := x \rightarrow x - \frac{f\left(x\right)}{f'\left(x\right)}$$

$$x \rightarrow x - \frac{f\left(x\right)}{\dfrac{d}{dx}f\left(x\right)} \tag{5.39}$$

$g\left(1.1\right)$

Error, (in g) invalid input: diff received 1.1, which is not valid for its 2nd argument

2. This can be done using the *NewtonsMethod* command in the $with\left(Student\left[Calculus1\right]\right)$ package as shown below. The *output = plot* option is used to obtain the graph and the *output = sequence* is used to find the sequence containing the 15 iterations calculated.

$$f := x \rightarrow x^5 + x - 5$$

$$x \rightarrow x^5 + x - 5 \tag{5.40}$$

$with\left(Student\left[Calculus1\right]\right):$

$NewtonsMethod\left(f\left(x\right),1.1, iterations = 15, output = plot\right)$

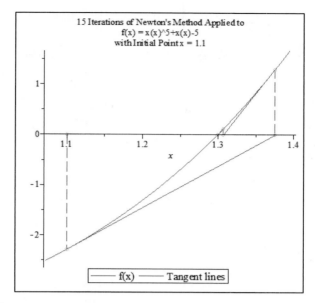

$$NewtonsMethod\left(f\left(x\right),1.1,iterations=15,output=sequence\right)$$

1.1, 1.375162550, 1.306682555, 1.299233535, 1.299152802, 1.299152792, **(5.41)**
1.299152792, 1.299152792, 1.299152792, 1.299152792, 1.299152792,
1.299152792, 1.299152792, 1.299152792, 1.299152792

If the options are not mentions the final iteration is put as an answer as shown below.

$$NewtonsMethod\left(f\left(x\right),1.1,iterations=15\right)$$

1.299152792   **(5.42)**

Related Rates

Example 7: A lighthouse is located at 120 meters away from the closest point O on a shoreline and its light source makes 1 revolution per second. How fast is the beam of light moving along the shoreline when it is 60 meters from O?

Solution: Let us first draw a picture. Let lighthouse be located at L. Suppose that the angle that the beam makes is $\theta(t)$ and distance of the beam from the shore is $h(t)$. We will use the *plot*, *textplot* and *display* commands as shown below. Do not forget to activate the $with(plots)$ package.

$with\left(plots\right):$

$s1:=plot\left(\left\{\left[\left[0,0\right],\left[3,3\right]\right]\left[\left[3,0\right],\left[3,3\right]\right],\left[\left[0,0\right],\left[3,0\right]\right]\right\},thickness=3,axes=none\right)$

$PLOT\left(...\right)$ **(5.43)**

$s2 := textplot\left(\left[1.5,-2,`120`\right],\left[3.3,1.5,`h(t)`\right],\left[0.3,0.15,`\theta(t)`\right],\left[-0.1,0,`L`\right],\left[3.1,0,`O`\right],\left[3.1,3.1,`A`\right]\right)$

$$PLOT\left(...\right) \tag{5.44}$$

$display\left(s1,s2\right)$

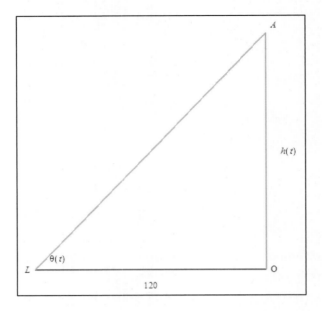

From the above sketch $\tan\left(\theta\right) = \dfrac{h(t)}{120}$ so that $h(t) = 120\tan\left(\theta(t)\right)$.

$h := t \rightarrow 120 \cdot \tan\left(\theta(t)\right)$

$$t \rightarrow 120\tan\left(\theta(t)\right) \tag{5.45}$$

$hrate := diff\left(h(t),t\right)$

$$t \rightarrow 120\left(1 + \tan\left(\theta(t)\right)^2\right)\left(\frac{d}{dt}\theta(t)\right) \tag{5.46}$$

Since the light source moves at 1 revolution per second, we have that (the angles are in radians) $\dfrac{d}{dt}\theta(t) = 2\pi$. At the instant that $h(t) = 60$ meters, $\tan\left(\theta(t)\right) = \dfrac{60}{120} = \dfrac{1}{2}$. Let us now substitute these values in the formula for *hrate*.

$$subs\left(\left\{\tan\left(\theta(t)\right) = \frac{1}{2}, \frac{d}{dt}\theta(t) = 2\cdot\pi, hrate\right\}\right)$$

$$300\pi \tag{5.47}$$

Hence the rate at which the beam is moving along the shoreline at the instant the beam is 60 meters from O is 300π meters per second.

Example 8: An insect moves along a curve $y = \dfrac{\sqrt{x^3 + 2x + 4\sin(x)}}{x^2 + 2}$. If its horizontal coordinate is changing at a rate of 5 ft/sec, at what rate is its vertical coordinate changing at the instant when $y = \dfrac{2}{3}$ feet?

Solution: We will enter the curve as a *Maple* function and plot it.

$$f := x \to \frac{\sqrt{x^3 + 2x + 4\sin(x)}}{x^2 + 2}$$

$$x \to \frac{\sqrt{x^3 + 2x + 4\sin(x)}}{x^2 + 2} \tag{5.48}$$

$$plot\left(f(x), x = 0..5\right)$$

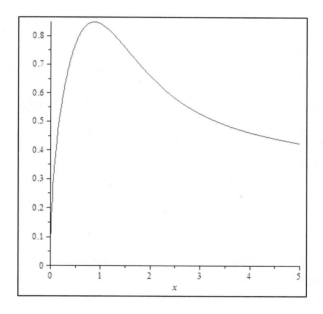

Since x is a function of t (time), f is also a function of t. In this case, *Maple* cannot compute derivatives using the chain rule by taking derivatives of both sides of the expression with respect to t (See example 5 in this chapter where we needed a rewrite command). If we take the derivative operator D, it works in an equivalent fashion to taking the derivative with respect to t (time) of both f and x providing we consider $D(x) = \dfrac{dx}{dt}$. Note that $D(\sin(x)) = \cos(x)D(x)$.

$$simplify\left(D\left(\frac{\sqrt{x^3+2\cdot x+4\cdot \sin(x)}}{x^2+2}\right)\right)$$

$$-\frac{1}{2}\frac{D(x)\,x^4-4\,D(x)-4\,D(\sin(x))\,x^2-8\,D(\sin(x))+16\,D(x)\,x\sin(x)}{\sqrt{x^3+2\cdot x+4\cdot \sin(x)}\,(x^2+2)^2} \qquad \textbf{(5.49)}$$

Let us now substitute the fact that $\dfrac{dx}{dt}=5$ ft/sec and $D(\sin(x))=\cos(x)D(x)$ in (3.56).

$$subs\left(\frac{dx}{dt}=5, D(\sin(x))=\cos(x)D(x),\ (\textbf{5.49})\right)$$

$$-\frac{1}{2}\frac{5\,x^4-20-4\left(\cos(x).5\right)x^2-8\cos(x).5+80\,x\sin(x)}{\sqrt{x^3+2\cdot x+4\cdot \sin(x)}\,(x^2+2)^2} \qquad \textbf{(5.50)}$$

To evaluate the above expression, we will find the value of x when $f(x)=\dfrac{2}{3}$.

$$evalf\left(solve\left(f(x)=\frac{2}{3},x\right)\right)$$

$$0.3268312752 \qquad \textbf{(5.51)}$$

Let us now substitute this value of x in (5.50)

$$subs\left(x=0.3268312752,\ (\textbf{5.50})\right)$$

$$4.126713513 \qquad \textbf{(5.52)}$$

Therefore, the vertical coordinate is increasing at a rate of 4.126713513 ft/sec. Note that the rate is increasing as it is positive.

Optimization

Example 9: Find the minimum distance from the point (-2, 3) to the graph of the function $f(x)=e^x\sin(x)$.

Solution: We will begin by entering f as a *Maple* function. We will then plot the function along with the point (-2, 3).

$$f:=x\rightarrow e^x\cdot \sin(x)$$

$$x\rightarrow e^x\sin(x) \qquad \textbf{(5.53)}$$

$$with(plots):$$

$s1 := plot\left(f(x), x = -3.5 .. 3.3\right)$

$$PLOT\,(...) \qquad\qquad \textbf{(5.54)}$$

$s2 := pointplot\left([-2,3], color = blue, symbol = solidcircle, symbolsize = 15\right)$

$$PLOT\,(...) \qquad\qquad \textbf{(5.55)}$$

$display\,(s1, s2)$

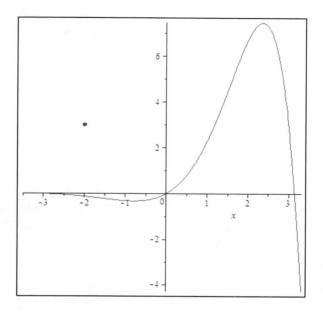

The distance between any point (x, y) on the curve f and the point (-2, 3) is given by $\sqrt{(x-(-2))^2 + (y-3)^2}$. We will replace y by $f(x) = e^x \sin(x)$ in this distance formula.

$$subs\left(y = e^x \cdot \sin(x), \sqrt{(x-(-2))^2 + (y-3)^2}\right)$$

$$\sqrt{x^2 + 4x + 13 + \left(e^x\right)^2 \sin(x)^2 - 6\,e^x \sin(x)} \qquad\qquad \textbf{(5.56)}$$

$g := unapply\left((\textbf{5.56}), x\right)$

$$x \rightarrow \sqrt{x^2 + 4x + 13 + \left(e^x\right)^2 \sin(x)^2 - 6\,e^x \sin(x)} \qquad\qquad \textbf{(5.57)}$$

We will now plot this function to see where the minimum occurs.

$plot\left(g(x), x = -4 .. 1.5\right)$

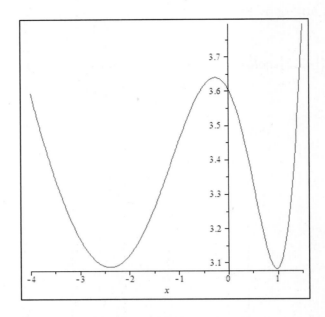

From the graph above, it seems that the minimum value occurs when x is in the interval [-3, -2] or in [0.5, 1.5]. Let us verify this using the *CriticalPoints* command. Do not forget to activate the $with\left(Student\left[Calculus1\right]\right)$ package.

$with\left(Student\left[Calculus1\right]\right):$

$c := CriticalPoints\left(g\left(x\right), numeric\right)$

$$\left[-2.394649061, -0.2626436025, 0.9742928616, 2.419021486,\right. \tag{5.58}$$

$$\left.2.971003513, 5.497910763, -6.288728266, 8.639380131, 9.424535729\right]$$

$g\left(c[1]\right); g\left(c[2]\right); g\left(c[3]\right); g\left(c[4]\right); g\left(c[5]\right); g\left(c[6]\right); g\left(c[7]\right); g\left(c[8]\right); g\left(c[9]\right);$

3.087292311
3.640913416
3.082150026
6.257057488
4.980805654
175.8008354
8.288742403
39992.043537
11.42453577 (5.59)

Based on the calculations the minimum distance of the point (-2, 3) to the function $f\left(x\right) = e^x \sin\left(x\right)$ is 3.082150026. This value occurs when $x = 0.974298616$. note that the point -2.394649061 is almost at the same distance from the point (-2, 3).

Example 10: Find the dimensions of the largest rectangle that can be inscribed in the region enclosed by the curve $f(x) = \dfrac{9-3x}{5+2x}$ and the first quadrant, where two sides of the rectangle are on the coordinate axes. Note that the picture is generated in *Maple*.

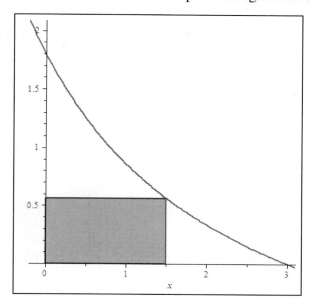

Solution: We will begin by entering f as a *Maple* function.

$f := x \to \dfrac{9-3 \cdot x}{5+2 \cdot x}$

$$x \to \dfrac{9-3\,x}{5+2\,x} \qquad\qquad \textbf{(5.60)}$$

Let the width of the rectangle be w and the height be h. Then $h = f(w)$ and $0 \le w \le 3$ (why?). We will now enter the area $A = wh = wf(w)$ as a function of w and graph it.

$A := w \to w \cdot f(w)$

$$w \to w\, f(w) \qquad\qquad \textbf{(5.61)}$$

$A(w)$

$$x \to \dfrac{w(9-3\,w)}{5+2\,w} \qquad\qquad \textbf{(5.62)}$$

$plot(A(w), w = 0..3)$

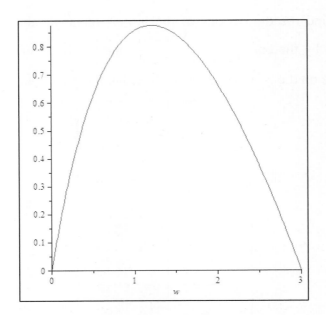

From the graph above, it seems that the maximum value occurs when x is in the interval [1, 2]. We will verify this using the *CriticalPoints* command. Do not forget to activate the $with\left(Student\left[Calculus1\right]\right)$ package.

$with\left(Student\left[Calculus1\right]\right)$:

$c := CriticalPoints\left(A(w), numeric\right)$

$$\left[-6.208099244, -2.49999999998080, 1.208099244\right] \qquad \textbf{(5.63)}$$

Note that the only value in the interval [0, 3] is 1.208099244. We will evaluate $f\left(c[3]\right)$ and $A\left(c[3]\right)$.

$f\left(c[3]\right)$

$$0.7248595459 \qquad \textbf{(5.64)}$$

$A\left(c[3]\right)$

$$0.8757022694 \qquad \textbf{(5.65)}$$

Hence the maximum area is 0.8757022694 and the dimensions of the rectangle are 1.208099244 by 0.7248595459.

Exercises

1. Find the domain and real zeros of the rational function $f(x) = \dfrac{4x^2 + 19x - 30}{x^3 + 9x^2 + 22x + 24}$. Determine if the Extreme Value Theorem can be applied to this function over the interval $[-5, 1]$. If so, find the maximum and minimum value of the function f in this interval.

2. Find the domain and real zeros of the rational function $f(x) = \dfrac{2x^3 + 11x^2 - 31x - 180}{2x^2 - 3x - 54}$. Determine if the Extreme Value Theorem can be applied to this function over the interval $[-1, 3]$. If so, find the maximum and minimum value of the function f in this interval.

3. Find the domain and real zeros of the polynomial function $f(x) = 3x^2 - 10x - 112$. Determine if the Extreme Value Theorem can be applied to this function over the interval $[-1, 1]$. If so, find the maximum and minimum value of the function f in this interval.

4. Find the domain the function $f(x) = 2 - \ln(5 - 3x - x^2)$. Determine if the Extreme Value Theorem can be applied to this function over the interval $[-3, 1]$. If so, find the maximum and minimum value of the function f in this interval.

5. Find the domain the function $f(x) = 5e^x + 6e^{-x}$. Determine if the Extreme Value Theorem can be applied to this function over the interval $[-2, 3]$. If so, find the maximum and minimum value of the function f in this interval.

6. Determine if the Rolle's Theorem can be applied to the function $f(x) = 6x^4 - 37x^3 - 429x^2 + 1462x + 1848$ over the interval $[-1, 4]$. If so, find the value of c in the open interval $(-1, 4)$ such that $f'(c) = 0$.

7. Determine if the Rolle's Theorem can be applied to the function $f(x) = 7x^3 - 19x^2 - 122x + 164$ over the interval $[-4, 2]$. If so, find the value of c in the open interval $(-4, 2)$ such that $f'(c) = 0$.

8.   Determine if the Rolle's Theorem can be applied to the function $f(x) = x^3 + x^2 - 33x - 7$ over the interval $[-7, 3]$. If so, find the value of c in the open interval $(-7, 3)$ such that $f'(c) = 0$.

9.   Determine if the Rolle's Theorem can be applied to the function $f(x) = x^2 \sin(x^2)$ over the interval $\left[-\dfrac{\pi}{2}, \dfrac{\pi}{2} \right]$. If so, find the value of c in the open interval $\left(-\dfrac{\pi}{2}, \dfrac{\pi}{2} \right)$ such that $f'(c) = 0$.

10.   Let $f(x) = \ln(|x|)$. Show that $\left(-\dfrac{1}{2}, \dfrac{1}{2} \right)$. Find the value of c in the open interval $\left(-\dfrac{1}{2}, \dfrac{1}{2} \right)$ such that $f'(c) = 0$. Does this contradict Rolle's Theorem?

11.   Find point(s) c (if they exist) that satisfy the Mean Value Theorem for the function $f(x) = 3x^4 - 14x^3 + 8x^2 - x - 12$ over the interval $[-2, 4]$. Find the equation(s) of the tangent line(s) at these point(s). Graph the function along with these tangent lines and the secant lines through the points $(-2, f(-2))$ and $(4, f(4))$.

12.   Find point(s) c (if they exist) that satisfy the Mean Value Theorem for the function $f(x) = 8x^4 - 12x^3 - 216x^2$ over the interval $[-2, 3]$. Find the equation(s) of the tangent line(s) at these point(s). Graph the function along with these tangent lines and the secant lines through the points $(-2, f(-2))$ and $(3, f(3))$.

13.   Find point(s) c (if they exist) that satisfy the Mean Value Theorem for the function $f(x) = 3x^4 - 23x^3 - 19x^2 + 203x + 196$ over the interval $[-3, 5]$. Find the equation(s) of the tangent line(s) at these point(s). Graph the function along with these tangent lines and the secant lines through the points $(-3, f(-3))$ and $(5, f(5))$.

14.   Find point(s) c (if they exist) that satisfy the Mean Value Theorem for the function $f(x) = -x^3 \cos(4x)$ over the interval $[-1, 0]$. Find the equation(s) of the tangent line(s) at these point(s). Graph the function along with these tangent lines and the secant lines through the points $(-1, f(-1))$ and $(0, f(0))$.

15.   Find point(s) c (if they exist) that satisfy the Mean Value Theorem for the function $f(x) = 7\sin(2x) - 2\cos(3x)$ over the interval $[-\pi, \pi]$. Find the

equation(s) of the tangent line(s) at these point(s). Graph the function along with these tangent lines and the secant lines through the points $(-\pi, f(-\pi))$ and $(\pi, f(\pi))$.

16. Find point(s) c (if they exist) that satisfy the Mean Value Theorem for the function $f(x) = \arcsin(2x-3) - 2x$ over the interval $\left[\dfrac{7}{6}, \dfrac{5}{3}\right]$. Find the equation(s) of the tangent line(s) at these point(s). Graph the function along with these tangent lines and the secant lines through the points $\left(\dfrac{7}{6}, f\left(\dfrac{7}{6}\right)\right)$ and $\left(\dfrac{5}{3}, f\left(\dfrac{5}{3}\right)\right)$.

17. Find all the real zeroes of $f(x) = 5x^{10} - 3x^5 + 2x^3 - 175$. Write a *Maple* procedure to find these roots using ten iterations in Newton's method (approximate the roots to six decimal places).

18. Find all the real zeroes of $f(x) = x^3 - \sin(x) * \cos(2x) - 1$. Write a *Maple* procedure to find these roots using ten iterations in Newton's method (approximate the roots to six decimal places).

19. Find all the real zeroes of $f(x) = \sin(x^2) - \cos(x^2)$. Write a *Maple* procedure to find these roots using ten iterations in Newton's method (approximate the roots to six decimal places).

20. Find all the real zeroes of $f(x) = \tan(x^2) - x - 2$. Write a *Maple* procedure to find these roots using ten iterations in Newton's method (approximate the roots to six decimal places).

21. Estimate the cube $\sqrt[3]{19}$ by Newton's method. (Hint: Solve $x^3 - 19 = 0$)

22. The cost for the government to support research to find a cure for x percent of all the diseases in a country is denoted by $C(x)$, where $C(x) = \dfrac{27x}{100 - x}$ and x is in millions of dollars.
    (a) Find the cost to support research for 10% of the diseases.
    (b) Find the cost to support research for 50% of the diseases.
    (c) What happens if we wanted to support research for 100% of the diseases? Explain.

23. Air is let out of a spherical balloon at a rate of 3 cubic centimeters per minute. Find the rate at which the radius of the balloon is decreasing when the diameter is 11 cm.

24. A filled cylindrical tank with diameter 12 meters is being drained of water at a rate of 2 cubic meters per second. At what rate is the height of the water decreasing? How long does it take for the tank to empty?

25. Two cars move away from the same point on perpendicular paths. One car travels at 55 mph whilst the other car travels at 33 mph. Find the rate at which the distance between the cars is changing in 1.2 hours.

26. The hour hand on a clock is one meter and the minute hand is 0.7 meters. What is rate of change of the distance between the tips of the hands at the instant when it reads 2:35 p.m.?

27. The sides of an equilateral triangle are changing at 0.2 meters per second. At what rate is its area increasing at the instant when the area is 9.76 square meters?

28. Two sides of a triangle are 7 meters and 9 meters long and the angle between them is increasing at the rate of 0.013 radians per second. Find the rate at which the area is increasing when the angle between these sides is 1.15 radians. Geometrically describe what occurs when the angle between these fixed sides is π radians and state the rate of change of the area at this instant.

29. A 4 foot boy walks away from a 9 foot lamppost at 2 ft/sec. At what rate is his shadow changing in length?

30. A star shaped bug travels along a curve $y = x^4 \sin^2\left(\dfrac{3}{x^2}\right)$. The horizontal speed of the bug changes at a rate of 0.23 centimeters per second. Find the change in its vertical speed at the instant when $x = 0.923$ radians. The path of the bug is sketched below.

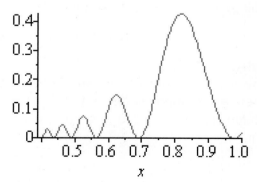

31. Another bug is moving on a path described by the curve $y = x^4 \sin^2\left(\dfrac{3}{x^2}\right)$. As the bug passes the point $\left(\sqrt{\dfrac{6}{\pi}}, \dfrac{36}{\pi^2}\right)$, its horizontal speed changes at a rate of 2 centimeters per second. How fast is the distance between the bug and the point $\left(1, \sin^2(3)\right)$ changing at this instant?

32. A street light is located at the top of a 25 foot pole. A lady walks away from the pole at a rate of 3 feet per second along a straight line. How fast is the tip of her shadow moving when she is 39 feet from the pole?

33. It costs \$900 for the local community cooperative to obtain permission to manufacture 1000 items. The cost in dollars for producing x items is $C(x) = x^2 - 5x + 900$. Find the minimum average cost (The average cost is $A(x) = \dfrac{C(x)}{x}$).

34. An ice cream company wants to make cones to hold 12 oz of Italian ices. Each cone is made from a disc of paper where a sector had been cut. The company wants to minimize the amount of paper used. What should be the radius and height of the cone?

35. In an electrical circuit, the power delivered to a load resistance R from a source with voltage E and source resistance r is given by $P = \left(\dfrac{V}{R+r}\right)^2 R$. For which value of R, do we have a maximum power?

36. Students were assigned to make open storage boxes out of rectangular white boards of size 20 cm by 14 cm. At each corner of the board squares of the same size will be cut. The resulting rectangular borders will then be folded to form the box. What should be the size of the squares that will maximize the volume of the box?

37. A manufacturing company produces ½ liter circular cylindrical cans. The company wants to minimize the production cost by using the least amount of material. What should be the dimensions of the can?

38. Three towns A, B and C form a right triangle. Towns A and B are 12 miles apart. A car starts driving from town A towards town B at 55 miles per hour. At the same time, a second car starts driving from town C towards town A at 30 miles per hour. What will be the shortest distance between the two cars?

39. Find the point on the line $7x + 3y = 6$ that is closest to the point $(2,7)$.

40. Find the point on the curve $xy = 8$ that is closest to the point $(0,3)$. Show that this is equivalent to solving the equation $y^4 - 3y^3 - 64 = 0$.

41. Find the point on the curve $y = x \ln x$ that is closest to the point $\left(\frac{1}{2}, \frac{1}{2}\right)$.

42. Find the point on the curve $y = x^2 \sin\left(\frac{1}{x}\right)$ that is closest to the point $\left(\frac{\pi}{122}, \frac{1}{5}\right)$.

    Execute the following code by loading the appropriate package in *Maple* to get an idea of what occurs.

    $$plot\left(\left(x^2 \cdot \sin\left(\frac{1}{x}\right) - 0.2\right)^2 + \left(x - \frac{\pi}{122}\right)^2, x = 0..0.07\right)$$

43. What is the minimum possible perimeter for a rectangle with area $1789 \ ft^2$.

44. What is the minimum possible perimeter for an equilateral triangle with area $1789 \ ft^2$.

45. A piece of wire 100 meters long is used to form a circle and a square. Find their maximum total area.

46. A piece of wire 100 meters long is used to form a circle and a square. Find their minimum total area.

47. A piece of wire 100 meters long is used to form a circle and an equilateral triangle. Find their minimum total area.

48. Find the dimensions of the largest rectangle with base on the x- axis and upper sides on the curve $y = 51 - 3x^2$.

49. Find the largest rectangle that can be inscribed in an equilateral triangle of sides with unit length.

50. Find the largest rectangle that can be inscribed in the curve $4x^2 + y^2 - 8x + 6y = 3$.

51.   Find the biggest rectangle that can be inscribed in a right triangle with legs of length 5 and 6 if the sides of the rectangle are parallel to the legs of the triangle.

52.   Minimize the surface area of a closed cylindrical can with volume 200 m^3. Can you maximize the surface area of this can? Carefully justify your answer.

53.   An explorer must cross a circular lake with radius 3 miles. She can row from her starting point A to the point B and then walk along the arc BC to get to point C. If she rows at a constant rate of 3 mph and walks at a constant rate of 5 mph, what is the best way for her to proceed to get to the other side of the lake in the shortest time? See the picture below:

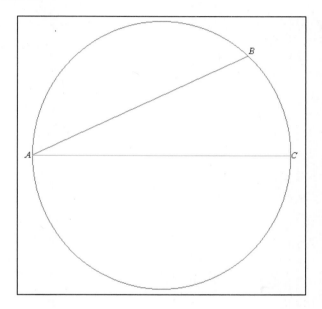

54.*   Exploration Problem: Generate a path shown in problem 31 and animate the movement of the star shaped bug along with the path. Write the *Maple* code to show the value of the bug's vertical speed at every instant in the interval $0.4 \le x \le 1$. Estimate the distance travelled by this bug.

55.   Below is a list of some commands used in this chapter. Describe the significance of each command. Can you find examples where each command is used?

(a) $display\left(p1, p2, ..., pn\right)$

(b) $ExtremePoints\left(f\left(x\right), x = a .. b, numeric\right)$

(c) $MeanValueTheorem\left(f\left(x\right), x = a .. b\right)$

(d) $NewtonsMethod\left(f\left(x\right), c, iterations = n, output = sequence\right)$

(e) $RollesTheorem\left(f\left(x\right), x = a .. b, output = points\right)$

      (f) $Tangent\left(f\left(x\right), x = c, a\mathrel{..}b, output = line\right)$

      (g) $with\left(plots\right)$

      (h) $with\left(Student\left[Calculus1\right]\right)$

56.   Can you list new *Maple* commands that were used in this chapter but were not listed in exercise 55?

Chapter 6   Integration

The object of this chapter is to use Riemann sums to evaluate definite integrals. Different partitions such as right sums, left sums, mid sums and other variations will be used. Indefinite integrals will be computed using the *Expression* palette and the *Clickable Calculus* feature. *Maple* will be used to vary different parameters to achieve better results. We will also use *Maple* to estimate integrals numerically via the trapezoidal rule and Simpson's rule.

Riemann Sums

Definition: Let f be any function. The function F is called the *antiderivative* of the function f if $F'(x) = f(x)$.

Definition: Let f be any function defined for $a \le x \le b$. Partition the interval $[a,b]$ into n subintervals of equal length. Let the end point of these subintervals be $x_0(=a), x_1, x_2, \ldots x_n(=b)$. Then the length of each subinterval $[x_{i-1}, x_i]$ is $\dfrac{b-a}{n}$ for $i = 1$, 2, .. n. Let x_i^* be any point in the subinterval $[x_{i-1}, x_i]$. The definite integral of f from a to b, denoted by $\displaystyle\int_a^b f(x)\,dx$, is defined as

$$\int_a^b f(x)\,dx = \lim_{n \to \infty} \sum_{i=1}^n f(x_i^*)(x_i - x_{i-1})$$

provided this limit exists. In this case f is said to be *integrable over the interval* $[a,b]$.

Remark: The sum in the definition of the definite integral is called a *Riemann Sum*.

Definition: Let f be any function defined for $a \le x \le b$. Partition the interval $[a,b]$ into n subintervals of equal length. Let the end point of these subintervals be $x_0(=a), x_1, x_2, \ldots x_n(=b)$. We define the sums $L(n)$ *left sum for n partitions*, $R(n)$ *right sum for n partitions*, and $M(n)$ *middle sum or midpoint sum for n partitions* as:

$$L(n) = \sum_{i=1}^n f(x_{i-1})(x_i - x_{i-1})$$

$$R(n) = \sum_{i=1}^n f(x_i)(x_i - x_{i-1})$$

$$M(n) = \sum_{i=1}^n f\left(\frac{x_i + x_{i-1}}{2}\right)(x_i - x_{i-1})$$

Caution: The answers for the approximations for all the sums may vary from those mentioned in this chapter after 6 significant digits.

Example 1: Estimate the value of $\int_0^1 (5-3x^2)\,dx$ using the left Riemann sum and the right Riemann sum for 20, 50, 100, 500, 1000, 5000, and 10000 partitions. Find a formula for the left Riemann sum $L(n)$ and the right Riemann sum $R(n)$ when the number of partitions is n. Evaluate $\lim_{n\to\infty} L(n)$ and $\lim_{n\to\infty} R(n)$. Use these numbers to evaluate $\int_0^1 (5-3x^2)\,dx$. Verify yours answers using *Maple* commands.

Solution: We will enter the integrand $f(x)=5-3x^2$ as a *Maple* function and graph it.

$f := x \to 5-3x^2$

$$x \to 5-3x^2 \qquad\qquad \textbf{(6.1)}$$

$plot\big(f(x),\, x=0..1,\, y=0..5\big)$

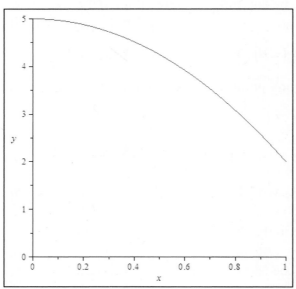

The function $f(x)$ is nonnegative in the interval [0, 1]. In this case the definite integral is the area of the region bounded by the curve $y=5-3x^2$, the x-axis, the y-axis, and the line $x=1$. This area can be approximated by constructing several rectangles, finding the area of each rectangle and then finding the sum of these areas. In other words, in this case the Riemann sums can be interpreted as the sum of the areas of these approximating rectangles. The *ApproximateInt* command with the options of *method = left* and *output = plot* sketches rectangles in a manner that the height of each rectangle is the value of the function at the left endpoint of each rectangle. The default method is the *midpoint* Riemann sum. The default number of partitions (which give rise to the rectangles) is 10. The number of partitions needs to be specified in other cases. The graph below has 20 such rectangles. The width of each rectangle is $\dfrac{1-0}{20} = \dfrac{1}{20}$. We will need to activate the $with\big(Student\left[Calculus1\right]\big)$ package.

$with\big(Student\left[Calculus1\right]\big):$

$ApproximateInt\big(f(x),\, x = 0..1,\, method = left,\, output = plot,\, partition = 20\big)$

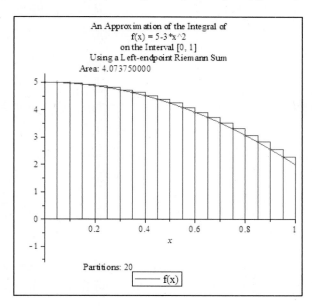

The sum of the areas of the 20 rectangles as sketched above is 4.073750000. However, if we are only interested in finding the total area of these rectangles, in other words the left Riemann sum, then we use the *RiemannSum* command with the option of *method = left* and without the option *output = value*. Note that the default option is *output = value*. This gives the answer in the form of a fraction in this case.

$ApproximateInt\big(f(x),\, x = 0..1,\, method = left,\, partition = 20\big)$

$$\frac{3259}{800}$$

(6.2)

$ApproximateInt\big(f(x),\, x = 0..1,\, method = left,\, output = value,\, partition = 20\big)$

$$\frac{3259}{800}$$

(6.3)

We will now use the *evalf* option to get a floating-point approximation of the answer. Note that the *numeric* option is not available in this case.

$evalf\big(ApproximateInt\big(f(x),\, x = 0..1,\, method = left,\, partition = 20\big)\big)$

$$4.073750000$$

(6.4)

Using the same function $f(x)$, observe what happens when the number of rectangles is increased to 50, 100, 500, 1000, 5000, and 10000.

$$evalf\left(ApproximateInt\left(f(x), x = 0..1, method = left, partition = 50\right)\right)$$

$$4.0029800000 \qquad\qquad\qquad \textbf{(6.5)}$$

$$evalf\left(ApproximateInt\left(f(x), x = 0..1, method = left, partition = 100\right)\right)$$

$$4.014950000 \qquad\qquad\qquad \textbf{(6.6)}$$

$$evalf\left(ApproximateInt\left(f(x), x = 0..1, method = left, partition = 500\right)\right)$$

$$4.002998000 \qquad\qquad\qquad \textbf{(6.7)}$$

$$evalf\left(ApproximateInt\left(f(x), x = 0..1, method = left, partition = 1000\right)\right)$$

$$4.001499500 \qquad\qquad\qquad \textbf{(6.8)}$$

$$evalf\left(ApproximateInt\left(f(x), x = 0..1, method = left, partition = 5000\right)\right)$$

$$4.000299980 \qquad\qquad\qquad \textbf{(6.9)}$$

$$evalf\left(ApproximateInt\left(f(x), x = 0..1, method = left, partition = 10000\right)\right)$$

$$4.000149995 \qquad\qquad\qquad \textbf{(6.10)}$$

Notice that as the number of rectangles increases, the values of these sums get closer and closer to 4. We will let $L(n)$ be the left sum with n partitions and evaluate the limit $\lim_{n\to\infty} L(n)$.

$$L := n \rightarrow ApproximateInt\left(f(x), x = 0..1, method = left, partition = n\right)$$
$$n \rightarrow Student : -Calculus1 : -ApproximateInt\left(f(x), x = 0..1, method = left, partition = n\right)$$

$$\textbf{(6.11)}$$

$$L(n)$$

$$\frac{\sum_{i=0}^{n-1}\left(5 - \dfrac{3i^2}{n^2}\right)}{n}$$

$$\textbf{(6.12)}$$

$$\lim_{n\to\infty} L(n)$$

$$4 \qquad\qquad\qquad \textbf{(6.13)}$$

Hence $\lim_{n\to\infty} L(n) = 4$. The same procedure will be repeated to find the right Riemann sum. The *ApproximateInt* command with the option of *method = right*. The option with *output = plot* sketches rectangles in a manner that the height of each rectangle is the value of the function at the right endpoint of each rectangle. The graph below has 20 such rectangles. The width of each rectangle is again $\dfrac{1-0}{20} = \dfrac{1}{20}$. We do not need to activate the $with\left(Student\left[Calculus1\right]\right)$ package as it has been done so earlier for the left Riemann sums.

$ApproximateInt\left(f\left(x\right), x = 0 .. 1, method = right, output = plot, partition = 20\right)$

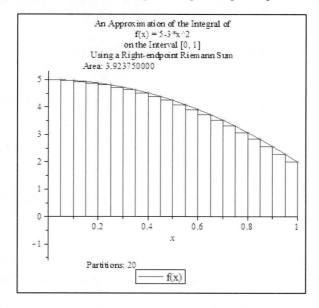

The sum of the areas of the 20 rectangles as sketched above is 3.923750000. However, if we are only interested in finding the total area of these rectangles, in other words the right Riemann sum, then we use the $ApproximateInt$ command with the option of $method = right$ and without the option $output = value$. Note that the default option is $output = value$. This gives the answer in the form of a fraction.

$ApproximateInt\left(f\left(x\right), x = 0 .. 1, method = right, partition = 20\right)$

$$\frac{3139}{800}$$

(6.14)

$ApproximateInt\left(f\left(x\right), x = 0 .. 1, method = right, output = value, partition = 20\right)$

$$\frac{3139}{800}$$

(6.15)

We will again use the $evalf$ option to get a floating-point approximation of the answer.

$evalf\left(ApproximateInt\left(f\left(x\right), x = 0 .. 1, method = right, partition = 20\right)\right)$

$$3.923750000$$

(6.16)

Using the same function $f(x)$, observe what happens when the number of rectangles is increased to 50, 100, 500, 1000, 5000, and 10000.

$$evalf\left(ApproximateInt\left(f(x), x = 0 ..1, method = right, partition = 50\right)\right)$$
$$3.969800000 \tag{6.17}$$

$$evalf\left(ApproximateInt\left(f(x), x = 0 ..1, method = right, partition = 100\right)\right)$$
$$3.984950000 \tag{6.18}$$

$$evalf\left(ApproximateInt\left(f(x), x = 0 ..1, method = right, partition = 500\right)\right)$$
$$3.996998000 \tag{6.19}$$

$$evalf\left(ApproximateInt\left(f(x), x = 0 .\text{-}1, method = right, partition = 1000\right)\right)$$
$$3.998499500 \tag{6.20}$$

$$evalf\left(ApproximateInt\left(f(x), x = 0 ..1, method = right, partition = 5000\right)\right)$$
$$3.999699980 \tag{6.21}$$

$$evalf\left(ApproximateInt\left(f(x), x = 0 ..1, method = right, partition = 10000\right)\right)$$
$$3.999849995 \tag{6.22}$$

Notice that as the number of rectangles increases, the value of these sums get closer and closer to four. We will let $R(n)$ be the left sum with n partitions and evaluate the limit $\lim_{n\to\infty} R(n)$.

$$R := n \to ApproximateInt\left(f(x), x = 0 ..1, method = right, partition = n\right)$$
$$n \to Student:-Calculus1:-ApproximateInt\left(f(x), x = 0 ..1, method = right, partition = n\right) \tag{6.23}$$

$$R(n)$$

$$\frac{\sum_{i=1}^{n}\left(5 - \frac{3i^2}{n^2}\right)}{n} \tag{6.24}$$

$$\lim_{n\to\infty} R(n)$$

$$4 \tag{6.25}$$

Hence $\lim_{n\to\infty} R(n) = 4$. Since $\lim_{n\to\infty} L(n) = 4 = \lim_{n\to\infty} R(n)$, we see that $\int_0^1 (5-3x^2)\,dx = 4$.

This can be verified by clicking on $\int_a^b f\,dx$ available in the *Expression* palette. Insert 0 as a is shaded, then use the Tab key and insert 1 in place of b, use the Tab key and insert $f(x)$ in place of f and then use the Tab key once more and insert x in place of x and then press the Enter key.

$$\int_0^1 f(x)\,dx$$

4

(6.26)

Remarks:

1. We can vary the heights of the rectangles. The *method* option gives us different possibilities. For example, in the option with *method* = *random* the height of each rectangle is the value of the function at a random point in each subinterval, and in the one with *method* = *lower* the height of each rectangle is the smallest value of the function in each subinterval. The sum for *n* rectangles cannot be calculated using these options. However, with the option *method* = *midpoint*, where the height of each rectangle is the value of the function at the midpoint of each subinterval, one can calculate the sum for *n* rectangles.

2. The option *output* = *animation* gives an animation of the rectangles where the number of rectangles increases. The following plot will give an animation (the *method* = *random* option is used here). When the option *method* = *random* is used, each subinterval is randomly subdivided. The default setting again has 10 rectangles. The default number of frames is 6. Furthermore, the number of partitions in each frame is twice the number of partitions in the previous frame and the partition of the subinterval is obtained by dividing the each subinterval in the previous frame into half. This is also the default setting. The partitions for the frames are 10, 20, 40, 80, 160, and 320.

ApproximateInt $\left(f(x),\, x = 0\,..\,1,\, method = random,\, output = animation \right)$

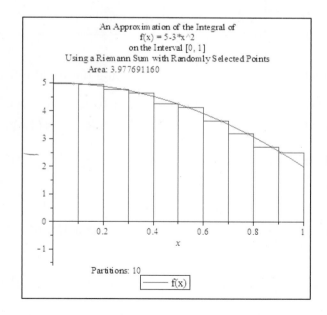

Right click on the graph, select *Animation* and then select *Play*. Notice that the area changes as the number of rectangles changes. One can also view this frame by frame.

One can change the number of frames and the way each subinterval is divided. In the graphs below the number of frames is 7 and the refinement for each subinterval is random. That means that each subinterval is not divided into half to get the partitions for the next frame. The division for each subinterval is random. The default subdivision for each rectangle is *refinement = halve*. However, the number of partitions in each frame is still twice the number of partitions in the previous frame. The second graph below is that the third frame.

$$ApproximateInt\left(f\left(x\right), x = 0..1, method = random, output = animation, iteration = 7,$$

$$refinement = random\right)$$

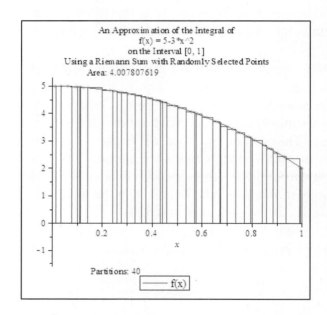

Example 2: Estimate the value of $\int_{-2}^{1} xe^{-x^2}dx$ using the middle Riemann sum for 15, 20, 50, 100, 500, 1000, and 5000 partitions. Find a formula for the middle Riemann sum $M(n)$ when the number of partitions is n. Estimate $\lim_{n\to\infty} M(n)$, using a floating-point approximation. Use these numbers to estimate $\int_{-2}^{1} xe^{-x^2}dx$. Verify yours answers using *Maple* commands.

Solution: We will enter the integrand $f(x) = xe^{-x^2}$ as a *Maple* function and graph it.

$$f := x \rightarrow x \cdot e^{-x^2}$$

$$x \rightarrow x\,e^{-x^2} \tag{6.27}$$

$$plot\left(f\left(x\right), x = -2..1\right)$$

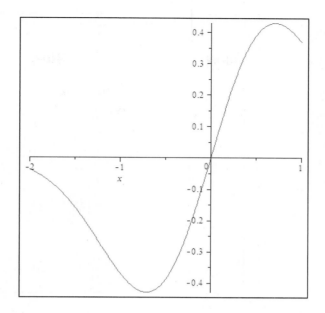

The function $f(x)$ is nonpositive in the interval [-2, 0] and nonnegative in the interval [0, 1]. Notice that the area A_1 bounded by the curve $y = xe^{-x^2}$, the x-axis, the y-axis, and the line $x = -2$ is more than the area A_2 bounded by the curve $y = xe^{-x^2}$, the x-axis, the y-axis, and the line $x = 1$. In this case the definite integral $\int_{-2}^{1} xe^{-x^2}\,dx = A_2 - A_1$. As in Example 1 we will use Riemann sums to find the value of the integral. In this case, we will use the *ApproximateInt* command with the option of *method = midpoint* (though the default option is *method = midpoint*, we will insert it so that we do not forget it at other instances) and *output = plot* that sketches rectangles in a manner that the height of each rectangle is the value of the function at the midpoint of each rectangle. The graph below has 15 such rectangles. The width of each rectangle is $\dfrac{1-(-2)}{15} = \dfrac{1}{5}$. Do not forget to activate the $with\left(Student\left[Calculus1\right]\right)$ package.

$with\left(Student\left[Calculus1\right]\right):$

$ApproximateInt\left(f(x), x = -2..1, method = midpoint, output = plot, partition = 15\right)$

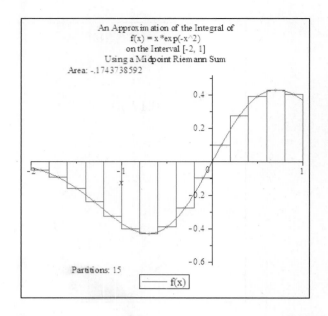

The sum of the areas of the 15 rectangles as sketched above is -0.1743738592. However, if we are only interested in finding the total area of these rectangles, in other words the middle Riemann sum, then we use the *ApproximateInt* command with the option of *method* = *midpoint* and then the *evalf* option to get a floating-point approximation of the answer.

$ApproximateInt\left(f\left(x\right), x = -2..1, method = midpoint, partition = 15\right)$

$$-\frac{19}{50}\,e^{-\frac{361}{100}} - \frac{17}{50}\,e^{-\frac{289}{100}} - \frac{3}{10}\,e^{-\frac{9}{4}} - \frac{13}{50}\,e^{-\frac{169}{100}} - \frac{11}{50}\,e^{-\frac{121}{100}} \qquad \textbf{(6.28)}$$

$evalf\left(ApproximateInt\left(f\left(x\right), x = -2..1, method = midpoint, partition = 15\right)\right)$

$$-0.1743738592 \qquad \textbf{(6.29)}$$

Using the same function $f\left(x\right)$, observe what happens when the number of rectangles is increased to 20, 50, 100, 500, 1000, and 5000.

$evalf\left(ApproximateInt\left(f\left(x\right), x = -2..1, method = midpoint, partition = 20\right)\right)$

$$-0.1745545038 \qquad \textbf{(6.30)}$$

$evalf\left(ApproximateInt\left(f\left(x\right), x = -2..1, method = midpoint, partition = 50\right)\right)$

$$-0.1747458815 \qquad \textbf{(6.31)}$$

$evalf\left(ApproximateInt\left(f\left(x\right), x = -2..1, method = midpoint, partition = 100\right)\right)$

$$-0.1747729093 \qquad \textbf{(6.32)}$$

$evalf\left(ApproximateInt\left(f\left(x\right), x = -2..1, method = midpoint, partition = 500\right)\right)$

$$-0.1747815416 \qquad \textbf{(6.33)}$$

$$evalf\left(ApproximateInt\left(f\left(x\right), x=-2..1, method=midpoint, partition=1000\right)\right)$$
$$-0.1747818101 \tag{6.34}$$

$$evalf\left(ApproximateInt\left(f\left(x\right), x=-2..1, method=midpoint, partition=5000\right)\right)$$
$$-0.1747818978 \tag{6.35}$$

Notice that as the number of rectangles increases, the value of this sum gets closer and closer to -0.174782. We will let $M\left(n\right)$ be the left sum with n partitions and evaluate the limit $\lim\limits_{n\to\infty} M\left(n\right)$.

$$M := n \to evalf\left(ApproximateInt\left(f\left(x\right), x=-2..1, method=midpoint, partition=n\right)\right)$$
$$n \to Student:-Calculus1:-ApproximateInt\left(f\left(x\right), x=0..1, method=left, partition=n\right) \tag{6.36}$$

$M\left(n\right)$

$$\frac{3\left(\sum_{i=0}^{n-1}\left(-2+\frac{3\left(i+\frac{1}{2}\right)}{n}\right)e^{-\left(-2+\frac{3\left(i+\frac{1}{2}\right)}{n}\right)^2}\right)}{n} \tag{6.37}$$

$\lim\limits_{n\to\infty} M\left(n\right)$

$$\frac{1}{2}e^{-4} - \frac{1}{2}e^{-1} \tag{6.38}$$

$evalf\left(\left(\mathbf{6.38}\right)\right)$

$$-0.1747819012 \tag{6.39}$$

Hence $\lim\limits_{n\to\infty} M\left(n\right) = -0.1747819012$. Therefore, we see that $\int_{-2}^{1} xe^{-x^2}dx = -0.1747819012$. This can be verified by using *Maple* commands as done in Example 1.

$\int_{-2}^{1} xe^{-x^2}dx$

$$\frac{1}{2}e^{-4} - \frac{1}{2}e^{-1} \tag{6.40}$$

$evalf\left(\left(\mathbf{6.40}\right)\right)$

$$-0.1747819012 \tag{6.41}$$

Remarks: Just as in Example 1 we can give an animation for this approximation. We have used the option $method=midpoint$ for this animation. In this case we have the option refinement = 0.3. Each subinterval is divided into two subintervals so that the ratio of the lengths of the two subintervals is 3:7. In fact one could choose any number in the open interval (0, 1). Once again the number of partitions in each frame is twice the number of partitions in the previous frame. The pictures below are of the first two frames.

$$ApproximateInt\left(f(x),\, x=-2..1,\, method=random,\, output=animation,\, partiton=15,\right.$$
$$\left. iterations=9,\, refinement=0.3\right)$$

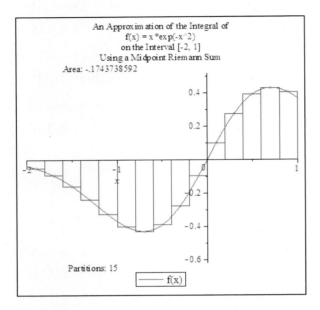
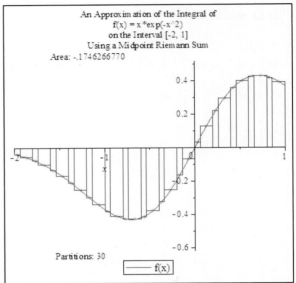

In the next chapter we will show the reader a technique on how to evaluate $\int_{-2}^{1} xe^{-x^2} dx$ step-by-step.

Example 3: Graph the function $f(x) = \sin(x)$ in the interval [1, 2].

(a)   For any integer $n > 0$, construct a partition of [1, 2] in the following manner:
$$x_0 = 1,\, x_1 = 2^{1/n},\, x_2 = 2^{2/n},\, ...,\, x_n = 2^{n/n} = 2$$
Plot the graph with this partition for $n > 0$ for the left Riemann sum and for $n > 0$ for the right Riemann sum.

(b)   Use *Maple* commands to find the left Riemann sum and the right Riemann sum for the partition:
$$x_0 = 1,\, x_1 = 2^{1/n},\, x_2 = 2^{2/n},\, ...,\, x_n = 2^{n/n} = 2$$
When $n = 5, 10, 20, 50, 100, 500, 1000,$ and 10000 .

(c)   Use the definition of the left Riemann sum $L(n)$ and the right Riemann sum $R(n)$ and the partition
$$x_0 = 1,\, x_1 = 2^{1/n},\, x_2 = 2^{2/n},\, ...,\, x_n = 2^{n/n} = 2$$

of [1, 2] to find a general formula for $L(n)$ and $R(n)$.

(d)   Evaluate $\lim_{n \to \infty} L(n)$ and $\lim_{n \to \infty} R(n)$. Use these numbers to evaluate $\int_1^2 \sin(x)\,dx$.

(e)   Verify yours answers using *Maple* commands.

Solution: We will enter the integrand $f(x) = \sin(x)$ as a *Maple* function and graph it.

$f := x \to \sin(x)$

$$x \to \sin(x) \tag{6.42}$$

$plot\left(f(x),\, x = 1..2,\, y = 0.8..1 \right)$

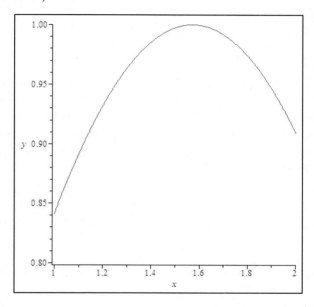

(a)   Note that when $n = 5$, the points in the partition are $x_0 = 1$, $x_1 = 2^{1/5}$, $x_2 = 2^{2/5}$, $x_3 = 2^{3/5}$, $x_4 = 2^{4/5}$, $x_5 = 2^{5/5} = 2$. Instead of using the *partition* = *n* command we will now use the command $partition = \left[seq\left(2^{\frac{i}{5}},\, i = 0..5 \right) \right]$. Do not forget to activate the $with\left(Student[Calculus1] \right)$ package.

$with\left(Student[Calculus1] \right):$

$ApproximateInt\left(f(x),\, x = 1..2,\, method = left,\, output = plot,\, partition = \left[seq\left(2^{\frac{i}{5}},\, i = 0..5 \right) \right] \right)$

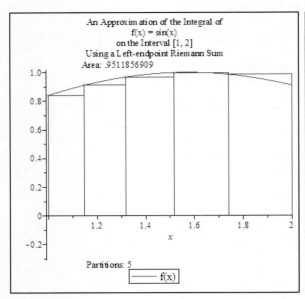
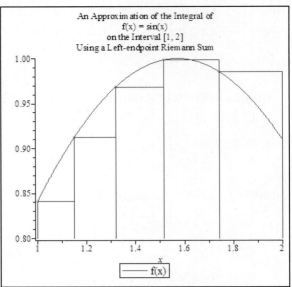

The graph is the one on the left side. To obtain the picture on the right, right click on the plot, select *Axis*, select *Properties…*, select *Vertical*, and change *Range min.* to 0.8 and *Range max.* to 1.0. Similarly we can draw the graph for the right Riemann sums. In this case we let $n = 7$.

$$ApproximateInt\left(f\left(x \right), x = 1..2, method = right, output = plot, partition = \left[seq\left(2^{\frac{i}{7}}, i = 0..7 \right) \right] \right)$$

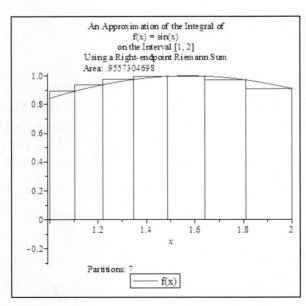
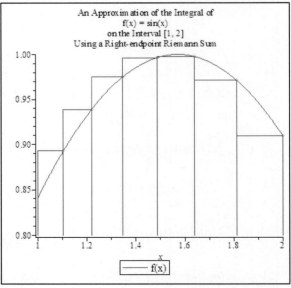

(b)   We will now use *Maple* commands to find the left Riemann sum for the partition
$$x_0 = 1, \; x_1 = 2^{1/n}, \; x_2 = 2^{2/n}, ..., x_n = 2^{n/n} = 2$$

when $n = 5, 10, 20, 50, 100, 500, 1000,$ and 10000. As in part (a) instead of using the *partition* $= n$ command we will now use the $partition = \left[seq\left(2^{\frac{i}{5}}, i = 0..5 \right) \right]$.

We do not need to activate the $with\left(Student\left[Calculus1 \right] \right)$ package since we already did it for part (a).

$$evalf\left(ApproximateInt\left(f(x), x = 1..2, method = left, partition = \left[seq\left(2^{\frac{i}{5}}, i = 0..5 \right) \right] \right) \right)$$

$$0.951185691 \tag{6.43}$$

$$evalf\left(ApproximateInt\left(f(x), x = 1..2, method = left, partition = \left[seq\left(2^{\frac{i}{10}}, i = 0..10 \right) \right] \right) \right)$$

$$0.954803217 \tag{6.44}$$

$$evalf\left(ApproximateInt\left(f(x), x = 1..2, method = left, partition = \left[seq\left(2^{\frac{i}{20}}, i = 0..20 \right) \right] \right) \right)$$

$$0.9558622294 \tag{6.45}$$

$$evalf\left(ApproximateInt\left(f(x), x = 1..2, method = left, partition = \left[seq\left(2^{\frac{i}{50}}, i = 0..50 \right) \right] \right) \right)$$

$$0.9562696104 \tag{6.46}$$

$$evalf\left(ApproximateInt\left(f(x), x = 1..2, method = left, partition = \left[seq\left(2^{\frac{i}{100}}, i = 0..100 \right) \right] \right) \right)$$

$$0.956368468 \tag{6.47}$$

$$evalf\left(ApproximateInt\left(f(x), x = 1..2, method = left, partition = \left[seq\left(2^{\frac{i}{500}}, i = 0..500 \right) \right] \right) \right)$$

$$0.956434459 \tag{6.48}$$

$$evalf\left(ApproximateInt\left(f(x), x = 1..2, method = left, partition = \left[seq\left(2^{\frac{i}{1000}}, i = 0..1000 \right) \right] \right) \right)$$

$$0.956441895 \tag{6.49}$$

$$evalf\left(ApproximateInt\left(f(x), x = 1..2, method = left, partition = \left[seq\left(2^{\frac{i}{10000}}, i = 0..10000 \right) \right] \right) \right)$$

$$0.9564489440 \tag{6.50}$$

We will now use *Maple* commands to find the right Riemann sum for the partition
$$x_0 = 1, \ x_1 = 2^{1/n}, \ x_2 = 2^{2/n}, ..., x_n = 2^{n/n} = 2$$
when $n = 5, 10, 20, 50, 100, 500, 1000,$ and 10000.

$$evalf\left(ApproximateInt\left(f(x), x=1..2, method=right, partition=\left[seq\left(2^{\frac{i}{5}}, i=0..5\right)\right]\right)\right)$$

$$0.954543373 \tag{6.51}$$

$$evalf\left(ApproximateInt\left(f(x), x=1..2, method=right, partition=\left[seq\left(2^{\frac{i}{10}}, i=0..10\right)\right]\right)\right)$$

$$0.956297964 \tag{6.52}$$

$$evalf\left(ApproximateInt\left(f(x), x=1..2, method=right, partition=\left[seq\left(2^{\frac{i}{20}}, i=0..20\right)\right]\right)\right)$$

$$0.9565864811 \tag{6.53}$$

$$evalf\left(ApproximateInt\left(f(x), x=1..2, method=right, partition=\left[seq\left(2^{\frac{i}{50}}, i=0..50\right)\right]\right)\right)$$

$$0.9565567204 \tag{6.54}$$

$$evalf\left(ApproximateInt\left(f(x), x=1..2, method=right, partition=\left[seq\left(2^{\frac{i}{100}}, i=0..100\right)\right]\right)\right)$$

$$0.956511832 \tag{6.55}$$

$$evalf\left(ApproximateInt\left(f(x), x=1..2, method=right, partition=\left[seq\left(2^{\frac{i}{500}}, i=0..500\right)\right]\right)\right)$$

$$0.956463100 \tag{6.56}$$

$$evalf\left(ApproximateInt\left(f(x), x=1..2, method=right, partition=\left[seq\left(2^{\frac{i}{1000}}, i=0..1000\right)\right]\right)\right)$$

$$0.9564561848 \tag{6.57}$$

$$evalf\left(ApproximateInt\left(f(x), x=1..2, method=right, partition=\left[seq\left(2^{\frac{i}{10000}}, i=0..10000\right)\right]\right)\right)$$

$$0.956449352 \tag{6.58}$$

(c)   We will now use the definition of the left Riemann sum and the partition

$$x_0 = 1, \ x_1 = 2^{1/n}, \ x_2 = 2^{2/n}, \ ..., \ x_n = 2^{n/n} = 2$$

of [1, 2] to find a general formula for $L(n)$. Recall that we cannot get a general formula using *Maple* commands. This will be seen in the remark after this example.

$$L := n \rightarrow \sum_{i=0}^{n-1}\left(f\left(2^{\frac{i}{n}}\right)\cdot\left(2^{\frac{i+1}{n}} - 2^{\frac{i}{n}}\right)\right)$$

$$n \to \sum_{i=0}^{n-1}\left(f\left(2^{\frac{i}{n}} \right) \cdot \left(2^{\frac{i+1}{n}} - 2^{\frac{i}{n}} \right) \right) \tag{6.59}$$

$L(n)$

$$\sum_{i=0}^{n-1}\left(\sin\left(2^{\frac{i}{n}} \right) \cdot \left(2^{\frac{i+1}{n}} - 2^{\frac{i}{n}} \right) \right) \tag{6.60}$$

$\lim\limits_{n\to\infty} L(n)$

$$\cos(1) - \cos(2) \tag{6.61}$$

$evalf\left((\mathbf{6.61}) \right)$

$$0.9564491424 \tag{6.62}$$

Hence we see that $\lim\limits_{n\to\infty} L(n) = 0.9564491424$. We will now repeat the same process for the right Riemann sum and the partition
$$x_0 = 1,\ x_1 = 2^{1/n},\ x_2 = 2^{2/n},\ \dots,\ x_n = 2^{n/n} = 2$$
of $[1, 2]$ to find a general formula for $R(n)$. Recall that as in the case of $L(n)$, we cannot get a general formula using *Maple* commands.

$$R := n \to \sum_{i=1}^{n}\left(f\left(2^{\frac{i}{n}} \right) \cdot \left(2^{\frac{i}{n}} - 2^{\frac{i-1}{n}} \right) \right)$$

$$n \to \sum_{i=1}^{n}\left(f\left(2^{\frac{i}{n}} \right) \cdot \left(2^{\frac{i}{n}} - 2^{\frac{i-1}{n}} \right) \right) \tag{6.63}$$

$R(n)$

$$\sum_{i=1}^{n}\left(\sin\left(2^{\frac{i}{n}} \right) \cdot \left(2^{\frac{i}{n}} - 2^{\frac{i-1}{n}} \right) \right) \tag{6.64}$$

$\lim\limits_{n\to\infty} R(n)$

$$\cos(1) - \cos(2) \tag{6.65}$$

$evalf\left((\mathbf{6.65}) \right)$

$$0.9564491424 \tag{6.66}$$

Hence we see that $\lim\limits_{n\to\infty} R(n) = 0.9564491424$. Therefore, we can conclude that
$$\int_{1}^{2} \sin(x)\,dx = 0.9564491424.$$

(e)   We will now verify our answers using *Maple* commands.

$$\int_1^2 \sin(x)\,dx$$

$$\cos(1) - \cos(2) \qquad\qquad\qquad \textbf{(6.67)}$$

$$evalf\left((\textbf{6.67})\right)$$

$$0.9564491424 \qquad\qquad\qquad \textbf{(6.68)}$$

Remarks: If we use *Maple* commands to find the left (or for that matter the right) Riemann sum in Example 3 we would get the following message.

$$L1 := n \rightarrow ApproximateInt\left(f(x), x = 1..2, method = left, partition = \left[seq\left(2^{\frac{i}{n}}, i = 0..10000 \right) \right] \right)$$

$$n \rightarrow Student : -Calculus1 : -ApproximateInt\left(f(x), x = 1..2, method = left, \right.$$
$$\left. partition = \left[seq\left(2^{\frac{i}{n}}, i = 0..10000 \right) \right] \right) \qquad \textbf{(6.69)}$$

$$L1(n)$$

Error, (in L1) unable to execute seq

Simpson's Rule and the Trapezoidal Rule

Definition: Let f be any function defined for $a \le x \le b$. Partition the interval $[a,b]$ into n subintervals of equal length. Let the end point of these subintervals be $x_0 (= a), x_1, x_2, ... x_n (= b)$. We define the Trapezoidal sum

$$T(n) = \frac{b-a}{2n}\left(f(x_0) + 2f(x_1) + 2f(x_2) + 2f(x_3) + 2f(x_4).... + 2f(x_{n-2}) + 2f(x_{n-1}) + f(x_n) \right)$$

Note: Let f be a function satisfying $\left| f''(x) \right| \le C$ for $a \le x \le b$. If E_M and E_T denote the errors in the Midpoint and the Trapezoidal Rules, respectively, then $\left| E_M \right| \le \dfrac{C(b-a)^3}{24n^2}$

and $\left| E_T \right| \le \dfrac{C(b-a)^3}{12n^2}$.

Definition: Let f be any function defined for $a \le x \le b$. Partition the interval $[a,b]$ into n subintervals of equal length where n is an even number. Let the end point of these subintervals be $x_0 (= a), x_1, x_2, ... x_n (= b)$. We define the Simpson sum as

$$S(n) = \frac{b-a}{3n}\left(f(x_0) + 4f(x_1) + 2f(x_2) + 4f(x_3) + 2f(x_4).... + 2f(x_{n-2}) + 4f(x_{n-1}) + f(x_n) \right)$$

Note: Let f be a function satisfying $\left|f^{(4)}(x)\right| \le K$ for $a \le x \le b$. If E_S is the error in

Simpson's Rule then $\left|E_S\right| \le \dfrac{K(b-a)^5}{180n^4}$.

Example 4: Graph the function $f(x) = \sin(x^2)$ in the interval $\left[0, \dfrac{\pi}{2}\right]$.

(a)   Find the values of left Riemann sum, the right Riemann sum, the midpoint Riemann sum, the sum using the Trapezoidal rule and the sum using Simpson's rule for 5 partitions of the interval $\left[0, \dfrac{\pi}{2}\right]$ for $\int_0^{\frac{\pi}{2}} \sin(x^2)\,dx$. Approximate your answer to 10 significant digits.

(b)   Use *Maple* commands to find errors involved in the approximations of part (a).

Solution: We will enter the integrand $f(x) = \sin(x^2)$ as a *Maple* function and graph it.

$f := x \rightarrow \sin(x^2)$

$$x \rightarrow \sin(x^2) \qquad\qquad\qquad \textbf{(6.70)}$$

$plot\left(f(x), x = 0..\dfrac{\pi}{2}, y = 0..1\right)$

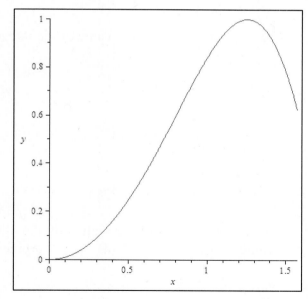

(a)   We will now use *Maple* commands to find the values of $L(5)$, $R(5)$, $M(5)$, $T(5)$, and $S(5)$. We will use the *ApproximateInt* command which is the approximation of the integral. Among its options are those that allow us to find approximations using the Riemann sums, the Trapezoidal rule, and Simpson's rule. Do not forget to activate the $with\left(Student\left[Calculus1\right]\right)$ package.

$with\left(Student\left[Calculus1\right]\right):$

$evalf\left(ApproximateInt\left(f\left(x\right), x=0..\frac{\pi}{2}, method=left, partition=5\right)\right)$

$$0.7097135060 \qquad\qquad \textbf{(6.71)}$$

$evalf\left(ApproximateInt\left(f\left(x\right), x=0..\frac{\pi}{2}, method=right, partition=5\right)\right)$

$$0.9058324389 \qquad\qquad \textbf{(6.72)}$$

$evalf\left(ApproximateInt\left(f\left(x\right), x=0..\frac{\pi}{2}, method=midpoint, partition=5\right)\right)$

$$0.8383462273 \qquad\qquad \textbf{(6.73)}$$

$evalf\left(ApproximateInt\left(f\left(x\right), x=0..\frac{\pi}{2}, method=trapezoid, partition=5\right)\right)$

$$0.8077729725 \qquad\qquad \textbf{(6.74)}$$

$evalf\left(ApproximateInt\left(f\left(x\right), x=0..\frac{\pi}{2}, method=simpson, partition=5\right)\right)$

$$0.8281551423 \qquad\qquad \textbf{(6.75)}$$

(b) To find the error we first need to find the value of the integral $\int_0^{\frac{\pi}{2}} \sin\left(x^2\right)dx$.

$\int_0^{\frac{\pi}{2}} \sin\left(x^2\right)dx$

$$\frac{1}{2}\text{FresnelS}\left(\frac{1}{2}\sqrt{2}\sqrt{\pi}\right)\sqrt{2}\sqrt{\pi} \qquad\qquad \textbf{(6.76)}$$

$evalf\left(\left(6.76\right)\right)$

$$0.8281163285 \qquad\qquad \textbf{(6.77)}$$

Hence the errors in approximating the value of the integral are obtained by subtracting the values obtained in part (a) of this problem from 0.8281163285. They are

$0.8281163285 - 0.7097135060$

$$0.1184028225 \qquad\qquad \textbf{(6.78)}$$

Hence, the error in approximating the value of the integral using the left Riemann sum is 0.1184028225.

$$0.8281163285 - 0.9058324389$$

$$-0.0777161104 \tag{6.79}$$

Hence, the error in approximating the value of the integral using the right Riemann sum is -0.0777161104.

$$0.8281163285 - 0.8383462273$$

$$-0.0102298988 \tag{6.80}$$

Hence, the error in approximating the value of the integral using the midpoint Riemann sum is -0.0102298988.

$$0.8281163285 - 0.8077729725$$

$$0.0203435275 \tag{6.81}$$

Hence, the error in approximating the value of the integral using the Trapezoidal rule is 0.0203435275.

$$0.8281163285 - 0.8281551423$$

$$-0.0000388138 \tag{6.82}$$

Hence, the error in approximating the value of the integral using the Simpson's rule is -0.0000388138.

Example 5: Consider the integral $\int_0^{\frac{\pi}{2}} \sin(x^2) dx$.

(a)   Find the number of partitions n of the interval $\left[0, \frac{\pi}{2}\right]$ that guarantee that the Simpson's rule approximation to $\int_0^{\frac{\pi}{2}} \sin(x^2) dx$ is accurate to within 0.000001.

(b)   Find the number of partitions n of the interval $\left[0, \frac{\pi}{2}\right]$ that guarantee that the Trapezoidal rule and the midpoint rule approximation to $\int_0^{\frac{\pi}{2}} \sin(x^2) dx$ is accurate to within 0.000001.

Solution: We will once again enter the integrand $f(x) = \sin(x^2)$ as a *Maple* function.

$$f := x \rightarrow \sin(x^2)$$

$$x \rightarrow \sin(x^2) \tag{6.83}$$

(a)   Recall that if $\left| f^{(4)}(x) \right| \leq K$ for every value of x in the interval $\left[0, \dfrac{\pi}{2} \right]$, then the

error involved in using Simpson's rule is $\left| error \right| \leq \dfrac{K \left(\dfrac{\pi}{2} \right)^5}{180 n^4}$. We will now use

Maple commands to find $f^{(4)}(x)$ and plot this function.

$f^{(4)}(x)$

$$16 \sin\left(x^2\right) x^4 - 48 \cos\left(x^2\right) x^2 - 12 \sin\left(x^2\right) \qquad \textbf{(6.84)}$$

$plot\left(f^{(4)}(x), x = 0 .. \dfrac{\pi}{2} \right)$

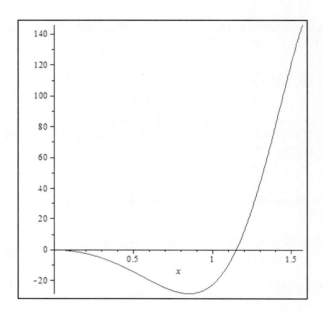

From the graph above the maximum value of $\left| f^{(4)}(x) \right|$ occurs at the point

$\left(\dfrac{\pi}{2}, f^{(4)}\left(\dfrac{\pi}{2} \right) \right)$ when $x = \dfrac{\pi}{2}$. Let us evaluate $f^{(4)}\left(\dfrac{\pi}{2} \right)$

$f^{(4)}\left(\dfrac{\pi}{2} \right)$

$$16 \sin\left(\dfrac{1}{4} \pi^2 \right) \pi^4 - 48 \cos\left(\dfrac{1}{4} \pi^2 \right) \pi^2 - 12 \sin\left(\dfrac{1}{4} \pi^2 \right) \qquad \textbf{(6.85)}$$

evalf $((\textbf{6.85}))$

$$145.8410156 \qquad \textbf{(6.86)}$$

Therefore, $\left|f^{(4)}(x)\right| \leq 145.8410156$ for every value of x in the interval $\left[0, \dfrac{\pi}{2}\right]$.

We will now solve the equation $0.000001 = \dfrac{145.8410156\left(\dfrac{\pi}{2}\right)^5}{180n^4}$ for nonnegative

values of n.

$$fsolve\left(0.000001 = \dfrac{145.8410156 \cdot \left(\dfrac{\pi}{2}\right)^5}{180 \cdot n^4}, n, n = 0..\infty\right)$$

$$52.75961771 \tag{6.87}$$

Hence n must be 53 or larger. Let us check our answer by actually calculating the error.

$$evalf\left(\int_0^{\frac{\pi}{2}} \sin\left(x^2\right)dx - ApproximateInt\left(f(x), x = 0..\dfrac{\pi}{2}, method = simpson, partition = 53\right)\right)$$

$$-5.1 \, 10^{-9} \tag{6.88}$$

Therefore the error is -5.1×10^{-9} which is indeed less than 0.000001.

(b)   Recall that if $\left|f''(x)\right| \leq C$ for every value of x in the interval $\left[0, \dfrac{\pi}{2}\right]$, then the

error involved in using the Trapezoidal Rule is $\left|error\right| \leq \dfrac{C\left(\dfrac{\pi}{2}\right)^3}{12n^2}$. We will now

use *Maple* commands to find $f''(x)$ and plot this function.

$f''(x)$

$$-4\sin\left(x^2\right)x^2 + 2\cos\left(x^2\right) \tag{6.89}$$

$$plot\left(f''(x), x = 0..\dfrac{\pi}{2}\right)$$

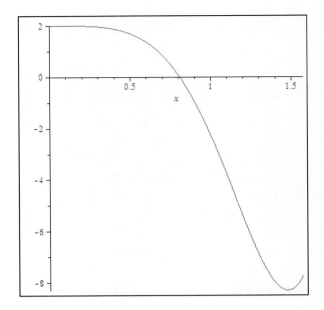

From the graph above the maximum value of $\left|f''(x)\right|$ occurs at a value of x between 1 and 1.5. Note that the maximum value of $\left|f''(x)\right|$ is the same as the minimum value of $f''(x)$. Let us find this value.

$$evalf\left(minimize\left(f''(x), x = 0 .. \frac{\pi}{2}\right)\right)$$
$$-7.723681775 \qquad\qquad\qquad \textbf{(6.90)}$$

Therefore, $\left|f''(x)\right| \le 7.723681775$ for every value of x in the interval $\left[0, \dfrac{\pi}{2}\right]$. We

will now solve the equation $0.000001 = \dfrac{7.723681775\left(\dfrac{\pi}{2}\right)^3}{12n^2}$ for nonnegative

values of n.

$$fsolve\left(0.000001 = \frac{7.723681775 \cdot \left(\dfrac{\pi}{2}\right)^3}{12 \cdot n^2}, n, n = 0 .. \infty\right)$$
$$1579.433621 \qquad\qquad\qquad \textbf{(6.91)}$$

Hence n must be 1580 or larger. Let us check our answer by actually calculating the error.

$$evalf\left(\int_0^{\frac{\pi}{2}} \sin\left(x^2\right) dx - ApproximateInt\left(f(x), x = 0 .. \frac{\pi}{2}, method = trapezoid, partition = 1580 \right) \right)$$

$$2.008\,10^{-7} \tag{6.92}$$

Therefore the error is 2.008×10^{-9} which is indeed less than 0.000001.

Remark: The actual number of partitions required for Simpson's rule is really 13 and for the Trapezoidal rule is 711. This is verified below.

$$evalf\left(\int_0^{\frac{\pi}{2}} \sin\left(x^2\right) dx - ApproximateInt\left(f(x), x = 0 .. \frac{\pi}{2}, method = simpson, partition = 13 \right) \right)$$

$$-9.125\,10^{-7} \tag{6.93}$$

$$evalf\left(\int_0^{\frac{\pi}{2}} \sin\left(x^2\right) dx - ApproximateInt\left(f(x), x = 0 .. \frac{\pi}{2}, method = trapezoid, partition = 711 \right) \right)$$

$$9.987\,10^{-7} \tag{6.94}$$

Chapter 6 Integration

Exercises

1. Estimate the value of $\int_0^7 (4-x)\,dx$ using the left Riemann sum and the right Riemann sum for 20, 50, 100, 500, 1000, 5000, and 10000 partitions. Find a formula for the left Riemann sum $L(n)$ and the right Riemann sum $R(n)$ when the number of partitions is n. Evaluate $\lim_{n\to\infty} L(n)$ and $\lim_{n\to\infty} R(n)$. Use these numbers to evaluate $\int_0^7 (4-x)\,dx$. Verify yours answers using *Maple* commands.

2. Estimate the value of $\int_{-1}^2 (2x+3)\,dx$ using the left Riemann sum and the right Riemann sum for 20, 50, 100, 500, 1000, 5000, and 10000 partitions. Find a formula for the left Riemann sum $L(n)$ and the right Riemann sum $R(n)$ when the number of partitions is n. Evaluate $\lim_{n\to\infty} L(n)$ and $\lim_{n\to\infty} R(n)$. Use these numbers to evaluate $\int_{-1}^2 (2x+3)\,dx$. Verify yours answers using *Maple* commands.

3. Estimate the value of $\int_{-1}^1 (3x^2 - 4x + 7)\,dx$ using the left Riemann sum and the right Riemann sum for 20, 50, 100, 500, 1000, and 5000 partitions. Find a formula for the left Riemann sum $L(n)$ and the right Riemann sum $R(n)$ when the number of partitions is n. Evaluate $\lim_{n\to\infty} L(n)$ and $\lim_{n\to\infty} R(n)$. Use these numbers to evaluate $\int_{-1}^1 (3x^2 - 4x + 7)\,dx$. Verify yours answers using *Maple* commands.

4. Estimate the value of $\int_{-2}^1 (2x^4 - 32)\,dx$ using the left Riemann sum and the right Riemann sum for 20, 50, 100, 500, 1000, and 5000 partitions. Find a formula for the left Riemann sum $L(n)$ and the right Riemann sum $R(n)$ when the number of partitions is n. Evaluate $\lim_{n\to\infty} L(n)$ and $\lim_{n\to\infty} R(n)$. Use these numbers to evaluate $\int_{-2}^1 (2x^4 - 32)\,dx$. Verify yours answers using *Maple* commands.

5. Estimate the value of $\int_0^1 (12x^3 - 9x^2 - 234x - 15)\,dx$ using the left Riemann sum and the right Riemann sum for 20, 50, 100, 500, 1000, and 5000 partitions. Find a formula for the left Riemann sum $L(n)$ and the right Riemann sum $R(n)$ when the number of partitions is n. Evaluate $\lim_{n\to\infty} L(n)$ and $\lim_{n\to\infty} R(n)$. Use these

numbers to evaluate $\int_0^1 \left(12x^3 - 9x^2 - 234x - 15 \right) dx$. Verify yours answers using *Maple* commands.

6. Estimate the value of $\int_0^\pi \cos(x)\, dx$ using the left Riemann sum and the right Riemann sum for 20, 50, 100, 500, 1000, and 5000 partitions. Find a formula for the left Riemann sum $L(n)$ and the right Riemann sum $R(n)$ when the number of partitions is n. Evaluate $\lim_{n\to\infty} L(n)$ and $\lim_{n\to\infty} R(n)$. Use these numbers to evaluate $\int_0^\pi \cos(x)\, dx$. Verify yours answers using *Maple* commands.

7. Estimate the value of $\int_0^1 \dfrac{x}{1+x^7}\, dx$ using the left Riemann sum and the right Riemann sum for 20, 50, 100, 500, and 1000 partitions. Find a formula for the left Riemann sum $L(n)$ and the right Riemann sum $R(n)$ when the number of partitions is n. Can you evaluate $\lim_{n\to\infty} L(n)$ and $\lim_{n\to\infty} R(n)$? Can you use these numbers to evaluate $\int_0^1 \dfrac{x}{1+x^7}\, dx$? Can you verify yours answers using *Maple* commands? What happens when you enter the command $\int_{0.0}^{1.0} \dfrac{x}{1+x^7}\, dx$? Does that give you an estimated answer for $\int_0^1 \dfrac{x}{1+x^7}\, dx$?

8. Estimate the value of $\int_{-1}^1 \dfrac{x^2}{1+x^8}\, dx$ using the left Riemann sum and the right Riemann sum for 20, 50, 100, 500, and 1000 partitions. Find a formula for the left Riemann sum $L(n)$ and the right Riemann sum $R(n)$ when the number of partitions is n. Can you evaluate $\lim_{n\to\infty} L(n)$ and $\lim_{n\to\infty} R(n)$? Can you use these numbers to evaluate $\int_{-1}^1 \dfrac{x^2}{1+x^8}\, dx$? Can you verify yours answers using *Maple* commands? What happens when you enter the command $\int_{-1.0}^{1.0} \dfrac{x^2}{1+x^8}\, dx$? Does that give you an estimated answer for $\int_{-1}^1 \dfrac{x^2}{1+x^8}\, dx$?

9. Graph the function $f(x) = \dfrac{1}{x+1}$ in the interval $[0,1]$.

(a) Find the values of left Riemann sum, the right Riemann sum, the midpoint Riemann sum, the sum using the Trapezoidal rule and the sum using Simpson's rule for 5 partitions of the interval $[0,1]$ for $\int_0^1 \dfrac{1}{x+1} dx$. Approximate your answer to 10 significant digits.

(b)   Use *Maple* commands to find errors involved in the approximations of part (a).

10.   Consider the integral $\int_0^1 \dfrac{1}{x+1} dx$.

(a)   Find the number of partitions n of the interval $[0,1]$ that guarantee that the Simpson's rule approximation to $\int_0^1 \dfrac{1}{x+1} dx$ is accurate to within 0.000001.

(b)   Find the number of partitions n of the interval $[0,1]$ that guarantee that the Trapezoidal rule and the midpoint rule approximation to $\int_0^1 \dfrac{1}{x+1} dx$ is accurate to within 0.000001.

11.   Graph the function $f(x) = \ln(x)$ in the interval $\left[e, 2e^2\right]$.

(a)   Find the values of left Riemann sum, the right Riemann sum, the midpoint Riemann sum, the sum using the Trapezoidal rule and the sum using Simpson's rule for 5 partitions of the interval $\left[e, 2e^2\right]$ for $\int_e^{2e^2} \ln(x) dx$. Approximate your answer to 10 significant digits.

(b)   Use *Maple* commands to find errors involved in the approximations of part (a).

12.   Consider the integral $\int_e^{2e^2} \ln(x) dx$.

(a)   Find the number of partitions n of the interval $\left[e, 2e^2\right]$ that guarantee that the Simpson's rule approximation to $\int_e^{2e^2} \ln(x) dx$ is accurate to within 0.000001.

(b)   Find the number of partitions n of the interval $\left[e, 2e^2\right]$ that guarantee that the Trapezoidal rule and the midpoint rule approximation to $\int_e^{2e^2} \ln(x) dx$ is accurate to within 0.000001.

13.   Graph the function $f(x) = \sec(x)$ in the interval $\left[0, \dfrac{\pi}{3}\right]$.

(a)   Find the values of left Riemann sum, the right Riemann sum, the midpoint Riemann sum, the sum using the Trapezoidal rule and the sum using Simpson's rule for 5 partitions of the interval $\left[0, \dfrac{\pi}{3}\right]$ for $\int_0^{\frac{\pi}{3}} \sec(x) dx$. Approximate your answer to 10 significant digits.

(b)   Use *Maple* commands to find errors involved in the approximations of part (a).

14. Consider the integral $\int_0^{\frac{\pi}{3}} \sec(x)\,dx$.

(a) Find the number of partitions n of the interval $\left[0, \dfrac{\pi}{3}\right]$ that guarantee that the

 Simpson's rule approximation to $\int_0^{\frac{\pi}{3}} \sec(x)\,dx$ is accurate to within 0.000001.

(b) Find the number of partitions n of the interval $\left[0, \dfrac{\pi}{3}\right]$ that guarantee that the

 Trapezoidal rule and the midpoint rule approximation to $\int_0^{\frac{\pi}{3}} \sec(x)\,dx$ is accurate
 to within 0.000001.

15. Let $f(x) = \dfrac{1}{x}$. In this problem the value of $\ln(3)$ will be estimated using
 Riemann sums.

(a) Write a *Maple* procedure to generate the left and right Riemann sums for n equal
 partitions of the interval $[1, 3]$ for the function $f(x)$. Run your procedure for
 1000, 10000, and 60000 partitions.

(b) For any integer $n > 0$, construct a partition of $[1, 3]$ in the following manner:
$$x_0 = 1,\ x_1 = 3^{1/n},\ x_2 = 3^{2/n},\ \dots,\ x_n = 3^{n/n} = 3.$$
 Modify your procedure in part (a) to find the left and right Riemann sums for
 $f(x)$ using this partition. Run your procedure for 1000, 10000, and 60000
 partitions.

(c) For part (b) show that the left sum $L(n)$ is given by $L(n) = n\left(3^{1/n} - 1\right)$ and the

 right sum $R(n)$ is given by $R(n) = n\left(1 - 3^{-1/n}\right)$.

(d) Use that fact that $\int_1^3 \dfrac{1}{x}\,dx = \ln(3)$ and your results from parts (a), (b), and (c) to
 find an estimate for ln3.

16. Below is a list of some *Maple* commands used in this chapter. Describe the
 significance of each command. Can you find examples where each command is
 used?

 (a) $ApproximateInt\left(f(x), x = a \mathbin{..} b, method = left, partition = n\right)$

 (b) $ApproximateInt\left(f(x), x = a \mathbin{..} b, method = midpoint, output = integral,\right.$
 $\left. partition = n\right)$

 (c) $ApproximateInt\left(f(x), x = a \mathbin{..} b, method = random, output = animation,\right.$
 $\left. partition = n, iterations = m, refinement = k\right)$

(d) $ApproximateInt\left(f\left(x\right), x = a\mathbin{..}b, method = right, output = integral,\right.$

$\left.partition = n\right)$

(e) $with\left(Student\left[Calculus1\right]\right)$

17.   Can you list some *Maple* commands that were used in this chapter but were not listed in exercise 16?

Chapter 7   Integration Techniques

The object of this chapter is to use different integration techniques to evaluate both definite and indefinite integrals. Integrals will be computed using the *Expression* palette and the *Clickable Calculus* feature. *Maple* will be used to vary different parameters to achieve better results. Different methods of integration will be reinforced using step-by-step procedures.

Definite and Indefinite Integrals

Fundamental Theorem of Calculus Part I: Let f be a continuous function on the closed interval $[a,b]$. Define the function g by $g(x) = \int_a^x f(t)\,dt$ for $a \le x \le b$. Then g is continuous on $[a,b]$, differentiable on (a,b) and $g'(x) = f(x)$ for every value of x in (a,b).

Fundamental Theorem of Calculus Part II: Let f be a continuous function on the closed interval $[a,b]$. Then $\int_a^b f(x)\,dx$ exists and $\int_a^b f(x)\,dx = F(b) - F(a)$ where F is any antiderivative of f.

Example 1: Use *Maple* to find $g'(x)$ and $g''(x)$ for each of the following and verify that these answers follow from the Fundamental Theorem of Calculus I.

(a) $g(x) = \int_1^x t\,dt$

(b) $g(x) = \int_1^x x^2 t\,dt$

Solution: We will evaluate the integrals using the *Expression* palette.

(a)   In order to enter $g(x) = \int_1^x t\,dt$, click on $\int_a^b f\,dx$ in the *Expression* palette (after entering $g := x \rightarrow$). Insert 1 in place of a, use the $\boxed{Tab}$ key to insert x in place of b, again use the $\boxed{Tab}$ key to insert t in place of f and then use the $\boxed{Tab}$ key once more and insert t in place of x and then press the $\boxed{Enter}$ key to get the answer shown below.

$$g := x \rightarrow \int_1^x t\,dt$$

$$x \rightarrow \int_1^x t\,dt \qquad\qquad (7.1)$$

In this case $f(t) = t$, which is a continuous function. Hence by Fundamental Theorem of Calculus Part I we see that $g'(x) = x$ so that $g''(x) = 1$. We will verify this using *Maple* commands.

$g'(x)$

$$x \qquad \qquad \textbf{(7.2)}$$

$g''(x)$

$$1 \qquad \qquad \textbf{(7.3)}$$

(b) We will enter $g(x) = \int_1^x x^2 t \, dt$ using a similar procedure as in part (a). Do not forget to write the integrand as a product $x^2 \cdot t$.

$$g := x \rightarrow \int_1^x x^2 \cdot t \, dt$$

$$x \rightarrow \int_1^x x^2 \, t \, dt \qquad \qquad \textbf{(7.4)}$$

Since $g(x) = \int_1^x x^2 t \, dt = x^2 \int_1^x t \, dt$, we see that (by the product rule) $g'(x) = \left(x^2\right)\left(\int_1^x t \, dt\right)' + \left(\int_1^x t \, dt\right)\left(x^2\right)'$. We will now use Fundamental Theorem of Calculus Parts I and II to obtain $g'(x) = \left(x^2\right)(x) + \frac{1}{2}\left(x^2 - 1\right)(2x) = 2x^3 - x$. Therefore, $g''(x) = 6x^2 - 1$. We will verify this using *Maple* commands (we will need to use the *simplify* command to simplify $g'(x)$).

$g'(x)$

$$x\left(x^2 - 1\right) + x^3 \qquad \qquad \textbf{(7.5)}$$

simplify
$=$

$$2\,x^3 - x \qquad \qquad \textbf{(7.6)}$$

$g''(x)$

$$6\,x^2 - 1 \qquad \qquad \textbf{(7.7)}$$

Note that can use *Maple* to find $g(x)$.

$g(x)$

$$\frac{1}{2}x^2\left(x^2-1\right)$$

(7.8)

Properties of the definite integral:

I. Let f and g be continuous on the closed interval $[a,b]$ and let k be any constant. Then

1. $\displaystyle\int_a^b f(x)\,dx = -\int_b^a f(x)\,dx$

2. $\displaystyle\int_a^a f(x)\,dx = 0$

3. $\displaystyle\int_a^b k\,dx = k(b-a)$

4. $\displaystyle\int_a^b k\,f(x)\,dx = k\int_a^b f(x)\,dx$

5. $\displaystyle\int_a^b \left[f(x)+g(x)\right]dx = \int_a^b f(x)\,dx + \int_a^b g(x)\,dx$

6. $\displaystyle\int_a^b \left[f(x)-g(x)\right]dx = \int_a^b f(x)\,dx - \int_a^b g(x)\,dx$

7. $\displaystyle\int_a^b f(x)\,dx = \int_a^c f(x)\,dx + \int_c^b f(x)\,dx$, where c is any number in the interval $[a,b]$.

8. If $f(x)\geq 0$ for every value of x in the closed interval $[a,b]$, then $\displaystyle\int_a^b f(x)\,dx \geq 0$.

9. If $f(x)\geq g(x)$ for every value of x in the closed interval $[a,b]$, then $\displaystyle\int_a^b f(x)\,dx \geq \int_a^b g(x)\,dx$

10. If $m\leq f(x)\leq M$ for every value of x in the closed interval $[a,b]$, then $m(b-a)\leq \displaystyle\int_a^b f(x)\,dx \leq M(b-a)$.

II. Let f be continuous on the closed interval $[-a,a]$. Then

1. If f is an even function, then $\displaystyle\int_{-a}^a f(x)\,dx = 2\int_0^a f(x)\,dx$.

2. If f is an odd function, then $\displaystyle\int_{-a}^a f(x)\,dx = 0$.

Notation: Using the convention from the Fundamental Theorem of Calculus Part II we will use the notation $\int f(x)\,dx$ for an antiderivative of f and call it the indefinite integral.

Remark: The definite integral $\int_a^b f(x)\,dx$ is a number. The indefinite integral $\int f(x)\,dx$ is a function of x that is valid only on an interval.

Properties of the indefinite integral:

I. Let f be a function and let k be any constant. Then

1. $\int k\,f(x)\,dx = k\int f(x)\,dx$

2. $\int\left[f(x)+g(x)\right]dx = \int f(x)\,dx + \int g(x)\,dx$

3. $\int\left[f(x)-g(x)\right]dx = \int f(x)\,dx - \int g(x)\,dx$

II. The following is a list of formulas for indefinite integrals. Note that C is the constant of integration.

1. $\int k\,dx = kx + C$, where k is a constant

2. $\int x^n\,dx = \dfrac{x^{n+1}}{n+1} + C$, $n \neq -1$

3. $\int \dfrac{1}{x}\,dx = \ln|x| + C$

4. $\int e^x\,dx = e^x + C$

5. $\int a^x\,dx = \dfrac{a^x}{\ln(a)} + C$, $a > 0, a \neq 1$

6. $\int \sin(x)\,dx = -\cos(x) + C$

7. $\int \cos(x)\,dx = \sin(x) + C$

8. $\int \sec^2(x)\,dx = \tan(x) + C$

9. $\int \csc^2(x)\,dx = -\cot(x) + C$

10. $\int \tan(x)\sec(x)\,dx = \sec(x) + C$

11. $\int \cot(x)\csc(x)\,dx = -\csc(x) + C$

12. $\int \dfrac{1}{\sqrt{1-x^2}}\,dx = \arcsin(x) + C$

13. $\int \dfrac{1}{1+x^2}\,dx = \arctan(x) + C$

14. $\int \dfrac{1}{x\sqrt{x^2-1}}\,dx = \operatorname{arc\,sec}(x) + C$

15. $\int \tan(x)\,dx = \ln|\sec(x)| + C$

16. $\int \cot(x)\,dx = \ln|\sin(x)| + C$

17. $\int \sec(x)\,dx = \ln|\sec(x) + \tan(x)| + C$

18. $\quad \int \csc(x)\,dx = \ln\left|\csc(x) - \cot(x)\right| + C$

In our first example we will use $\int f\,dx$ in the *Expression* palette to evaluate different indefinite integrals to get used to using *Maple* commands. Observe that when evaluating an indefinite integral, *Maple* omits the constant of integration.

Example 2: Use the *Expression* palette to evaluate the integrals

(a) $\quad \int \dfrac{1}{x}\,dx$,

(b) $\quad \int \tan(x)\,dx$,

(c) $\quad \int \csc(x)\,dx$,

(d) $\quad \int_0^1 \dfrac{1}{1+x^2}\,dx$, and

(e) $\quad \int \dfrac{1}{x\sqrt{x^2-1}}\,dx$.

Solution: We will evaluate the integrals using the *Expression* palette.

(a) Click on $\int f\,dx$. Insert $\dfrac{1}{x}$ in place of f, then use the $\boxed{\text{Tab}}$ key and insert x in place of x and lastly press the $\boxed{\text{Enter}}$ key to get the answer shown below.

$$\int \frac{1}{x}\,dx$$

$$\ln(x) \qquad\qquad\qquad\qquad \textbf{(7.9)}$$

Note that instead of the answer for $\int \dfrac{1}{x}\,dx$ as $\ln|x|$, *Maple* gives an output of

$$\int \frac{1}{x}\,dx = \ln(x).$$

(b) To evaluate the integral $\int \tan(x)\,dx$, insert $\tan(x)$ in place of f and use the same steps as in part (a).

$$\int \tan(x)\,dx$$

$$-\ln(\cos(x)) \qquad\qquad\qquad \textbf{(7.10)}$$

Hence $\int \tan(x)\,dx = -\ln(\cos(x))$, using *Maple* commands. Verify that this answer is the same as the one given in formula 15 in the list of formulas for indefinite integrals (without the absolute value).

(c)   To evaluate the integral $\int \csc(x)\,dx$, insert $\csc(x)$ in place of f and use the same steps as in part (a).

$\int \csc(x)\,dx$

$$-\ln\big(\csc(x)+\cot(x)\big) \tag{7.11}$$

Hence $\int \csc(x)\,dx = -\ln\big(\csc(x)+\cot(x)\big)$, using *Maple* commands. Is this answer the same as the one given in formula 18 for the list of indefinite integrals?

(d)   In order to evaluate $\int_0^1 \dfrac{1}{1+x^2}\,dx$, click on $\int_a^b f\,dx$ in the *Expression* palette. Insert 0 in place of a, use the $\boxed{Tab}$ key to insert 1 in place of b, again use the $\boxed{Tab}$ key to insert $\dfrac{1}{1+x^2}$ in place of f and then use the $\boxed{Tab}$ key once more and insert x in place of x and then press the $\boxed{Enter}$ key to get the answer shown below.

$\int_0^1 \dfrac{1}{1+x^2}\,dx$

$$\frac{1}{4}\pi \tag{7.12}$$

Hence $\int_0^1 \dfrac{1}{1+x^2}\,dx = \dfrac{1}{4}\pi$, using *Maple* commands.

(e)   To evaluate the integral $\int \dfrac{1}{x\sqrt{x^2-1}}\,dx$, insert $\dfrac{1}{x\sqrt{x^2-1}}$ in place of f and use the same steps as in part (a).

$\int \dfrac{1}{x\sqrt{x^2-1}}\,dx$

$$-\arctan\left(\frac{1}{\sqrt{x^2-1}}\right) \tag{7.13}$$

Hence $\int \dfrac{1}{x\sqrt{x^2-1}}\,dx = -\arctan\left(\dfrac{1}{\sqrt{-1+x^2}}\right)$, using *Maple* commands. Is this the same as formula 14 in the list of indefinite integrals?

Remarks: We could use the right click command to evaluate these integrals. For example, in order to integrate $\int \frac{1}{x} \, dx$, enter $\frac{1}{x}$, right click on the expression, select *Integrate*, and select *x*.

$$\frac{1}{x} \xrightarrow{\text{integrate w.r.t. x}} \ln(x)$$

Step-by-step Integration

The inert form of the integral, *Int*, will be used to perform a step-by-step evaluation of the integral. We will need to activate the $with\big(Student[Calculus1]\big)$ package. We will execute the command *restart* at the beginning of each problem.

Below is a list of different rules of integration recognized by *Maple*. We will illustrate these rules, when performing a step-by-step approach, with numerous examples throughout this chapter.

Rules for step-by-step integration

Let f and g be two functions such that $\int f(x) \, dx$ and $\int g(x) \, dx$ exist. Let k be any constant. The table below gives the rules for integration and the *Maple* rules used for step-by-step integration. We will omit the constant on integration in the table below in keeping with the rules for *Maple*. Although these rules are stated for indefinite integrals, the same rules can be used to evaluate definite integrals.

Rule for integration	*Maple* rule for step-by-step integration
$\int k \cdot f(x) \, dx = k \cdot \int f(x) \, dx$, where k is a constant	*constantmultiple*
$\int \big[f(x) + g(x) \big] \, dx = \int f(x) \, dx + \int g(x) \, dx$	*sum*
$\int \big[f(x) - g(x) \big] \, dx = \int f(x) \, dx - \int g(x) \, dx$	*difference*
$\int k \, dx = kx$, where k is a constant	*constant*
$\int x^n \, dx = \dfrac{x^{n+1}}{n+1}, \, n \neq -1$	*power*
$\int \frac{1}{x} \, dx = \ln(x)$	*power*
$\int e^x \, dx = e^x$	exp
$\int \sin(x) \, dx = -\cos(x)$	sin
$\int \cos(x) \, dx = \sin(x)$	cos

Rule for integration	*Maple* rule for step-by-step integration
$\int \tan(x)\,dx = -\ln(\cos(x))$	tan
$\int \cot(x)\,dx = \ln(\sin(x))$	cot
$\int \sec(x)\,dx = \ln(\sec(x)+\tan(x))$	sec
$\int \sec(x)\,dx = -\ln(\csc(x)+\cot(x))$	csc
Substitution rule	*change*
Integration by parts	*parts*
Integrand is expressed as a sum of partial fractions	*partialfractions*
Flips the limits of integration	*flip*
Changes the substituted variable back in terms of the original variable	*revert*
Changes part of the integrand using a different expression	*rewrite*
algebraically solves an equation for an integral that appears more than once	*solve*

Remark: Note that even though there are rules *sin* and *cos* for step-by-step integration, there is no rule for a number of the other indefinite integrals. For example, *Maple* does not have a rule called *csc* for the integral $\int \csc(x)\cot(x)\,dx = -\csc(x)$ when performing a step-by-step evaluation of the integral. This integral is evaluated in example 3 using a step-by-step procedure.

Substitution Rule

The Substitution method:

(a)   Let $w = g(x)$ be a differentiable function. Let f be continuous on the range of g. Then

$$\int (f \circ g)(x)\, g'(x)\, dx = \int f(w)\, dw$$

(b)   Let $w = g(x)$ be a differentiable function whose range is an interval. Suppose that the derivative g' is a continuous function. Let f be continuous on the range of g. Then

$$\int_a^b (f \circ g)(x)\, g'(x)\, dx = \int_{g(a)}^{g(b)} f(w)\, dw$$

We will use w to denote the substituted variable so that we do not confuse it with u used while using the integration by parts method.

Since *Maple* does not have a rule called *csc* when performing a step-by-step evaluation of the integral we will integrate $\int \csc(x)\cot(x)\,dx$ in example 3 using a substitution method.

Example 3: Use the step-by-step procedure in *Maple* to integrate $\int \csc(x)\cot(x)\,dx$.

Solution: We will begin by executing the *restart* command. Activate the $with\big(Student[Calculus1]\big)$ package and set $infolevel\big[Student[Calculus1]\big]:=1$. We will also use the inert form of the integral which is the *Int* command. This command will return the integral unevaluated.

restart

$with\big(Student[Calculus1]\big):$

$infolevel\big[Student[Calculus1]\big]:=1:$

$Rule\big[\ \big]\big(Int(\csc(x)\cot(x),x)\big)$

```
Creating problem #1
```

$$\int \csc(x)\cot(x)\,dx = \int \csc(x)\cot(x)\,dx \qquad\qquad \textbf{(7.14)}$$

We will now make a change of variable $w = \csc(x)$ so that $dw = -\csc(x)\cot(x)\,dx$ or $-dw = \csc(x)\cot(x)\,dx$. We will use the *change* command here. Notice that the substitution routine is given below the command. The integral on the right side is the one where we have made the substitution. We will be using the labels for the commands as was the case for step-by-step limits and differentiation.

$Rule\big[change, w = \csc(x)\big]\big(\textbf{(7.14)}\big)$

```
Applying substitution x = arccsc(w), w = csc(x) with dx = -
1/(w)^(2)/(1-1/w^2)^(1/2)*dw, dw = -csc(x)*cot(x)*dx
```

$$\int \csc(x)\cot(x)\,dx = \int (-1)\,dw \qquad\qquad \textbf{(7.15)}$$

Notice that the substitution routine is given below the command. We will integrate the constant -1 using the *constant* command here.

$Rule[constant]\big(\textbf{(7.15)}\big)$

$$\int \csc(x)\cot(x)\,dx = -w \qquad\qquad \textbf{(7.16)}$$

We will change w in terms of x by using the *revert* command here.

$Rule[revert]\big(\textbf{(7.16)}\big)$

Chapter 7 Integration Techniques

Reverting substitution using w = csc(x)

$$\int \csc(x)\cot(x)\,dx = -\csc(x) \qquad\qquad \textbf{(7.17)}$$

Observe that the *revert* routine is given below the command. Finally we can show the step-by-step computation of this integral using the *ShowSteps* command. Again the commands that appear on the left side are features available in version 13 and not in the earlier versions.

ShowSteps( )

$$\int \csc(x)\cot(x)\,dx$$

$$= \int (-1)\,dw \qquad \left[change, w = \csc(x)\right] \qquad\qquad \textbf{(7.18)}$$

$$= -w \qquad\qquad\qquad \left[constant\right]$$

$$= -\csc(x) \qquad\qquad\quad \left[revert\right]$$

Example 4: Use *Maple* commands to integrate $\int x^2\left(2-5x^3\right)^7 dx$. Show your work step-by-step. Are the answers obtained using *Maple* commands and the step-by-step procedure the same? Give reasons for your answer.

Solution:

Using the *Expression* palette

We will integrate $\int x^2\left(2-5x^3\right)^7 dx$ by clicking on $\int f\,dx$ in the *Expression* palette.

restart

$$\int x^2 \cdot \left(2-5\cdot x^3\right)^7 dx$$

$$-\frac{78125}{24}x^{24} + \frac{31250}{3}x^{21} - \frac{43750}{3}x^{18} + \frac{35000}{3}x^{15} - \frac{17500}{3}x^{12} + \frac{5600}{3}x^9 - \frac{1120}{3}x^6 + \frac{128}{3}x^3$$

$$\textbf{(7.19)}$$

Using a step-by-step approach

We will solve this problem using a step-by-step and a change of variable. Do not forget to activate the $with\left(Student\left[Calculus1\right]\right)$ package.

restart

$with\left(Student\left[Calculus1\right]\right):$

$infolevel\left[Student\left[Calculus1\right]\right]:=1:$

$Rule\left[\ \right]\left(Int\left(x^2\cdot\left(2-5\cdot x^3\right)^7,x\right)\right)$

Creating problem #1

$$\int x^2 \left(2-5\,x^3\right)^7 dx = \int x^2 \left(2-5\,x^3\right)^7 dx \qquad (7.20)$$

We will now make a change of variable $w = 2 - 5 \cdot x^3$ so that $dw = -15x^2 dx$ or $-\dfrac{1}{15} dw = x^2 dx$. The *change* command will be used here. Observe that the substitution routine is given below the command. The integral is on the left side and the right side is the one where we have made the substitution.

$Rule\left[change, w = 2 - 5 \cdot x^3\right]((\textbf{7.20}))$

```
Applying  substitution  x  =  1/5*(-25*w+50)^(1/3),  w  =  2-5*x^3
with dx = -5/3/(-25*w+50)^(2/3)dw,  dw = -15*x^2*dx
```

$$\int x^2 \left(2-5\,x^3\right)^7 dx = \int \left(-\frac{1}{15} w^7\right) dw \qquad (7.21)$$

We will factor $-\dfrac{1}{15}$ outside the integral by using the *constantmultiple* command.

$Rule[constantmultiple]((\textbf{7.21}))$

$$\int x^2 \left(2-5\,x^3\right)^7 dx = -\frac{1}{15} \int w^7\, dw \qquad (7.22)$$

The *power* command will be used to find $-\dfrac{1}{15} \int w^7\, dw$.

$Rule[power]((\textbf{7.22}))$

$$\int x^2 \left(2-5\,x^3\right)^7 dx = -\frac{1}{120} w^8 \qquad (7.23)$$

We will change w in terms of x by using the *revert* command here.

$Rule[revert]((\textbf{7.23}))$

```
Reverting substitution using w = 2-3*x^3
```

$$\int x^2 \left(2-5\,x^3\right)^7 dx = -\frac{32}{15} + \frac{128}{3} x^3 - \frac{1120}{3} x^6 + \frac{5600}{3} x^9 - \frac{17500}{3} x^{12} + \frac{35000}{3} x^{15} - \frac{43750}{3} x^{18}$$
$$+ \frac{31250}{3} x^{21} - \frac{78125}{24} x^{24} \qquad (7.24)$$

Remarks: Notice that we have a constant $-\dfrac{32}{15}$ which is a part of this answer but was not a part of the answer when we evaluated the integral using the *Expression Palette*. The reason for this is that the constant $-\dfrac{32}{15}$ appears when we expand

$-\dfrac{1}{120}w^8 = -\dfrac{1}{120}\left(2-3\cdot x^3\right)^8$. This constant is absorbed with the constant of integration when we integrate directly using *Maple* commands. We can factor the right-hand side of the above expression (number **(7.24)**) by using the *factor* command.

factor $\big((\mathbf{7.24})\big)$

$$-\frac{1}{120}\left(-2+5\,x^3\right)^8 \tag{7.25}$$

Observe that we do not get the same answer when we try to use the same command for equation **(7.19)**.

factor $\big((\mathbf{7.19})\big)$

$$-\frac{1}{24}x^3\left(5\,x^3-4\right)\left(25\,x^6-20\,x^3+8\right)\left(625\,x^{12}-1000\,x^9+600\,x^6-160\,x^3+32\right) \tag{7.26}$$

The reader could show the step-by-step computation of this integral using the *ShowSteps* command. Check the number of steps required to integrate $\int x^2\left(2-5x^3\right)^7\,dx$ using the *ShowSolution* command.

Integration by Parts

Integration by Parts:
Let f and g be differentiable functions. If $u = f(x)$ and $v = g(x)$, then $du = f'(x)dx$ and $dv = g'(x)dx$ and

$$\int u\,dv = uv - \int v\,du$$

The next example uses integration by parts. When using the *parts* command it is important to indicate u and v (in that order), not u and dv.

Example 5: Use the step-by-step procedure in *Maple* to integrate $\int x\cos(x)\,dx$.

Solution: Notice that this problem is one where one uses the integration by parts method. We will use *Maple* commands to solve this problem step-by-step using a change of variable. We will again need to activate the *with*$\big($*Student*$\big[$*Calculus1*$\big]\big)$ package and set *infolevel*$\big[$*Student*$\big[$*Calculus1*$\big]\big] := 1$. We will also use the inert form of the integral which is the *Int* command. This command will return the integral unevaluated.

restart

$with\big(Student[Calculus1]\big):$

$infolevel\big[Student[Calculus1]\big]:=1:$

$Rule[\ ]\big(Int\big(x\cdot\cos(x),x\big)\big)$
Creating problem #1

$$\int x\cos(x)\,dx = \int x\cos(x)\,dx \qquad\qquad\qquad (7.27)$$

We will now use integration by parts with $u=x$ and $dv=\cos(x)\,dx$ so that $du=dx$ and $v=\sin(x)$. We will use the *parts* command. In this case we need to specify both u and v (not u and dv).

$Rule\big[parts,x,\sin(x)\big]\big((\mathbf{7.27})\big)$

$$\int x\cos(x)\,dx = x\sin(x)-\left(\int\sin(x)\,dx\right) \qquad\qquad (7.28)$$

Unlike the previous example, no routine appeared below. We will now complete the problem by integrating $\sin(x)$.

$Rule[\sin]\big((\mathbf{7.28})\big)$

$$\int x\cos(x)\,dx = \cos(x)+x\sin(x) \qquad\qquad\qquad (7.29)$$

Hence $\int x\cos(x)\,dx = \cos(x)+x\sin(x)$.

The following example uses both the substitution and the integration by parts methods. This example is a definite integral.

Example 6: Use the step-by-step procedure in *Maple* to integrate $\int_{-1}^{0}x^3 e^{x^2}\,dx$.

Solution: We will use *Maple* commands to solve this problem step-by-step. Do not forget to activate the $with\big(Student[Calculus1]\big)$ package.

$restart$

$with\big(Student[Calculus1]\big):$

$infolevel\big[Student[Calculus1]\big]:=1:$

$Rule[\ ]\big(Int\big(x^3\cdot e^{x^2},x=-1..0\big)\big)$
Creating problem #1

$$\int_{-1}^{0} x^3 \, e^{x^2} \, dx = \int_{-1}^{0} x^3 \, e^{x^2} \, dx \qquad (7.30)$$

First observe that the integrand can be rewritten as $x^3 \cdot e^{x^2} = x^2 \cdot e^{x^2} \cdot x$. We will make a change of variable $w = x^2$ so that $dw = 2x \, dx$ or $\frac{1}{2} dw = x \, dx$. We will use the *change* command here. Notice that the substitution routine is given below the command. The integral is on the left side and the right side is the one where we have made the substitution.

$Rule\left[change, w = x^2 \right]\left(\left(\textbf{7.30}\right)\right)$

```
Applying substitution  x  =  w^(1/2),  w  =  x^2  with  dx  =
1/2/w^(1/2)*dw,  dw  =  2*x*dx
```

$$\int_{-1}^{0} x^3 \, e^{x^2} \, dx = \int_{1}^{0} \frac{1}{2} e^{w} \, w \, dw \qquad (7.31)$$

Notice that the limits of integration change automatically. Let us now flip the limits of integration by using the *flip* command here.

$Rule\left[flip \right]\left(\left(\textbf{7.31}\right)\right)$

$$\int_{-1}^{0} x^3 \, e^{x^2} \, dx = -\left(\int_{0}^{1} \frac{1}{2} e^{w} \, w \, dw \right) \qquad (7.32)$$

We will factor $\frac{1}{2}$ outside the integral by using the *constantmultiple* command here.

$Rule\left[constantmultiple \right]\left(\left(\textbf{7.32}\right)\right)$

$$\int_{-1}^{0} x^3 \, e^{x^2} \, dx = -\frac{1}{2} \int_{0}^{1} e^{w} \, w \, dw \qquad (7.33)$$

Now we will use the integration by parts method with $u = w$ and $dv = e^{w} dw$ so that $du = dw$ and $v = e^{w}$. As was the case with the previous example we need to specify both u and v (and not u and dv). Use the Common Symbols Palette when typing e.

$Rule\left[parts, w, e^{w} \right]\left(\left(\textbf{7.33}\right)\right)$

$$\int_{-1}^{0} x^3 \, e^{x^2} \, dx = -\frac{1}{2} e + \frac{1}{2} \int_{0}^{1} e^{w} \, dw \qquad (7.34)$$

Next we will use the *exp* command to integrate $\int e^{w} \, dx$.

$Rule\left[\exp \right]\left(\left(\textbf{7.34}\right)\right)$

$$\int_{-1}^{0} x^3 \, e^{x^2} \, dx = -\frac{1}{2} \tag{7.35}$$

Hence $\int_{-1}^{0} x^3 e^{x^2} \, dx = -\frac{1}{2}$. Finally we can show the step-by-step computation of this integral using the *ShowSteps* command.

ShowSteps( )

$$\int_{-1}^{0} x^3 \, e^{x^2} \, dx$$

$$= \int_{1}^{0} \frac{1}{2} e^{w} \, w \, dw \qquad \left[change, w = x^2 \right]$$

$$= -\left(\int_{0}^{1} \frac{1}{2} e^{w} \, w \, dw \right) \qquad \left[flip \right]$$

$$\tag{7.36}$$

$$= -\frac{1}{2} \int_{0}^{1} e^{w} \, w \, dw \qquad \left[constant\,multiple \right]$$

$$= -\frac{1}{2} e + \frac{1}{2} \int_{0}^{1} e^{w} \, dw \qquad \left[parts, w, e^{w} \right]$$

$$= -\frac{1}{2} \qquad \left[exp \right]$$

Trigonometric Integrals

<u>Rule for evaluating</u> $\int \sin^m (x) \cos^n (x) \, dx$:

(a)   If n is an odd integer ($n = 2k + 1$) let $w = \sin(x)$ so that $dw = \cos(x) dx$.

$$\int \sin^m (x) \cos^n (x) \, dx \; = \; \int \sin^m x \cos^{2k+1} x \, dx \; = \; \int \sin^m (x) \cos^{2k} (x) \cos(x) dx$$

$$= \; \int \sin^m (x) \left(1 - \sin^2 (x) \right)^{k} \cos(x) dx \; = \; \int w^m \left(1 - w^2 \right)^{k} dw$$

(b)   If m is an odd integer ($m = 2k + 1$) let $w = \cos(x)$ so that $dw = -\sin(x) dx$.

$$\int \sin^m (x) \cos^n (x) \, dx \; = \; \int \sin^{2k+1} (x) \cos^n (x) \, dx \; = \; \int \sin^{2k} (x) \cos^n (x) \sin(x) dx$$

$$= \; \int \cos^n (x) \left(1 - \cos^2 (x) \right)^{k} \sin(x) dx \; = \; -\int w^n \left(1 - w^2 \right)^{k} dw$$

(c)   If both m and n even integers then use the half-angle identities

$$\sin^2 (x) = \frac{1}{2} \left(1 - \cos(2x) \right) \text{ and } \cos^2 (x) = \frac{1}{2} \left(1 + \cos(2x) \right).$$

<u>Rule for evaluating</u> $\int \tan^m (x) \sec^n (x) \, dx$:

(a)   If n is an even integer ($n = 2k$)let $w = \tan(x)$ so that $dw = \sec^2 (x) dx$.

$$\int \tan^m(x)\sec^n(x)\,dx \;=\; \int \tan^m(x)\sec^{2k}(x)\,dx \;=\; \int \tan^m(x)\sec^{2k-2}(x)\sec^2(x)\,dx$$

$$= \int \tan^m(x)\left(1+\tan^2(x)\right)^{k-1}\sec^2(x)\,dx$$

$$= \int w^m\left(1+w^2\right)^{k-1}\,dw$$

(b)   If m is an odd integer ($m = 2k+1$) let $w = \sec(x)$ so that $dw = \sec(x)\tan(x)\,dx$.

$$\int \tan^m(x)\sec^n(x)\,dx \;=\; \int \tan^{2k+1}(x)\sec^n(x)\,dx$$

$$= \int \tan^{2k}(x)\sec^{n-1}(x)\tan(x)\sec(x)\,dx$$

$$= \int \left(\sec^2(x)-1\right)^k \sec^{n-1}(x)\tan(x)\sec(x)\,dx$$

$$= \int \left(w^2-1\right)^k w^{n-1}\,dw$$

<u>Rule for evaluating</u> $\int \sin(mx)\cos(nx)\,dx$:

$$\int \sin(mx)\cos(nx)\,dx = \frac{1}{2}\int\left[\sin(mx-nx)+\sin(mx+nx)\right]dx$$

<u>Rule for evaluating</u> $\int \sin(mx)\sin(nx)\,dx$:

$$\int \sin(mx)\sin(nx)\,dx = \frac{1}{2}\int\left[\cos(mx-nx)-\cos(mx+nx)\right]dx$$

<u>Rule for evaluating</u> $\int \cos(mx)\cos(nx)\,dx$:

$$\int \cos(mx)\cos(nx)\,dx = \frac{1}{2}\int\left[\cos(mx-nx)+\cos(mx+nx)\right]dx$$

In the following example the rule for $\int \sin(mx)\cos(nx)\,dx$ will be used. This gives rise to two different integrals each requiring a different substitution. We will evaluate one integral at a time because *Maple* cannot make two substitutions in one step.

Example 7: Use the step-by-step procedure in *Maple* to integrate $\int \sin(11x)\cos(25x)\,dx$.

Solution: We will use *Maple* commands to solve this problem step-by-step. Do not forget to activate the $with\left(Student\left[Calculus1\right]\right)$ package.

restart
$with\left(Student\left[Calculus1\right]\right):$
$infolevel\left[Student\left[Calculus1\right]\right]:=1:$
$Rule\left[\;\right]\left(Int\left(\sin(11\cdot x)\cdot\cos(25\cdot x),x\right)\right)$

Creating problem #1

$$\int \sin(11\,x)\cos(25\,x)\,dx = \int \sin(11\,x)\cos(25\,x)\,dx \qquad \textbf{(7.37)}$$

First observe that the integrand can be rewritten as $\sin(11x)\cos(25x) =$ $\frac{1}{2}\sin(36x)+\frac{1}{2}\sin(-14x) = \frac{1}{2}\sin(36x)-\frac{1}{2}\sin(14x)$ (as $\sin(-14x)=-\sin(14x)$).

$$Rule\left[rewrite, \sin(11\cdot x)\cos(25\cdot x)=\frac{1}{2}\cdot\sin(36\cdot x)-\frac{1}{2}\cdot\sin(14\cdot x)\right]((\textbf{7.37}))$$

$$\int \sin(11\,x)\cos(25\,x)\,dx = \int\left(\frac{1}{2}\sin(36\,x)-\frac{1}{2}\sin(14\,x)\right)dx \qquad \textbf{(7.38)}$$

Now we will write the right hand side as a difference of two integrals using the *difference* command.

$$Rule[difference]((\textbf{7.38}))$$

$$\int \sin(11\,x)\cos(25\,x)\,dx = \int\frac{1}{2}\sin(36\,x)\,dx - \left(\int\frac{1}{2}\sin(14\,x)\,dx\right) \qquad \textbf{(7.39)}$$

We will now work with the two integrals separately as there is a different substitution command for each of the two integrals. We will use the *ShowIncomplete* command to do so. The labels are of the form % + *Int* (since we are dealing with integrals) + integer (in this case 3 and 4).

$$ShowIncomplete((\textbf{7.39}))$$

$$\%Int3 = \int\frac{1}{2}\sin(36\,x)\,dx$$

$$\%Int4 = \int\frac{1}{2}\sin(14\,x)\,dx \qquad \textbf{(7.40)}$$

We will first work with the integral *%Int3*. The first operation that we will perform is to factor the constant $\frac{1}{2}$ from the integral. We will use the *constantmultiple* command here.

$$Rule[constantmultiple](\%Int3)$$

$$\int\frac{1}{2}\sin(36\,x)\,dx = \frac{1}{2}\int\sin(36\,x)\,dx \qquad \textbf{(7.41)}$$

Use the *change* command to make the substitution $w1 = 36x$ so that $dw1 = 36\,dx$ or $\frac{1}{36}dw1 = dx$.

$Rule[change, w1 = 36 \cdot x]((\textbf{7.41}))$

```
Applying   substitution   x   =   1/36*w,   w1   =   36*x   with   dx   =
1/36*dw1, dw1 = 36*dx
```

$$\int \frac{1}{2}\sin(36\,x)\,dx = \frac{1}{2}\int \frac{1}{36}\sin(w1)\,dw1 \qquad (\textbf{7.42})$$

The first operation that we will perform is to factor the constant $\frac{1}{36}$ from the integral. We will use the *constantmultiple* command here.

$Rule[constantmultiple]((\textbf{7.42}))$

$$\int \frac{1}{2}\sin(36\,x)\,dx = \frac{1}{72}\int \sin(w1)\,dw1 \qquad (\textbf{7.43})$$

The sine function will be integrated using the *sin* command.

$Rule[\sin]((\textbf{7.43}))$

$$\int \frac{1}{2}\sin(36\,x)\,dx = \frac{1}{72}\cos(w1) \qquad (\textbf{7.44})$$

The *revert* command get this integral in terms of x.

$Rule[revert]((\textbf{7.44}))$

```
Reverting substitution using w1 = 36*x
```

$$\int \sin(11\,x)\cos(25\,x)\,dx = -\frac{1}{72}\cos(36\,x) - \left(\int \frac{1}{2}\sin(14\,x)\,dx \right) \qquad (\textbf{7.45})$$

Notice that once we finished evaluating *%Int3* the value of this integral was automatically replaced in the equation labeled (6.42). We will now work with the integral *%Int4*. We will follow the same procedure as we did for *%Int3*. The change of variable here is $w2 = 14x$ so that $dw2 = 14\,dx$ or $\frac{1}{14}dw2 = dx$. We will use the *constantmultiple*, *change*, *sin*, *constantmultiple*, and *revert* commands, in that order, to evaluate this integral.

$Rule[constantmultiple](\%Int4)$

$$\int \frac{1}{2}\sin(14\,x)\,dx = \frac{1}{2}\int \sin(14\,x)\,dx \qquad (\textbf{7.46})$$

$Rule[change, w2 = 14 \cdot x]((\mathbf{7.46}))$

```
Applying  substitution  x  =  1/14*w,  w2  =  14*x  with  dx  =
1/14*dw2,  dw2  =  14*dx
```

$$\int \frac{1}{2} \sin(14\,x)\,dx = \frac{1}{2} \int \frac{1}{14} \sin(w2)\,dw2 \qquad\qquad (\mathbf{7.47})$$

$Rule[constantmultiple]((\mathbf{7.47}))$

$$\int \frac{1}{2} \sin(14\,x)\,dx = \frac{1}{28} \int \sin(w2)\,dw2 \qquad\qquad (\mathbf{7.48})$$

$Rule[\sin]((\mathbf{7.48}))$

$$\int \frac{1}{2} \sin(14\,x)\,dx = -\frac{1}{28} \cos(w2) \qquad\qquad (\mathbf{7.49})$$

$Rule[revert]((\mathbf{7.49}))$

```
Reverting substitution using w2 = 14*x
```

$$\int \sin(11\,x)\cos(25\,x)\,dx = -\frac{1}{72}\cos(36\,x) + \frac{1}{28}\cos(14\,x) \quad (\mathbf{7.50})$$

Hence $\int \sin(11\,x)\cos(25\,x)\,dx = -\frac{1}{72}\cos(36\,x) + \frac{1}{28}\cos(14\,x)$.

Example 8: Use *Maple* to integrate $\int \sin^3(x)\cos^{26}(x)\,dx$. Show your work step-by-step. Are the answers you obtained the same? Give reasons for your answer.

Solution:
Using the *Expression* palette
We will integrate $\int \sin^3(x)\cos^{26}(x)\,dx$ by clicking on $\int f\,dx$ in the *Expression* palette.

restart
$\int \sin(x)^3 \cdot \cos(x)^{26}\,dx$

$$-\frac{1}{29}\sin(x)^2\cos(x)^{27} - \frac{2}{783}\cos(x)^{27} \qquad\qquad (\mathbf{7.51})$$

Hence, using *Maple*, $\int \sin(x)^3 \cos(x)^{26}\,dx = -\frac{1}{29}\sin(x)^2\cos(x)^{27} - \frac{2}{783}\cos(x)^{27}$.

Using a step-by-step approach
We will use *Maple* commands to solve this problem step-by-step. Do not forget to activate the $with\left(Student\left[Calculus1\right]\right)$ package.

restart
$with\left(Student\left[Calculus1\right]\right):$

$infolevel\left[\,Student\left[\,Calculus1\right]\right]:=1:$

$Rule[\ ]\left(Int\left(\sin(x)^{3}\cdot\cos(x)^{26},x\right)\right)$

```
Creating problem #1
```

$$\int\sin(x)^{3}\cos(x)^{26}\,dx=\int\sin(x)^{3}\cos(x)^{26}\,dx \tag{7.52}$$

We will make a change of variable $w=\cos(x)$ so that $dw=-\sin(x)\,dx$ or $-dw=\sin(x)\,dx$. We will use the *change* command here.

$Rule\left[change,w=\cos(x)\right]\left((\mathbf{7.52})\right)$

```
Applying substitution x = arcos(w), w = cos(x) with dx = -
1/(1-w^2)^(1/2)*dw, dw = -sin(x)*dx
```

$$\int\sin(x)^{3}\cos(x)^{26}\,dx=\int\left(-w^{26}+w^{28}\right)dx \tag{7.53}$$

Now we will write the right hand side as a sum of two integrals using the *sum* command.

$Rule[sum]\left((\mathbf{7.53})\right)$

$$\int\sin(x)^{3}\cos(x)^{26}\,dx=\int\left(-w^{26}\right)dx+\int w^{28}\,dx \tag{7.54}$$

We will factor the negative sign from the first integral.

$Rule[constantmultiple]\left((\mathbf{7.54})\right)$

$$\int\sin(x)^{3}\cos(x)^{26}\,dx=-\left(\int w^{26}\,dx\right)+\int w^{28}\,dx \tag{7.55}$$

Let us now use the power rule twice, to evaluate the two integrals.

$Rule[power]\left((\mathbf{7.55})\right)$

$$\int\sin(x)^{3}\cos(x)^{26}\,dx=-\frac{1}{27}w^{27}+\int w^{28}\,dx \tag{7.56}$$

$Rule[power]\left((\mathbf{7.56})\right)$

$$\int\sin(x)^{3}\cos(x)^{26}\,dx=-\frac{1}{27}w^{27}+\frac{1}{29}w^{29} \tag{7.57}$$

The *revert* command get this integral in terms of x.

$Rule[revert]\left((\mathbf{7.57})\right)$

```
Reverting substitution using w = cos(x)
```

$$\int \sin(x)^3 \cos(x)^{26}\, dx = -\frac{1}{27}\cos(x)^{27} + \frac{1}{29}\cos(x)^{29} \qquad \textbf{(7.58)}$$

Hence $\int \sin(x)^3 \cos(x)^{26}\, dx = -\frac{1}{27}\cos(x)^{27} + \frac{1}{29}\cos(x)^{29}$, using a step-by-step approach. The reader can check that the answers in (7.51) and (7.58) are the same.

Example 9: Use the step-by-step procedure in *Maple* to integrate $\int \tan^4(x)\, dx$.

Solution: We will use *Maple* commands to solve this problem step-by-step. Do not forget to activate the $with\big(Student[Calculus1]\big)$ package.

restart
$with\big(Student[Calculus1]\big):$
$infolevel\big[Student[Calculus1]\big]:=1:$
$Rule[\ ]\Big(Int\big(\tan(x)^4,x\big)\Big)$
Creating problem #1

$$\int \tan(x)^4\, dx = \int \tan(x)^4\, dx \qquad \textbf{(7.59)}$$

Since $\tan^2(x) = \sec^2(x) - 1$, the integrand can be rewritten as $\tan^4(x) = \tan^2(x)\sec^2(x) - \tan^2(x)$.

$Rule\Big[rewrite, \tan(x)^4 = \tan(x)^2 \cdot \sec(x)^2 - \tan(x)^2\Big]\big((7.59)\big)$

$$\int \tan(x)^4\, dx = \int \Big(\tan(x)^2 \sec(x)^2 - \tan(x)^2\Big)\, dx \qquad \textbf{(7.60)}$$

Now we will write the right hand side as a difference of two integrals using the difference command.

$Rule[difference]\big((7.60)\big)$

$$\int \tan(x)^4\, dx = \int \tan(x)^2 \sec(x)^2\, dx - \left(\int \tan(x)^2\, dx\right) \qquad \textbf{(7.61)}$$

We will work with the two integrals separately as there is a different substitution command for each of the two integrals. We will use the *ShowIncomplete* command to do so. The labels are of the form % + *Int* (since we are dealing with integrals) + integer (in this case 3 and 4)

$ShowIncomplete\big((7.61)\big)$

$$\%Int3 = \int \tan(x)^2 \sec(x)^2 \, dx$$

$$\%Int4 = \int \tan(x)^2 \, dx \tag{7.62}$$

We will first work with the integral *%Int3*. We will make the change of variable $w1 = \tan(x)$ so that $dw1 = \sec^2(x)\,dx$. We will use the *change* command here.

$Rule\big[change, w1 = \tan(x)\big](\%Int3)$

```
Applying substitution  x  =  arctan(w1),  w1  =  tan(x)  with  dx  =
1/(w1^2+1)*dw1,  dw1  =  (1+tan(x)^2)*dx
```

$$\int \tan(x)^2 \sec(x)^2 \, dx = \int w1^2 \, dw1 \tag{7.63}$$

We will now use the power rule using the *power* command here.

$Rule[power]((\mathbf{7.63}))$

$$\int \tan(x)^2 \sec(x)^2 \, dx = \frac{1}{3} w1^3 \tag{7.64}$$

The *revert* command get this integral in terms of *x*.

$Rule[revert]((\mathbf{7.64}))$

```
Reverting  substitution  using  w1  =  tan(x)
```

$$\int \left(\tan(x)^2 \sec(x)^2 - \tan(x)^2 \right) dx = \frac{1}{3} \tan(x)^3 - \left(\int \tan(x)^2 \, dx \right) \tag{7.65}$$

As in Example 7, the value of *%Int3* was automatically replaced in the equation labeled (7.61). We will work with the integral *%Int4*. The integrand can be rewritten as $\tan^2(x) = \sec^2(x) - 1$.

$Rule\big[rewrite, \tan(x)^2 = \sec(x)^2 - 1\big](\%Int4)$

$$\int \tan(x)^2 \, dx = \int \left(\sec(x)^2 - 1 \right) dx \tag{7.66}$$

Now we will write the right hand side as a difference of two integrals using the difference command and work with the two integrals separately.

$Rule[difference]((\mathbf{7.66}))$

$$\int \tan(x)^2 \, dx = \int \sec(x)^2 \, dx + \int (-1) \, dx \tag{7.67}$$

$ShowIncomplete((\mathbf{7.67}))$

$$\% Int7 = \int \sec(x)^2 \, dx$$

$$\% Int8 = \int (-1) \, dx \qquad \textbf{(7.68)}$$

We will use the *change* command to make the substitution $w2 = \tan(x)$ $dw2 = \sec^2(x) \, dx$.

$Rule\big[change, w2 = \tan(x)\big](\% Int7)$

```
Applying substitution x = arctan(w2), w2 = tan(x) with dx =
1/(w2^2+1)*dw2, dw2 = (1+tan(x)^2)*dx
```

$$\int \sec(x)^2 \, dx = \int 1 \, dw2 \qquad \textbf{(7.69)}$$

We will now integrate the integrand 1 using the *constant* command here.

$Rule[constant]((\textbf{7.69}))$

$$\int \sec(x)^2 \, dx = w2 \qquad \textbf{(7.70)}$$

Now let us get this integral in terms of x. We will use the *revert* command here.

$Rule[revert]((\textbf{7.70}))$

```
Reverting substitution using w2 = tan(x)
```

$$\int \big(\sec(x)^2 - 1\big) \, dx = \tan(x) + \int (-1) \, dx \qquad \textbf{(7.71)}$$

We will now integrate the integrand (-1) using the *constant* command here.

$Rule[constant]((\textbf{7.71}))$

$$\int \tan(x)^4 \, dx = \frac{1}{3} \tan(x)^3 - \tan(x) + x \qquad \textbf{(7.72)}$$

Hence $\int \tan(x)^4 \, dx = \dfrac{1}{3} \tan(x)^3 - \tan(x) + x$.

Trigonometric Substitution

Let a be a positive real number. If the integrand contains expressions of the form $\sqrt{a^2 - x^2}$, $\sqrt{a^2 + x^2}$, or $\sqrt{x^2 - a^2}$ and a regular substitution strategy cannot be used we use trigonometric substitutions.

The integrand contains the expression	Trigonometric substitution	Trigonometric identity used
$\sqrt{a^2 - x^2}$	$x = a\sin(\theta),\ -\dfrac{\pi}{2} \le \theta \le \dfrac{\pi}{2}$	$1 - \sin^2(\theta) = \cos^2(\theta)$
$\sqrt{a^2 + x^2}$	$x = a\tan(\theta),\ -\dfrac{\pi}{2} < \theta < \dfrac{\pi}{2}$	$1 + \tan^2(\theta) = \sec^2(\theta)$
$\sqrt{x^2 - a^2}$	$x = a\sec(\theta),\ 0 \le \theta < \dfrac{\pi}{2}$ or $\pi \le \theta < \dfrac{3\pi}{2}$	$\sec^2(\theta) - 1 = \tan^2(\theta)$

Example 10: Use the step-by-step procedure in *Maple* to integrate $\displaystyle\int_{\frac{\sqrt{2}}{2}}^{1} \frac{1}{\sqrt{2 - x^2}}\,dx$.

Solution: We will use *Maple* commands to solve this problem step-by-step. Do not forget to activate the $with\big(Student[Calculus1]\big)$ package.

restart
$with\big(Student[Calculus1]\big):$

$infolevel\big[Student[Calculus1]\big] := 1:$

$Rule[\ ]\left(Int\left(\dfrac{1}{\sqrt{2 - x^2}}, x = \dfrac{\sqrt{2}}{2} .. 1 \right) \right)$

```
Creating problem #1
```

$$\int_{\frac{1}{2}\sqrt{2}}^{1} \frac{1}{\sqrt{2 - x^2}}\,dx = \int_{\frac{1}{2}\sqrt{2}}^{1} \frac{1}{\sqrt{2 - x^2}}\,dx \tag{7.73}$$

Since the denominator is of the form $\sqrt{2 - x^2} = \sqrt{\left(\sqrt{2}\right)^2 - x^2}$, we will make a trigonometric substitution $x = \sqrt{2}\sin(w)$. Hence $\sqrt{2 - x^2} = \sqrt{2\left(1 - \sin^2(w)\right)} = \sqrt{2}\cos(w)$ and $dx = \sqrt{2}\cos(w)\,dw$. We will use the *change* command here.

$Rule\left[change, x = \sqrt{2}\cdot\sin(w) \right]\big((\mathbf{7.73})\big)$

```
Applying    substitution    x   =   2^(1/2)*sin(w),    w   =
arcsin(1/2*x*2^(1/2))   with   dx   =   2^(1/2)*cos(w)*dw,   dw   =
2^(1/2)/(4-2*x^2)^(1/2)*dx
```

$$\int_{\frac{1}{2}\sqrt{2}}^{1} \frac{1}{\sqrt{2 - x^2}}\,dx = \int_{\frac{1}{6}\pi}^{\frac{1}{4}\pi} 1\,dw \tag{7.74}$$

Now we will the constant integrand using the command *constant*.

$Rule[constant]((7.74))$

$$\int_{\frac{1}{2}\sqrt{2}}^{1} \frac{1}{\sqrt{2-x^2}} \, dx = \frac{1}{12}\pi \tag{7.75}$$

Hence $\int_{\frac{1}{2}\sqrt{2}}^{1} \frac{1}{\sqrt{2-x^2}} \, dx = \frac{1}{12}\pi$.

Example 11: Use the step-by-step procedure in *Maple* to integrate $\int \sqrt{x^2+2x+2} \, dx$.

Solution: We will use *Maple* commands to solve this problem step-by-step. Do not forget to activate the $with\left(Student[Calculus1]\right)$ package.

$restart$
$with\left(Student[Calculus1]\right):$
$infolevel\left[Student[Calculus1]\right]:=1:$
$Rule[\ ]\left(Int\left(\sqrt{x^2+2x+2}, x\right)\right)$
```
Creating problem #1
```

$$\int \sqrt{x^2+2x+2} \, dx = \int \sqrt{x^2+2x+2} \, dx \tag{7.76}$$

Since $x^2+2x+2 = (x+1)^2+1$, we will use the *change* command to make the substitution $w1 = x+1$ so that $dw1 = dx$.

$Rule[change, w1 = x+1]((7.76))$
```
Applying substitution x = w1-1, with dx = dw1
```
$$\int \sqrt{x^2+2x+2} \, dx = \int \sqrt{w1^2+1} \, dw1 \tag{7.77}$$

The *change* command will be used to make the trigonometric substitution $w1 = \tan(w2)$ so that $dw1 = \sec^2(w2) \, dw2$.

$Rule\left[change, w1 = \tan(w2)\right]((7.77))$
```
Applying   substitution   w1  =  tan(w2),   w2  =  arctan(w1)  with
dw1  =  (1+tan(w2)^1)*dw2,  dw2  =  1/(w1^2+1)*dw1
```
$$\int \sqrt{x^2+2x+2} \, dx = \int \sec(w2)^3 \, dw2 \tag{7.78}$$

Since the integrand can be rewritten as $\sec^3(w2) = \sec(w2)\sec^2(w2)$, we will use the integration by parts method with $u = \sec(w2)$ and $dv = \sec(w2)^2\,dw2$ so that $du = \sec(w2)\tan(w2)\,dw2$ and $v = \tan(w2)$. As was the case with the previous example we need to specify both u and v (and not u and dv).

$Rule\left[\,parts,\sec(w2),\tan(w2)\,\right]((\mathbf{7.78}))$

$$\int\sqrt{x^2+2x+2}\;dx = \sec(w2)\tan(w2) - \left(\int\sec(w2)\tan(w2)^2\,dw2\right) \qquad (\mathbf{7.79})$$

Since $\tan^2(w2) = \sec^2(w2) - 1$, we will use the *rewrite* command.

$Rule\left[\,rewrite,\sec(w2)\cdot\left(\tan(w2)^2 - 1\right) = \sec^3(w2) - \sec(w2)\,\right]((\mathbf{7.79}))$

$$\int\sqrt{x^2+2x+2}\;dx = \sec(w2)\tan(w2) - \left(\int\left(\sec(w2)^3 - \sec(w2)\right)dw2\right) \qquad (\mathbf{7.80})$$

Now we will write the integral on the right hand side as a difference of two integrals using the difference command.

$Rule\left[\,difference\,\right]((\mathbf{7.80}))$

$$\int\sqrt{x^2+2x+2}\;dx = \sec(w2)\tan(w2) - \left(\int\sec(w2)^3\,dw2\right) + \int\sec(w2)\,dw2 \quad (\mathbf{7.81})$$

From equation (7.78) observe that $\int\sqrt{x^2+2x+2}\;dx = \int\sec(w2)^3\,dw2$. Hence we will now use the solve command that adds $\int\sec(w2)^3\,dw2$ and solves the integral $\int\sqrt{x^2+2x+2}\;dx$ (or $\int\sec(w2)^3\,dw2$).

$Rule\left[\,solve\,\right]((\mathbf{7.81}))$

$$\int\sqrt{x^2+2x+2}\;dx = \frac{1}{2}\sec(w2)\tan(w2) + \frac{1}{2}\int\sec(w2)\,dw2 \qquad (\mathbf{7.82})$$

We will now integrate $\sec(w2)$ by using the *sec* command.

$Rule\left[\,sec\,\right]((\mathbf{7.82}))$

$$\int\sqrt{x^2+2x+2}\;dx = \frac{1}{2}\sec(w2)\tan(w2) + \frac{1}{2}\ln\left(\sec(w2) + \tan(w2)\right) \qquad (\mathbf{7.83})$$

$Rule\left[\,revert\,\right]((\mathbf{7.83}))$

```
Reverting substitution using w2 = arctan(w1)
```

$$\int \sqrt{x^2+2x+2}\ dx = \frac{1}{2}\ \sqrt{w1^2+1}\ w1 + \frac{1}{2}\ln\left(\sqrt{w1^2+1}+w1\right) \qquad \textbf{(7.84)}$$

Rule[*revert*]((**7.84**))

```
Reverting substitution using w1 = x+1
```

$$\int \sqrt{x^2+2x+2}\ dx = \frac{1}{2}\ \sqrt{x^2+2x+2}\ x + \frac{1}{2}\ \sqrt{x^2+2x+2} + \frac{1}{2}\ln\left(\sqrt{x^2+2x+2}+x+1\right)\ \textbf{(7.85)}$$

Hence $\int \sqrt{x^2+2x+2}\ dx = \frac{1}{2}\ \sqrt{x^2+2x+2}\ x + \frac{1}{2}\ \sqrt{x^2+2x+2} + \frac{1}{2}\ln\left(\sqrt{x^2+2x+2}+x+1\right).$

Remark: If we integrate $\int \sqrt{x^2+2x+2}\ dx$ by clicking on $\int f\ dx$ in the *Expression* palette then we get

$$\int \sqrt{x^2+2x+2}\ dx$$

$$\frac{1}{4}\left(2\ x+2\right)\sqrt{x^2+2\ x+2} + \frac{1}{2}\ \text{arcsin}\ \text{h}\left(x+1\right) \qquad \textbf{(7.86)}$$

This is the same answer as the one obtained above since $\text{arcsin}\ \text{h}\left(x\right) = \ln\left(x+\sqrt{x^2+1}\right)$ so

that $\text{arcsin}\ \text{h}\left(x+1\right) = \ln\left(x+1+\sqrt{\left(x+1\right)^2+1}\right) = \ln\left(x+1+\sqrt{x^2+2x+2}\right).$

Integration of Rational Functions using Partial Fractions

The partial fraction decomposition method:

Integrating using partial fractions is used to integrate rational functions (quotients of polynomials) as a sum of different parts. These sums are familiar expressions that can be integrated. Divide the numerator by the denominator if the degree of the numerator is greater than or equal to the degree of the denominator. Try to factor the denominator and then express the rational function (without the "quotient" after long division) as a sum of partial fractions. The numerator of each partial fraction has a degree that is one less than the degree of the denominator. The number of coefficients in the numerator, to be determined, is the same as to the degree of the polynomial in the denominator. For

example, $\dfrac{K}{\left(x+a\right)\left(x+b\right)} = \dfrac{A}{\left(x+a\right)} + \dfrac{B}{\left(x+b\right)}, \qquad \dfrac{p\left(x\right)}{\left(ax^2+b\right)\left(cx^2+d\right)} = \dfrac{Ax+B}{ax^2+b} + \dfrac{Cx+D}{cx^2+d}.$

However, if the powers of the factors are more than one then the rational expression has to be broken as a sum of the repeated factors such as

$\dfrac{K}{\left(x+a\right)\left(x+b\right)^n} = \dfrac{A}{\left(x+a\right)} + \dfrac{B_1}{\left(x+b\right)} + \dfrac{B_2}{\left(x+b\right)^2} + ... \dfrac{B_n}{\left(x+b\right)^n}.$ Note that there is no point

using partial fraction decompositions if the denominator consists of only one polynomial expression raised to some power.

Remark We will give an example of finding the partial fraction decomposition of the rational expression $\dfrac{3x^2+20x+50}{x^3+10x^2+25x}$. Note that the denominator x^3+10x^2+25x can be factored as $x^3+10x^2+25x = x(x^2+10x+25) = x(x+5)^2$. Hence we can write the given rational expression as $\dfrac{3x^2+20x+50}{x^3+10x^2+25x} = \dfrac{A}{x}+\dfrac{B}{x+5}+\dfrac{C}{(x+5)^2}$. Multiply both sides by the LCD $= x(x+5)^2$ to obtain $3x^2+20x+50 = A(x+5)^2+Bx(x+5)+Cx = A(x+10x+25)+B(x^2+5x)+Cx$. Compare the constant terms to obtain $50=25A$ or $A=2$. Compare the coefficients of x^2 to obtain $3=A+B$ so that $B=3-A=1$. Compare the coefficients of x to obtain $20=10A+5B+C$ so that $C=20-10A-5B=-5$. Hence the partial fraction decomposition for $\dfrac{3x^2+20x+50}{x^3+10x^2+25x}$ is

$$\dfrac{2}{x}+\dfrac{1}{x+5}-\dfrac{5}{(x+5)^2}.$$

Example 12: Use the step-by-step procedure in *Maple* to integrate $\displaystyle\int \dfrac{x^2+17x+15}{4x^3-5x^2+24x-30}\,dx$.

Solution: We will use *Maple* commands to solve this problem step-by-step. Do not forget to activate the $with(Student[Calculus1])$ package.

restart

$with(Student[Calculus1])$:

$infolevel[Student[Calculus1]]:=1$:

$Rule[\ ]\left(Int\left(\dfrac{x^2+17\cdot x+15}{4\cdot x^3-5\cdot x^2+24\cdot x-30},x\right)\right)$

```
Creating problem #1
```

$$\int \dfrac{x^2+17\,x+15}{4\,x^3-5\,x^2+24\,x-30}\,dx = \int \dfrac{x^2+17\,x+15}{4\,x^3-5\,x^2+24\,x-30}\,dx \qquad (7.87)$$

We will first write the integrand as a sum of its partial fractions using the *partialfractions* command. Please note that the order in which the partial fractions appear may not be the same.

$Rule[partialfractions]((7.87))$

$$\int \frac{x^2+17\,x+15}{4\,x^3-5\,x^2+24\,x-30}\,dx = \int\left(\frac{5}{4\,x-5}+\frac{-x+3}{x^2+6}\right)dx \qquad \textbf{(7.88)}$$

Now we will write the right hand side as a sum of two integrals using the *sum* command.

$Rule\left[\,sum\,\right]\left((\textbf{7.88})\right)$

$$\int \frac{x^2+17\,x+15}{4\,x^3-5\,x^2+24\,x-30}\,dx = \int\frac{5}{4\,x-5}\,dx + \int\frac{-x+3}{x^2+6}\,dx \qquad \textbf{(7.89)}$$

Since the order in which the partial fractions appear is not the same, we will now work with the two integrals separately. We will use the *ShowIncomplete* command to do so.

$ShowIncomplete\left((\textbf{7.89})\right)$

$$\%Int3 = \int\frac{5}{4\,x-5}\,dx$$

$$\%Int4 = \int\frac{-x+3}{x^2+6}\,dx \qquad \textbf{(7.90)}$$

Please note that if the order in which the partial fractions appears is reversed, perform the commands for the integral that is denoted by *%Int4*, shown below, first followed by the commands for integrating *%Int3*. We will first work with the integral *%Int3*. The *change* command will be used to make a substitution $w1 = 4x-5$ so that $dw1 = 4\,dx$ or $\frac{1}{4}dw1 = dx$.

$Rule\left[\,change, w1 = 4\cdot x-5\,\right]\left(\%Int3\right)$

```
Applying substitution x = 1/4*w1+5/4, w1 = 4*x-5 with dx =
1/4*dw1, dw1 = 4*dx
```

$$\int\frac{5}{4\,x-5}\,dx = \int\frac{5}{4\,w1}\,dw1 \qquad \textbf{(7.91)}$$

Factor the constant $\frac{5}{4}$ from the integral using the *constantmultiple* command.

$Rule\left[\,constantmultiple\,\right]\left((\textbf{7.91})\right)$

$$\int\frac{5}{4\,x-5}\,dx = \frac{5}{4}\int\frac{1}{w1}\,dw1 \qquad \textbf{(7.92)}$$

Now let us integrate using *power* command here.

$Rule\left[\,power\,\right]\left((\textbf{7.92})\right)$

$$\int \frac{5}{4x-5} \, dx = \frac{5}{4} \ln(w1) \tag{7.93}$$

Use the *revert* command to get this integral in terms of *x*.

$Rule[revert]((7.93))$

```
Reverting substitution using w1 = 4*x-5
```

$$\int \frac{x^2 + 17x + 15}{4x^3 - 5x^2 + 24x - 30} \, dx = \frac{5}{4} \ln(4x-5) + \int \frac{-x+3}{x^2+6} \, dx \tag{7.94}$$

Notice that once we finished evaluating *%Int3* the value of this integral was automatically replaced in the equation labeled (7.89). We will now work with the integral *%Int4*. We will first rewrite $\dfrac{-x+3}{x^2+6}$ as $-\dfrac{x}{x^2+6} + \dfrac{3}{x^2+6}$ using the *rewrite* command.

$$Rule\left[rewrite, \frac{-x+3}{x^2+6} = -\frac{x}{x^2+6} + \frac{3}{x^2+6}\right](\%Int4)$$

$$\int \frac{-x+3}{x^2+6} \, dx = \int \left(-\frac{x}{x^2+6} + \frac{3}{x^2+6}\right) dx \tag{7.95}$$

Express the integral as the sum of two integrals using the *sum* command.

$Rule[sum]((7.95))$

$$\int \frac{-x+3}{x^2+6} \, dx = \int \left(-\frac{x}{x^2+6}\right) dx + \int \frac{3}{x^2+6} \, dx \tag{7.96}$$

We will now work with the two integrals separately. We will use the *ShowIncomplete* command to do so.

$ShowIncomplete((7.96))$

$$\%Int8 = \int \left(-\frac{x}{x^2+6}\right) dx$$

$$\%Int9 = \int \frac{3}{x^2+6} \, dx \tag{7.97}$$

We will use a substitution $w2 = x^2 + 6$ so that $dw2 = 2x \, dx$ or $\dfrac{1}{2} dw2 = x \, dx$ for the integral *%Int9*.

$$Rule\left[change, w2 = x^2 + 6\right](\%Int8)$$

```
Applying substitution x = (w2-6)^(1/2), w2 = x^2+6 with dx
= 1/2/(w2-6)^(1/2)*dw2, dw2 = 2*x*dx
```

$$\int \left(-\frac{x}{x^2+6} \right) dx = \int \left(-\frac{1}{2\,w2} \right) dw2 \tag{7.98}$$

Now factor $-\dfrac{1}{2}$ outside the integrand.

Rule$\left[constantmultiple \right]\left((\mathbf{7.98}) \right)$

$$\int \left(-\frac{x}{x^2+6} \right) dx = -\frac{1}{2} \int \frac{1}{w2}\, dw2 \tag{7.99}$$

The *power* command will be used to integrate $\displaystyle\int \frac{1}{w2}\, dw2$

Rule$\left[power \right]\left((\mathbf{7.99}) \right)$

$$\int \left(-\frac{x}{x^2+6} \right) dx = -\frac{1}{2} \ln \left(w2 \right) \tag{7.100}$$

Use the *revert* command to get this integral in terms of *x*.

Rule$\left[revert \right]\left((\mathbf{7.100}) \right)$

```
Reverting substitution using w2 = x^2+6
```

$$\int \frac{x^2+17\,x+15}{4\,x^3-5\,x^2+24\,x-30}\, dx = -\frac{1}{2} \ln \left(x^2+6 \right) + \int \frac{3}{x^2+6}\, dx \tag{7.101}$$

Use the *change* command to make the trigonometric substitution $x = \sqrt{6} \tan \left(w3 \right)$ so that $dx = \sqrt{6} \sec^2 \left(w3 \right) dw3$ in %Int9.

Rule$\left[change, x = \sqrt{6} \cdot \tan \left(w3 \right) \right]\left(\%Int9 \right)$

```
Applying    substitution    x   =   6^(1/2)*tan(w3),    w3   =
arctan(1/6*x*6^(1/2))  with  dx  =  6^(1/2)*(1+tan(w3)^2)*dw3,
dw3 = 1/6*6^(1/2)/(1/6*x^2+1)*dx
```

$$\int \frac{3}{x^2+6}\, dx = \int \frac{1}{2} \sqrt{6}\, dw3 \tag{7.102}$$

Integrate using the *constant* command.

Rule$\left[constant \right]\left((\mathbf{7.102}) \right)$

$$\int \frac{3}{x^2+6}\, dx = \frac{1}{2} \sqrt{6}\, w3 \tag{7.103}$$

Change the variable back to x.

Rule[*revert*]((**7.103**))

```
Reverting substitution using w3 = arctan(1/6*x*6^(1/2))
```

$$\int \frac{x^2 + 17\,x + 15}{4\,x^3 - 5\,x^2 + 24\,x - 30}\,dx = \frac{5}{4}\ln(4\,x - 5) - \frac{1}{2}\ln(x^2 + 6) + \frac{1}{2}\sqrt{6}\,\arctan\left(\frac{1}{6}\,x\,\sqrt{6}\right) \quad (7.104)$$

Hence $\int \dfrac{x^2 + 17\,x + 15}{4\,x^3 - 5\,x^2 + 24\,x - 30}\,dx = \dfrac{5}{4}\ln(4\,x - 5) - \dfrac{1}{2}\ln(x^2 + 6) + \dfrac{1}{2}\sqrt{6}\,\arctan\left(\dfrac{1}{6}\,x\,\sqrt{6}\right).$

Remarks: We will show a different method to solve the integral *%Int4*. Use the *change* command to make the trigonometric substitution $x = \sqrt{6}\tan(w2)$ so that $dx = \sqrt{6}\sec^2(w2)\,dw2$.

Rule$\left[\,change, x = \sqrt{6}\tan(w4)\,\right]$((**7.90**))

```
Applying    substitution    x    =    6^(1/2)*tan(w2),    w3    =
arctan(1/6*x*6^(1/2))  with  dx  =  6^(1/2)*(1+tan(w2)^2)*dw2,
dw3 = 1/6*6^(1/2)/(1/6*x^2+1)*dx
```

$$\int \frac{-x + 3}{x^2 + 6}\,dx = \int \left(-\frac{1}{6}\sqrt{6}\left(\sqrt{6}\tan(w4) - 3\right)\right)dw4 \quad (7.105)$$

We will first simplify the integrand above.

simplify((**7.105**))

$$\int \frac{-x + 3}{x^2 + 6}\,dx = \frac{1}{2}\int \left(-2\tan(w4) + \sqrt{6}\right)dw4 \quad (7.106)$$

We will use the *sum* command and then proceed to integrate the above integral.

Rule[*sum*]((**7.106**))

$$\int \frac{-x + 3}{x^2 + 6}\,dx = \frac{1}{2}\int \left(-2\tan(w4)\right)dw4 + \frac{1}{2}\int \sqrt{6}\,dw4 \quad (7.107)$$

Rule[*constantmultiple*]((**7.107**))

$$\int \frac{-x + 3}{x^2 + 6}\,dx = -\left(\int \tan(w4)\,dw4\right) + \frac{1}{2}\int \sqrt{6}\,dw4 \quad (7.108)$$

Rule[*tan*]((**7.108**))

$$\int \frac{-x + 3}{x^2 + 6}\,dx = \ln\left(\cos(w4)\right) + \frac{1}{2}\int \sqrt{6}\,dw4 \quad (7.109)$$

Rule[*constant*]((**7.109**))

$$\int \frac{-x+3}{x^2+6}\,dx = \frac{1}{2}\ln\left(\cos\left(w4\right)\right)+\frac{1}{2}\sqrt{6}\ w4 \tag{7.110}$$

Change the variable back to x.

Rule$[\textit{revert}]((\textbf{7.110}))$

```
Reverting substitution using w3 = arctan(1/6*x*6^(1/2))
```

$$\int \frac{-x+3}{x^2+6}\,dx = \ln\left(\frac{6}{\sqrt{6x^2+36}}\right)+\frac{1}{2}\sqrt{6}\ \arctan\left(\frac{1}{6}\,x\,\sqrt{6}\right) \tag{7.111}$$

Note that this answer is the same as that for *%Int4* in 7.104 (why?).

Chapter 7 Integration Techniques

Exercises

1. Use *Maple* to find $g'(x)$ and $g''(x)$ for each of the following and verify that these answers follow from the Fundamental Theorem of Calculus.

   (a) $\quad g(x) = \int_2^{x^3} \sin(t^4) dt$

   (b) $\quad g(x) = \dfrac{1}{x} \int_1^{x^2} \dfrac{1 + t^{\sin(t)}}{t} dt$

   (c) $\quad g(x) = \int_{x-1}^{x+1} \sqrt{\sin(t)} \, dt$

   (d) $\quad g(x) = \int_{1+x+x^2}^{\sin(x)} \dfrac{t^{\frac{5}{2}}}{\sqrt{t^3 + 2009}} dt$

2. Sketch graph of the function $g(x)$. Use *Maple* to find $g'(x)$ and $g''(x)$ for each of the following and verify that these answers follow from the Fundamental Theorem of Calculus.

   (a) $\quad g(x) = \int_0^x \dfrac{t}{\sqrt{9 + t^2}} dt$

   (b) $\quad g(x) = \int_0^x \sqrt{1 - t^2} \, dt$

3. Let $g(x) = \int_0^{-x} g(-t) dt$. Find the conditions under which the function g is defined. Use *Maple* to find $g'(x)$ and $g''(x)$ for each of the following and verify that these answers follow from the Fundamental Theorem of Calculus.

4. Use the step-by-step procedure in *Maple* to integrate $\int e^{e^x + x} \, dx$.

5. Use the step-by-step procedure in *Maple* to integrate $\int \dfrac{1}{x\sqrt{x^2 - 1}} dx$ using the following change of variable:

   (a) $\quad w = x^2 - 1$

   (b) $\quad w = \sqrt{x^2 - 1} - x$

6. Can you use the step-by-step procedure in *Maple* to integrate $\int \dfrac{1}{x\sqrt{x^2 - 1}} dx$ using the change of variable $x = \sec(w)$? Give reasons for your answer.

7- 20 Use the step-by-step procedure in *Maple* to integrate

7. $\quad \int \dfrac{1}{x^3 - 1} dx$.

8. $\displaystyle\int \frac{6x-26}{x^2+3x-10}\, dx$.

9. $\displaystyle\int \frac{x^2+1}{x^3-3x^2++3x-1}\, dx$.

10. $\displaystyle\int \frac{x-3}{x^2+1}\, dx$.

11. $\displaystyle\int \frac{1}{x^3-1}\, dx$.

12. $\displaystyle\int \frac{2x+9}{x^2(2x-1)^2(3x+8)^3}\, dx$.

13. $\displaystyle\int \frac{1}{x^4+1}\, dx$. (Hint: Write the denominator $x^4+1 = \left(x^2+1\right)^2-2x^2 = \left(x^2+\sqrt{2}x^2+1\right)\left(x^2-\sqrt{2}x^2+1\right)$ and then find the partial function decomposition of the integrand)

14. $\displaystyle\int \frac{1}{x^6+1}\, dx$. (Hint: Write the denominator $x^6+1 = \left(x^2\right)^3+1 = \left(x^2+1\right)\left(x^4-x^2+1\right) = \left(x^2+1\right)\left[\left(x^4+2x^2+1\right)-3x^2\right] = \left(x^2+1\right)\left[\left(x^2+1\right)^2-3x^2\right]= \left(x^2+1\right)\left(x^2+\sqrt{3}x^2+1\right)\left(x^2-\sqrt{3}x^2+1\right)$ and then find the partial function decomposition of the integrand)

15. $\displaystyle\int \frac{1}{\left(x^2+x+1\right)^2}\, dx$.

16. $\displaystyle\int \frac{1}{\left(x^2+x+1\right)^3}\, dx$.

17. $\displaystyle\int \frac{48}{\left(x^2+1\right)^2}\, dx$.

18. $\displaystyle\int \frac{1}{\left(x^2+3\right)^4}\,dx$.

19. $\displaystyle\int \frac{1}{x^5+x^3+x^2+1}\,dx$.

20. $\displaystyle\int \frac{1}{x^6+x^4+x^2+1}\,dx$.

21. Below is a list of some *Maple* commands used in this chapter. Describe the significance of each command. Can you find examples where each command is used?

    (a)   *factor* $\left(f\left(x\right)\right)$

    (b)   *infolevel* $\left[\,Student\left[Calculus1\right]\right] := n$

    (c)   *restart*

    (d)   *Rule* $\left[\;\right]\left(Int\left(f\left(x\right),x=a\,..\,b\right)\right)$

    (e)   *ShowSolution* $\left(Int\left(f\left(x\right),x\right)\right)$

    (f)   *ShowSteps* $\left(\;\right)$

    (g)   *with* $\left(Student\left[Calculus1\right]\right)$

22. Can you list new *Maple* commands that were used in this chapter but were not listed in exercise 21?

Chapter 8   Applications of Integration

In this chapter we will explore the applications of integration using *Maple* for areas between curves, volume of revolution, arc lengths, surfaces of revolution, and function averages.

Area Between Curves

Definition: Let f and g be two functions. Then the area of the region bounded by the curves $y = f(x)$, $y = g(x)$, $x = a$, and $x = b$ is $\int_a^b [f(x) - g(x)]\,dx$.

Example 1: Find the area bounded by the curve $f(x) = 2x^3 + 9x^2 - 23x - 66$, the x-axis, $x = -4$ and $x = 2$.

Solution: We will enter $f(x)$ as a *Maple* function and sketch the graph for the values of x between the values -4 and 2.

$f := x \rightarrow 2 \cdot x^3 + 9 \cdot x^2 - 23 \cdot x - 66$

$$x \rightarrow 2\,x^3 + 9\,x^2 - 23\,x - 66 \qquad\qquad \textbf{(8.1)}$$

$plot(f(x), x = -4..2)$

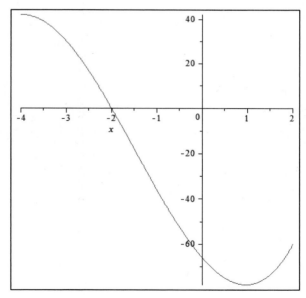

From the graph above, the function $f(x)$ intersects the x-axis at $x = -2$. This will be verified by using the *solve* command shown below.

$solve([\,f(x) = 0, -4 \le x \textbf{ and } x \le 2\,], x)$

Chapter 8 Applications of Integration

$$\{x=-2\} \tag{8.2}$$

The graph of the function is on or above the x-axis in the interval $[-4,-2]$ and is on or below the x-axis in the interval $[-2,2]$. Therefore, the integral $\int_{-4}^{-2} f(x)\,dx$ is positive and the integral $\int_{-2}^{2} f(x)\,dx$ is negative. Hence the required area can be evaluated as $\int_{-4}^{-2} f(x)\,dx + \left|\int_{-2}^{2} f(x)\,dx\right|$. We will calculate each integral separately using *Maple* commands and then calculate the area.

$$\int_{-4}^{-2} f(x)\,dx$$

$$54 \tag{8.3}$$

$$\int_{-2}^{2} f(x)\,dx$$

$$-216 \tag{8.4}$$

$$54+\left|-216\right|$$

$$270 \tag{8.5}$$

Therefore the area bounded by the curve $f(x)=2x^3+9x^2-23x-66$, the x-axis, $x=-4$ and $x=2$ is 270.

Example 2: Find the area bounded by the curves $y=5x^3+2x^2-23x-5$ and $y=2x^3+7x^2+31x-61$.

Solution: We will enter $y=5x^3+2x^2-23x-5$ and $y=2x^3+7x^2+31x-61$ as *Maple* functions (called $f(x)$ and $g(x)$ respectively) and find their points of intersection.

$$f:=x\rightarrow 5\cdot x^3+2\cdot x^2-23\cdot x-5$$

$$x\rightarrow 5x^3+2x^2-23x-5 \tag{8.6}$$

$$g:=x\rightarrow 2\cdot x^3+7\cdot x^2+31\cdot x-61$$

$$x\rightarrow 2x^3+7x^2+31x-61 \tag{8.7}$$

$$solve(f(x)=g(x),x)$$

$$1,-4,\frac{14}{3} \tag{8.8}$$

After finding the points of intersection, we graph the function in the interval $\left[-4,\frac{14}{3}\right]$.

The graphs of $f(x)$ and $g(x)$ are in red and blue respectively.

$$plot\left(\left[f(x)=g(x)\right],\, x=-4..\frac{14}{3},\, color=\left[red,blue\right]\right)$$

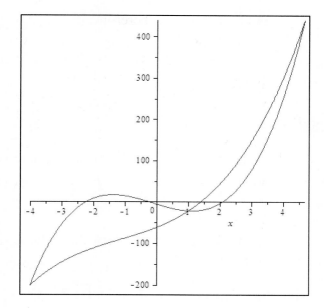

Notice that $f(x) \geq g(x)$ when x is in the interval $[-4,1]$ and $g(x) \geq f(x)$ when x is in the interval $\left[1, \dfrac{14}{3}\right]$. As a result the required area can be evaluated as

$$\int_{-4}^{1}\left(f(x)-g(x)\right)dx + \int_{1}^{\frac{14}{3}}\left(g(x)-f(x)\right)dx$$

$$\frac{89723}{162}$$

(8.9)

Therefore the area bounded by the curves $y = 5x^3 + 2x^2 - 23x - 5$ and $y = 2x^3 + 7x^2 + 31x - 61$ is $\dfrac{89723}{162}$.

Remarks: One could verify the above answer by evaluating the integrals separately and then adding them as shown below.

$$\int_{-4}^{1}\left(f(x)-g(x)\right)dx$$

$$\frac{4625}{12}$$

(8.10)

$$\int_1^{14/3}\left(g(x)-f(x)\right)dx$$

$$\frac{54571}{324} \tag{8.11}$$

$$\frac{4625}{12}+\frac{54571}{324}$$

$$\frac{89723}{162} \tag{8.12}$$

Volume Of Revolution

Definition: Let f and g be two functions. Then the volume of the solid obtained by rotating the region bounded by the curves $y=f(x)$, $y=g(x)$, $x=a$, and $x=b$ is called the *volume of revolution.*

Definition: Let f and g be two functions. Then the volume of the solid obtained by rotating the region bounded by the curves $y=f(x)$, $y=g(x)$, $x=a$, and $x=b$ about a horizontal line, using the *washer method*, is $\pi\int_a^b\left[R(x)^2-r(x)^2\right]dx$, where $R(x)$, known as the outer radius, is the larger distance from the axis of rotation and $r(x)$, called the inner radius, is the smaller distance of the region from the axis of rotation. In the event that $r(x)=0$, the method is called the *disc method.*

Definition: Let f and g be two functions. Then the volume of the solid obtained by rotating the region bounded by the curves $y=f(x)$, $y=g(x)$, $x=a$, and $x=b$ about a vertical line, using the *shell method*, is $2\pi\int_a^b p(x)h(x)dx$, where $p(x)$ is the horizontal distance from the axis of rotation and $h(x)$ is the vertical height of the solid.

Example 3: Find the volume of the solid obtained by rotating the region bounded by the curves $y=x^2$ and $y=1$ about (a) the x-axis, (b) the line $y=3$, and (c) the line $y=-4$.

Solution: We will first find the points where the curve $y=x^2$ intersects the line $y=1$.

$$solve\left(x^2=1,\,x\right)$$

$$1,-1 \tag{8.13}$$

Hence the points of intersection are (-1, 1) and (1, 1).

(a)   To find the volume of the solid obtained by rotating the region bounded by the curves $y = x^2$ and $y = 1$ about the x-axis we begin by sketching the graphs of $y = x^2$ and $y = 1$. These are in red and blue respectively.

$$plot\left(\left[x^2, 1\right], x = -1..1, color = \left[red, blue\right]\right)$$

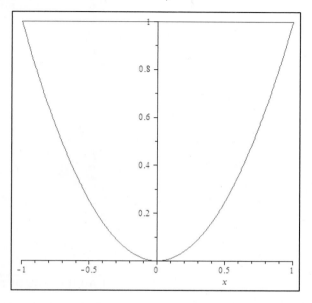

The volume is calculated using the washer method. Notice that the inner radius $r(x) = x^2$ and the outer radius $R(x) = 1$. In order to find the volume of revolution we need to activate the $with\left(Student\left[Calculus1\right]\right)$ package. The default axis of rotation is the x-axis.

$with\left(Student\left[Calculus1\right]\right):$
$VolumeOfRevolution\left(x^2, 1, x = -1..1\right)$

$$\frac{8}{5}\pi \tag{8.14}$$

Hence the volume of the solid obtained by rotating the region bounded by the curves $y = x^2$ and $y = 1$ about the x-axis is $\frac{8}{5}\pi$. If we are interested in finding the integral expression for the volume, then we need to add the $output = integral$ option as shown below.

$VolumeOfRevolution\left(x^2, 1, x = -1..1, output = integral\right)$

$$\int_{-1}^{1}\left(-\pi\left(x^4 - 1\right)\right)dx \tag{8.15}$$

In order to get the graph of the solid the option *output = plot* is added. We also add the option *scaling = constrained* .

VolumeOfRevolution $\left(x^2, 1, x = -1..1, output = plot, scaling = constrained \right)$

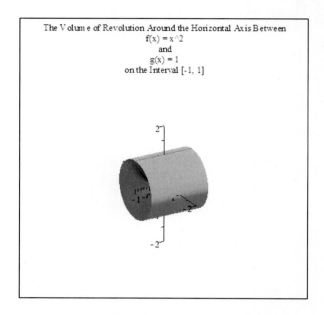

By adding more options *axes = box* and *scaling = constrained* the graph now looks like the one below.

VolumeOfRevolution $\left(x^2, 1, x = -1..1, output = plot, axes = box, scaling = constrained, \right.$

$\left. labels = \left[x, y, z \right] \right)$

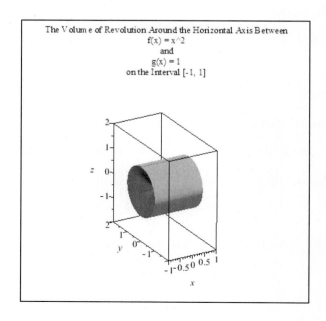

Get a different view of the solid by right clicking anywhere in the plot. One such view is shown below.

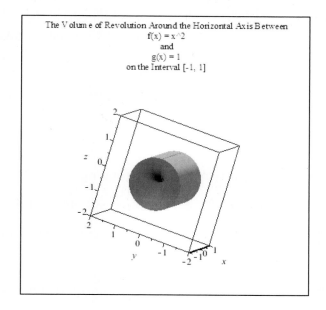

(b)   To find the volume of the solid obtained by rotating the region bounded by the curves $y = x^2$ and $y = 1$ about the line $y = 3$ we begin by sketching the graphs of $y = x^2$, $y = 1$, and the axis of rotation $y = 3$ as shown below. The graphs of $y = x^2$, $y = 1$, and $y = 3$ are in red, blue, and black respectively.

$$plot\left(\left[x^2, 1, 3\right], x = -1..1, color = \left[red, blue, black\right]\right)$$

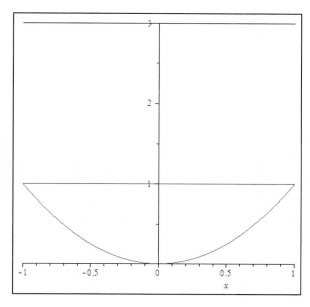

Chapter 8 Applications of Integration

The volume is calculated using the washer method. In this case the inner radius $r(x)=2$ and the outer radius $R(x)=3-x^2$. We do not need to activate the $with\left(Student\left[Calculus1\right]\right)$ package as we have done so in part (a). If the axis of rotation is parallel to the x-axis, its distance from the x-axis must be taken into account. In this case this "distance" is 3.

$$VolumeOfRevolution\left(x^2,1,\,x=-1..1,\,distancefromaxis=3\right)$$

$$\frac{32}{5}\pi \tag{8.16}$$

Hence the volume of the solid obtained by rotating the region bounded by the curves $y=x^2$ and $y=1$ about the line $y=3$ is $\frac{32}{5}\pi$. If we are interested in finding the integral expression for the volume, then we need to add the *output = integral* option as shown below.

$$VolumeOfRevolution\left(x^2,1,\,x=-1..1,\,output=integral,\,distancefromaxis=3\right)$$

$$\int_{-1}^{1}\pi\left|x^4-6x^2+5\right|dx \tag{8.17}$$

The graph of the solid is now given along with an alternate view.

$$VolumeOfRevolution\left(x^2,1,\,x=-1..1,\,distancefromaxis=3,\,output=plot,\,axes=box,\right.$$
$$\left.scaling=constrained,\,labels=\left[x,y,z\right]\right)$$

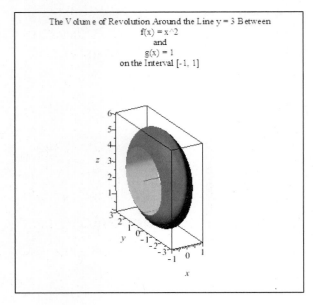
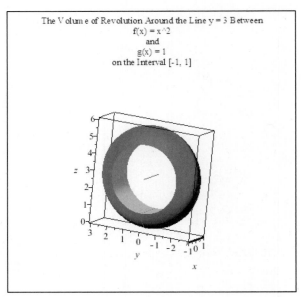

(c)   To find the volume of the solid obtained by rotating the region bounded by the curves $y = x^2$ and $y = 1$ about the line $y = -4$ we begin by sketching the graphs of $y = x^2$, $y = 1$ and the axis of rotation as shown below. The graphs of $y = x^2$, $y = 1$, and $y = -4$ are in red, blue, and black respectively.

$$plot\left(\left[x^2, 1, -4\right], x = -1..1, color = \left[red, blue, black\right]\right)$$

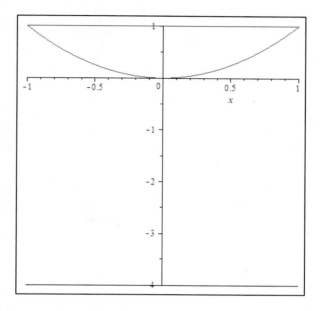

The volume is calculated using the washer method. In this case the inner radius $r(x) = 4 + x^2$ and the outer radius $R(x) = 5$. We do not need to activate the $with\left(Student\left[Calculus1\right]\right)$ package as we have done so in part (a). The default axis of rotation is the x-axis. However, if the axis of rotation is parallel to the x-axis, its distance from the x-axis must be taken into account. In this case this "distance" is -4 as the axis of rotation is the line $y = -4$.

$$VolumeOfRevolution\left(x^2, 1, x = -1..1, distancefromaxis = -4\right)$$

$$\frac{184}{15}\pi \tag{8.18}$$

Hence the volume of the solid obtained by rotating the region bounded by the curves $y = x^2$ and $y = 1$ about the line $y = -4$ is $\frac{184}{15}\pi$. If we are interested in finding the integral expression for the volume, then we need to add the *output = integral* option as shown below.

$$VolumeOfRevolution\left(x^2, 1, x = -1..1, output = integral, distancefromaxis = -4\right)$$

$$\int_{-1}^{1} \pi \left| x^4 + 8\,x^2 - 9 \right| dx \qquad\qquad (8.19)$$

The graph of the solid is given below.

$$VolumeOfRevolution\left(x^2, 1, x = -1..1, distancefromaxis = -4, output = plot, axes = box,\right.$$
$$\left.scaling = constrained, labels = [x, y, z]\right)$$

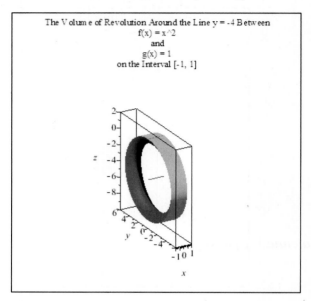

Example 4: Find the volume of the solid obtained by rotating the region bounded by the curves $y = \dfrac{1}{2 + x^7}$, $y = \dfrac{1}{3}$, $x = \dfrac{1}{2}$, and $x = 1$ about (a) the y-axis, (b) the line $x = 2$, and (c) the line $x = -\dfrac{1}{4}$.

Solution: Note that in all these cases we can find the volume using the shell method.

(a) To find the volume of the solid obtained by rotating the region bounded by the curves $y = \dfrac{1}{2 + x^7}$, $y = \dfrac{1}{3}$, $x = \dfrac{1}{2}$, and $x = 1$ about the y-axis we begin by sketching the graphs of $y = \dfrac{1}{2 + x^7}$ and $y = \dfrac{1}{3}$ for the values of x between $\dfrac{1}{2}$ and 1. These are in red and blue respectively.

$$plot\left(\left[\frac{1}{2 + x^7}, \frac{1}{3}\right], x = \frac{1}{2}..1, y = 0..\frac{1}{2}, color = [red, blue]\right)$$

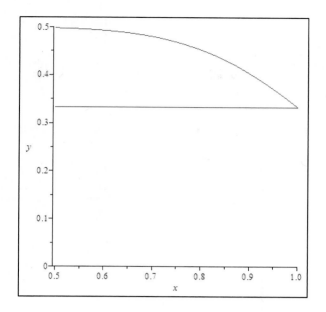

In this case the height is $h(x) = \dfrac{1}{2+x^7} - \dfrac{1}{3} = \dfrac{1-x^7}{3(2+x^7)}$ and the horizontal distance is $p(x) = x$. In order to find the volume of revolution we need to activate the $with(Student[Calculus1])$ package.

$with(Student[Calculus1]):$

$VolumeOfRevolution\left(\dfrac{1}{2+x^7}, \dfrac{1}{3}, x = \dfrac{1}{2}..1, axis = vertical\right)$

$$\int_{\frac{1}{2}}^{1} \left(-\frac{2}{3} - \frac{\pi x\left(-1+x^7\right)}{2+x^7}\right) dx \qquad\qquad (8.20)$$

The answer here is the same as that with the option *output = integral* seen below.

$VolumeOfRevolution\left(\dfrac{1}{2+x^7}, \dfrac{1}{3}, x = \dfrac{1}{2}..1, axis = vertical, output = integral\right)$

$$\int_{\frac{1}{2}}^{1} \left(-\frac{2}{3} - \frac{\pi x\left(-1+x^7\right)}{2+x^7}\right) dx \qquad\qquad (8.21)$$

However, we can get a floating point answer for the volume by setting the values of x to be from 0.5 to 1.

$VolumeOfRevolution\left(\dfrac{1}{2+x^7}, \dfrac{1}{3}, x = 0.5..1, axis = vertical\right)$

$$0.2551922724 \qquad\qquad \textbf{(8.22)}$$

Hence the volume of the solid is 0.2551922724. The graphs of the solid and with one with a different view are shown below.

$$VolumeOfRevolution\left(\frac{1}{2+x^7}, \frac{1}{3}, x = \frac{1}{2}..1, axis = vertical, output = plot, axes = box, \right.$$

$$\left. scaling = constrained, labels = [x, y, z]\right)$$

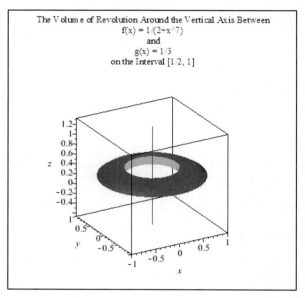

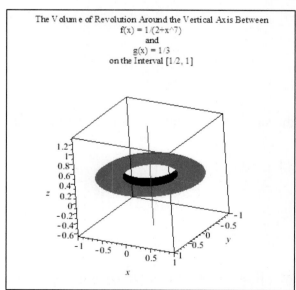

(b)   To find the volume of the solid obtained by rotating the region bounded by the curves $y = \dfrac{1}{2+x^7}$, $y = \dfrac{1}{3}$, $x = \dfrac{1}{2}$, and $x = 1$ about the line $x = 2$ we begin by sketching the graphs of $y = \dfrac{1}{2+x^7}$, $y = \dfrac{1}{3}$, and $x = 2$ using the *implicitplot*

command for the values of x between $\dfrac{1}{2}$ and 1. These are in red, blue, and black

respectively. But before doing so we must activate the *with(plots)* command.

with(plots) :

$$implicitplot\left(\left[y = \frac{1}{2+x^7}, y = \frac{1}{3}, x = 2\right], x = \frac{1}{2}..2, y = \frac{1}{3}..\frac{1}{2}, color = \left[red, blue, black\right]\right)$$

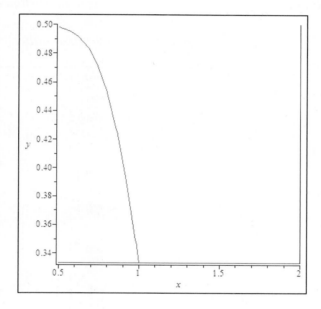

In this case the height is $h(x) = \dfrac{1}{2+x^7} - \dfrac{1}{3} = \dfrac{1-x^7}{3(2+x^7)}$ and the horizontal distance is

$p(x) = 2-x$. We do not need to activate the *with(Student[Calculus1])* package as we have done so in part (a). In this case we must take into account the "distance" of the axis of rotation from the y-axis which is 2.

$$VolumeOfRevolution\left(\frac{1}{2+x^7}, \frac{1}{3}, x = 0.5..1, axis = vertical, distancefromaxis = 2\right)$$

$$0.4795436933 \tag{8.23}$$

Hence the volume of the solid obtained by rotating the region bounded by the curves

$y = \dfrac{1}{2+x^7}$, $y = \dfrac{1}{3}$, $x = \dfrac{1}{2}$, and $x = 1$ about the line $x = 2$ is 0.4795436933. The integral

expression for the volume is given below.

Chapter 8 Applications of Integration

$$VolumeOfRevolution\left(\frac{1}{2+x^7}, \frac{1}{3}, x = \frac{1}{2}..1, axis = vertical, output = integral,\right.$$

$$\left. distancefromaxis = 2\right)$$

$$\int_{\frac{1}{2}}^{1}\left(\frac{2}{3}\frac{\pi(x-2)(-1+x^7)}{2+x^7}\right)dx \qquad \textbf{(8.24)}$$

The graph of the solid is given below.

$$VolumeOfRevolution\left(\frac{1}{2+x^7}, \frac{1}{3}, x = \frac{1}{2}..1, axis = vertical, distancefromaxis = 2, output = plot,\right.$$

$$\left. axes = box, scaling = constrained, labels = [x, y, z]\right)$$

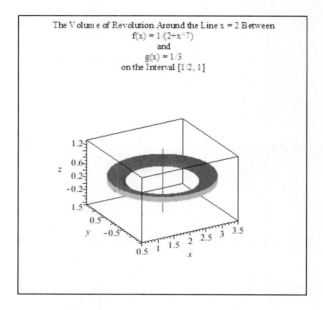

(c) To find the volume of the solid obtained by rotating the region bounded by the curves $y = \frac{1}{2+x^7}$, $y = \frac{1}{3}$, $x = \frac{1}{2}$, and $x = 1$ about the line $x = -\frac{1}{4}$ we begin by sketching the graphs of $y = \frac{1}{2+x^7}$, $y = \frac{1}{3}$, and $x = -\frac{1}{4}$ using the *implicitplot* command for the values of x between $\frac{1}{2}$ and 1. These are in red, blue, and black respectively. We do not need to activate the *with*(*plots*) command as it has been activated in (b).

$$implicitplot\left(\left[y=\frac{1}{2+x^7}, y=\frac{1}{3}, x=-\frac{1}{4}\right], x=-0.25\,..\,1.02, y=0.3\,..\,0.51,\right.$$

$$\left.color=\left[red, blue, black\right]\right)$$

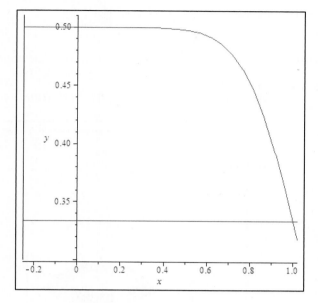

In this case the height is $h(x)=\dfrac{1}{2+x^7}-\dfrac{1}{3}=\dfrac{1-x^7}{3(2+x^7)}$ and the horizontal distance is

$p(x)=\dfrac{1}{4}+x$. We do not need to activate the $with\left(Student\left[Calculus1\right]\right)$ package as we

have done so in part (a). Here the "distance" of the axis of rotation from the y-axis is $-\dfrac{1}{4}$.

$$VolumeOfRevolution\left(\frac{1}{2+x^7}, \frac{1}{3}, x=0.5\,..\,1, axis=vertical, distancefromaxis=-\frac{1}{4}\right)$$
$$0.3470342681 \tag{8.25}$$

Hence the volume of the solid obtained by rotating the region bounded by the curves

$y=\dfrac{1}{2+x^7}$, $y=\dfrac{1}{3}$, $x=\dfrac{1}{2}$, and $x=1$ about the line $x=-\dfrac{1}{4}$ is 0.3470342681. The

integral expression for the volume is given below.

$$VolumeOfRevolution\left(\frac{1}{2+x^7}, \frac{1}{3}, x=\frac{1}{2}\,..\,1, axis=vertical, output=integral,\right.$$

$$\left.distancefromaxis=-\frac{1}{4}\right)$$

$$\int_{\frac{1}{2}}^{1} \left(-\frac{1}{6} \frac{\pi(4x+1)(-1+x^7)}{2+x^7} \right) dx \qquad \textbf{(8.26)}$$

The graph of the solid is given below.

$$VolumeOfRevolution\left(\frac{1}{2+x^7}, \frac{1}{3}, x = \frac{1}{2}..1, axis = vertical, distancefromaxis = -\frac{1}{4}, output = plot, \right.$$

$$\left. axes = box, scaling = constrained, labels = [x, y, z] \right)$$

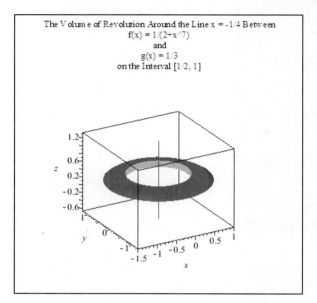

Arc Length

Definition: Let f be a function. Then the *arc length* for the curve $y = f(x)$ bounded by $x = a$, and $x = b$ is $\int_a^b \sqrt{1 + (f'(x))^2} \, dx$.

Example 5: Find the length of the curve $f(x) = 1 + \ln(\tan(x))$ from the point $\left(\frac{\pi}{6}, 1 + \ln\left(\frac{\sqrt{3}}{3} \right) \right)$ to the point $\left(\frac{\pi}{4}, 1 \right)$.

Solution: We will enter $f(x)$ as a *Maple* function and sketch the graph for the values of x between the values $\frac{\pi}{6}$ and $\frac{\pi}{4}$.

$$f := x \rightarrow 1 + \ln(\tan(x))$$

$$x \rightarrow 1 + \ln\left(\tan\left(x\right)\right) \tag{8.27}$$

$$plot\left(f\left(x\right), x = \frac{\pi}{6} .. \frac{\pi}{4}\right)$$

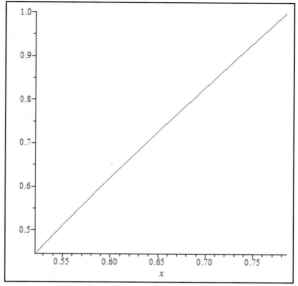

We will use the *ArcLength* command to calculate the required arc length. As in the previous two examples we will need to activate the *with*(*Student*[*Calculus1*]) package.

$$with\left(Student\left[Calculus1\right]\right):$$

$$ArcLength\left(f\left(x\right), x = \frac{\pi}{6} .. \frac{\pi}{4}\right)$$

$$\frac{1}{2}\ln\left(3\right) - \frac{1}{2}\ln\left(11 + 2\sqrt{19}\right) + \frac{1}{2}\arctan\left(\frac{1}{19}\sqrt{19}\right) + \frac{1}{4}\ln\left(9\sqrt{19} + 38\right) - \frac{1}{4}\ln\left(9\sqrt{19} - 38\right)$$

$$+ \frac{1}{2}\ln\left(5 + 2\sqrt{5}\right) - \frac{1}{4}\ln\left(\sqrt{5} + 2\right) + \frac{1}{4}\ln\left(-2 + \sqrt{5}\right) \tag{8.28}$$

$$\xrightarrow{\text{at 10 digits}}$$

$$0.6085868400 \tag{8.29}$$

Hence the arc length is 0.6085868400. The integral used to make this calculation is seen below. This includes the option *output = integral*.

$$ArcLength\left(f\left(x\right), x = \frac{\pi}{6} .. \frac{\pi}{4}, output = integral\right)$$

$$\int_{\frac{1}{6}\pi}^{\frac{1}{4}\pi} \frac{\sqrt{1 + 3\tan\left(x\right)^2 + \tan\left(x\right)^4}}{\tan\left(x\right)}\, dx \tag{8.30}$$

When the option *output = plot* is used we see three graphs – the one in red is that of the function, the one in blue is the integrand, and the one in green is the graph of the integral as a function of *x*. This has been shown below the graph.

$$ArcLength\left(f(x), x = \frac{\pi}{6} .. \frac{\pi}{4}, output = plot, numpoints = 20000 \right)$$

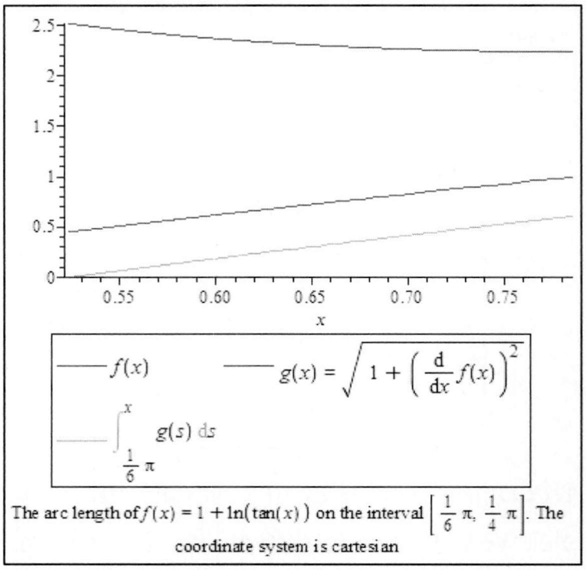

The arc length of $f(x) = 1 + \ln(\tan(x))$ on the interval $\left[\frac{1}{6}\pi, \frac{1}{4}\pi \right]$. The coordinate system is cartesian

Surface Area

Definition: Let f be a function. Then the *surface area* for solid found by rotating the region bounded by the curve $y = f(x)$, $x = a$, and $x = b$ about the line $y = k$ is

$$2\pi \int_a^b (f(x) - k)\sqrt{1 + (f'(x))^2} \, dx.$$

Example 6: Find the area of the surface obtained by rotating the curve $y = \sqrt{36 - x^2}$, $-2 \le x \le 2$, about (a) the *x*-axis, (b) the line $y = -3$, and (c) the *y*-axis.

Solution: We will enter $f(x)$ as a *Maple* function and sketch the graph for the values of *x* between the values -2 and 2. Since this is part of a circle (with radius 6) we will use the *scaling = constrained* option

$$f := x \rightarrow \sqrt{36 - x^2}$$

$$x \rightarrow \sqrt{36 - x^2} \qquad\qquad \textbf{(8.31)}$$

$$plot(f(x), x = -2..2, y = 5.6..6)$$

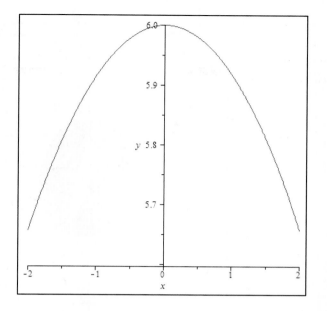

(a)   We will use the *SurfaceOfRevlolution* command to calculate the required area of the surface. As in the case of finding the volume of revolution, the default axis is the *x*-axis. Do not forget to activate the *with* (*Student* [*Calculus1*]) package.

$with \left(Student \left[Calculus1 \right] \right):$

$SurfaceOfRevolution \left(f \left(x \right), x = -2 .. 2 \right)$

$$48 \pi \qquad\qquad\qquad\qquad \textbf{(8.32)}$$

Hence the surface area is 48π. The integral used to make this calculation is seen below. This includes the use of the option *output = integral*.

$SurfaceOfRevolution \left(f \left(x \right), x = -2 .. 2, output = integral \right)$

$$\int_{-2}^{2} 12 \pi \, dx \qquad\qquad\qquad \textbf{(8.33)}$$

The graph of the surface along with its axis of rotation is seen below.

$$SurfaceOfRevolution \left(f \left(x \right), x = -2 .. 2, output = plot, axes = box, \right.$$

$$\left. scaling = constrained, labels = \left[x, y, z \right] \right)$$

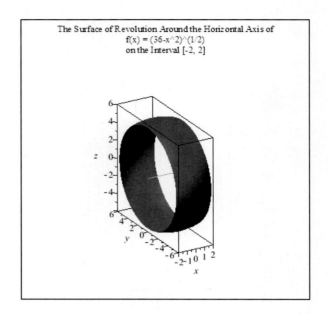

The Surface of Revolution Around the Horizontal Axis of
f(x) = (36-x^2)^(1/2)
on the Interval [-2, 2]

(b)   We will use the *SurfaceOfRevlolution* command to calculate the required surface area. As in the case of finding the volume of revolution, the default axis is the *x*-axis. We will not activate the $with\left(Student\left[Calculus1\right]\right)$ package as we have already activated it in part (a).

$$SurfaceOfRevolution\left(f\left(x\right), x=-2..2, distancefromaxis=-3\right)$$

$$72\,\pi\arcsin\left(\frac{1}{3}\right)+48\,\pi \tag{8.34}$$

Hence the surface area is $72\pi\arcsin\left(\dfrac{1}{3}\right)+48\pi$. As in part (a), the option *output = integral* gives the integral used to make this calculation. This is seen below.

$$SurfaceOfRevolution\left(f\left(x\right), x=-2..2, distancefromaxis=-3, output=integral\right)$$

$$\int_{-2}^{2}\frac{12\,\pi\left(\sqrt{36-x^2}+3\right)}{\sqrt{36-x^2}}\,dx \tag{8.35}$$

The graph of the surface along with its axis of rotation is seen below.

$$SurfaceOf\,Re\,volution\left(f\left(x\right), x=-2..2, distancefromaxis=-3,\right.$$

$$\left.output=plot, axes=box, labels=\left[x,y,z\right]\right)$$

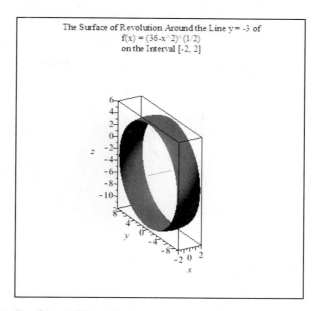

The Surface of Revolution Around the Line y = -3 of
f(x) = (36-x^2)^(1/2)
on the Interval [-2, 2]

(c)   We will use the *SurfaceOfRevlolution* command to calculate the required surface area. As in the case of finding the volume of revolution, the default axis is the *x*-axis. Since the axis of rotation is vertical, this needs to be specified. Do not forget to activate the *with*(*Student*[*Calculus1*]) package.

$with\big(Student[Calculus1]\big):$

$SurfaceOfRevolution\big(f(x), x = -2..2, axis = vertical\big)$

$$-96\,\pi\,\sqrt{2} + 144\,\pi \qquad\qquad \textbf{(8.36)}$$

Hence the surface area is $-96\pi\sqrt{2} + 144\pi$. As in part (a), the option *output = integral* gives the integral used to make this calculation. This is seen below.

$SurfaceOfRevolution\big(f(x), x = -2..2, axis = vertical, output = integral\big)$

$$\int_{-2}^{2} \frac{12\,\pi\,|x|}{\sqrt{36 - x^2}}\,dx \qquad\qquad \textbf{(8.37)}$$

The graph of the surface along with its axis of rotation is seen below.

$SurfaceOfRevolution\big(f(x), x = -2..2, axis = vertical, distancefromaxis = 1, output = plot,$

$$axes = box, scaling = constrained, labels = [x, y, z]\big)$$

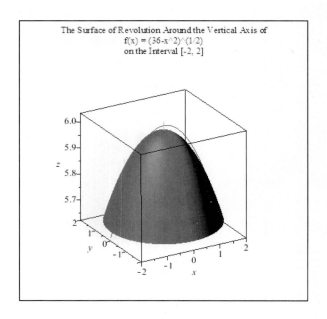

Average Value of a Function

Definition: The average value of a function $f(x)$ when x is in the interval $[a,b]$ is given by $\dfrac{1}{b-a} \displaystyle\int_a^b f(x)\,dx$.

Example 7: Find the average value of the function $f(x) = \arcsin(x)$ when x is in the interval $\left[\dfrac{1}{2}, 1\right]$.

Solution: We will enter $f(x)$ as a *Maple* function.

$f := x \rightarrow \arcsin(x)$

$$x \rightarrow \arcsin(x) \tag{8.38}$$

We will now use the *FuntionAverage* command to find the average value of the function. Do not forget to activate the $with\big(Student[Calculus1]\big)$ package.

$with\big(Student[Calculus1]\big):$

$FunctionAverage\left(f(x), x = \dfrac{1}{2}..1\right)$

$$\frac{5}{6}\pi - \sqrt{3} \tag{8.39}$$

Hence the average value of the function $f(x) = \arcsin(x)$ over the interval $\left[\dfrac{1}{2}, 1\right]$ is

$\dfrac{5}{6}\pi - \sqrt{3}$. The integral used to make this calculation is seen below. This includes the use of the option *output = integral*.

$$FunctionAverage\left(f(x),\, x = \frac{1}{2} .. 1,\, output = integral \right)$$

$$2\left(\int_{\frac{1}{2}}^{1} \left(\arcsin(x) \right) \, \mathrm{d}x \right) \qquad \textbf{(8.40)}$$

The plot of the function along with its average value over the interval $\left[\dfrac{1}{2}, 1\right]$ is given below. Note that this graph includes different options for the graph of the function (its thickness = 2) and for its average value (its graph is dotted with thickness 3)

$$FunctionAverage\left(f(x),\, x = \frac{1}{2} .. 1,\, output = plot,\, functionoptions = \left[thickness = 3 \right], \right.$$

$$\left. averageoptions = \left[linestyle = dot,\, thickness = 4 \right] \right)$$

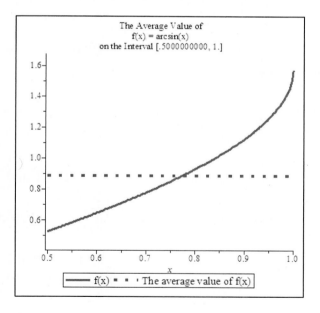

Chapter 8 Applications of Integration

Exercises

1. Find the area bounded by the curve $f(x) = 4x^3 + 7x^2 - 62x + 15$, the x-axis, $x = -6$ and $x = 6$.

2. Find the area bounded by the curve $f(x) = 4x^3 - 32x^2 - 173x + 1326$, the x-axis, $x = -2$ and $x = 8$.

3. Find the area bounded by the curve $f(x) = 6x^4 + 31x^3 - 387x^2 - 906x + 3016$, the x-axis, $x = -5$ and $x = 3$.

4. Find the area bounded by the curve $f(x) = \dfrac{\pi}{2 - x^7}$, the x-axis, $x = -1$ and $x = \dfrac{1}{2}$.

5.* Find the area bounded by the curve $f(x) = e^{-x^2}$, the x-axis, $x = -2$ and $x = 2$.

6. Find the area bounded by the curves $y = x^2 - 5$ and $y = 8 - x^2$.

7. Find the area bounded by the curves $y = 6x^3 + 7x^2 + 36x - 18$ and $y = 8x^3 + 4x^2 - 23x + 12$.

8. Find the area bounded by the curves $y = 9x^3 - 9x^2 - 42x + 95$ and $y = 5x^3 + 3x^2 + 19x - 10$.

9. Find the area bounded by the curves $y = 11x^4 + 4x^3 - 28x^2 - 5x + 13$ and $y = 5x^4 + 3x^3 + 22x^2 3x - 3$.

10. Find the area bounded by the curves $y = e^{2x} - 3e^x + 1$ and $y = 1$.

11.* Find the area bounded by the curves $y = \sin(x)$ and $y = \cos(x)$ where $0 \le x \le \pi$.

12. Find the volume of the solid obtained by rotating the region bounded by the curves $y = x^2 + 2$ and $y = 3x$ about (a) the x-axis and (b) the line $y = 6$.

13. Find the volume of the solid obtained by rotating the region bounded by the curves $y = x^2 + 2$ and $y = 3x$ about (a) the y-axis and (b) the line $x = 4$.

14. Find the volume of the solid obtained by rotating the region bounded by the curves $y = \cos(x)$ and $y = 1$, where $0 \le x \le 2\pi$, about (a) the line $y = -1$ and (b) the line $y = 3$.

15. Find the volume of the solid obtained by rotating the region bounded by the curves $y = \cos(x)$ and $y = 1$, where $0 \leq x \leq 2\pi$, about (a) the y-axis and (b) the line $x = -2$.

16. Find the volume of the solid obtained by rotating the region bounded by the curves $y = \dfrac{4x}{1+x^2}$ and $y = 1$ about (a) the x-axis and (b) the line $y = -1$.

17. Find the volume of the solid obtained by rotating the region bounded by the curves $y = \dfrac{4x}{1+x^2}$ and $y = 1$ about (a) the y-axis and (b) the line $x = 5$.

18. Find the volume of the solid obtained by rotating the region bounded by the curves $f(x) = e^{-x^2}$, $y = 0$, $x = 0$ and $x = 3$ about (a) the y-axis and (b) the line $y = 2$.

19. Find the volume of the solid obtained by rotating the region bounded by the curves $f(x) = e^{-x^2}$, $y = 0$, $x = 0$ and $x = 3$ about (a) the x-axis and (b) the line $x = -3$.

20. Find the volume of the solid obtained by rotating the region bounded by the curves $f(x) = e^{-x^2}$, $y = \dfrac{2}{5}$, $x = 0$ and $x = 3$ about (a) the y-axis and (b) the line $y = -1$.

21. Find the volume of the solid obtained by rotating the region bounded by the curves $f(x) = e^{-x^2}$, $y = \dfrac{2}{5}$, $x = 0$ and $x = 3$ about (a) the x-axis and (b) the line $x = 3$.

22. Find the length of the curve $f(x) = 2x^2 - 11x$ from the point $(-5, 105)$ to the point $(2, 14)$.

23. Find the length of the curve $f(x) = \left(\ln\left(\csc(x)\right)\right)^2$ from the point $\left(\dfrac{\pi}{6}, \left(\ln(2)\right)^2\right)$ to the point $\left(\dfrac{\pi}{2}, 0\right)$.

24. Find the length of the curve $f(x) = \sin(x)$ from the point $(0,0)$ to the point $\left(\dfrac{\pi}{3}, \dfrac{\sqrt{3}}{2} \right)$.

25. Find the area of the surface obtained by rotating the curve $y = 20 + 11x$, $-1 \le x \le 3$, about (a) the x-axis, (b) the line $y = 12$, and (c) the y-axis.

26.* Find the area of the surface obtained by rotating the curve $y = \ln(\sec(x))$, $\dfrac{\pi}{6} \le x \le \dfrac{\pi}{3}$, about (a) the x-axis, (b) the line $y = -3$, and (c) the y-axis.

27. Find the area of the surface obtained by rotating the curve $y = 3x^2$, $-1 \le x \le 1$, about (a) the x-axis, (b) the line $x = -2$, and (c) the y-axis.

28.* Find the area of the surface obtained by rotating the curve $y = e^{2x}$, $0 \le x \le 1$, about (a) the x-axis, (b) the line $x = 2$, and (c) the y-axis.

29. Find the average value of the function $f(x) = x^3 - 2x^2 - 5x + 7$ when x is in the interval $[-2,1]$.

30. Find the average value of the function $f(x) = \ln(x)$ when x is in the interval $[1,5]$.

31. Find the average value of the function $f(x) = \sec(x)$ when x is in the interval $\left[0, \dfrac{\pi}{4} \right]$.

32. * Find the average value of the function $f(x) = 2011e^{-x^2}$ when x is in the interval $[0, \ln(2)]$.

33. Below is a list of some *Maple* commands used in this chapter. Describe the significance of each command. Can you find examples where each command is used?

    (a) $ArcLength(f(x), x = a \mathbin{..} b, output = integral)$

    (b) $FunctionAverage(f(x), x = a \mathbin{..} b, output = plot)$

    (c) $implicitplot([implicit\ equation1, implicit\ equation2], x = a \mathbin{..} b, y = c \mathbin{..} d,$
$$color = [color1, color2])$$

(d) $SurfaceOfRevolution\big(f(x), x = a..b, axis = vertical\big)$

(e) $SurfaceOfRevolution\big(f(x), x = a..b, output = integral,$

$distancefromaxis = n\big)$

(f) $VolumeOfRevolution\big(f(x), g(x), x = a..b, distancefromaxis = n\big)$

(g) $with\big(plots\big)$

34.   Can you list new *Maple* commands that were used in this chapter but were not listed in exercise 33?

Chapter 9 Differential Equations

Applications of transcendental functions to solve simple differential equations, basic differential equations and families of solutions will be studied. *Maple* will be used to generate fields. The built-in *Maple* commands will be utilized to solve initial valued problems.

Definitions

Definition: A *differential equation* is an equation that contains a function and its derivative(s). An *ordinary differential equation* is a differential equation that involves a function with one variable. The *order* of a differential equation refers to the highest derivative present in the equation. The *degree* of a differential equation is the highest power of the highest derivative term.

Definition: A function is a *solution* of the differential equation if the function and its derivative(s) satisfy the equation.

Remarks:
1. The *general solution* of a differential equation is a solution that contains one or more constants. The number of constants is equal to the order of the differential equation. To find a specific solution, called *particular solution*, we use *initial conditions*. Choosing initial conditions means that we specify values of the solution and one or more of its derivatives at given points.

2. It is not always possible to find an explicit solution of a differential equation. Moreover, there is not a unique method that work for all equations. Many of the differential equations that arise in real life (physics, engineering, social science etc.) cannot be solved using general methods. For several of these equations, numerical methods with controlled errors are used. As we will discuss and observe throughout the different examples in this chapter, *Maple* is a powerful tool that provides solutions to different classes of differential equations.

Definition: *Quadrature equations* are differential equations of the form $\dfrac{dy}{dx} = f(x)$, where $y = \int f \, dx + C$ can be evaluated numerically by dividing the area under the curve $y = f(x)$ into squares.

First Order Differential Equations

Example 1: Consider the differential equation $\dfrac{dy}{dx} = x^3 - 5x\cos x$.

(a)   Find the general solution for this differential equation.

(b)   Find the particular solution to this differential equation that satisfies the initial condition $y(0) = 1$ and graph this function.

Solution:

(a)   In order to find the general solution to the differential equation, we will use the *Maple* command *dsolve*. We will set $infolevel[dsolve] := 3$ to get the information about the techniques or algorithms used to solve the differential equation.

$infolevel[dsolve] := 3:$

$ode := \dfrac{d}{dx} y(x) = x^3 - 5 \cdot x \cdot \cos(x)$

$$\frac{d}{dx} y(x) = x^3 - 5x\cos(x) \tag{9.1}$$

$dsolve(ode)$

```
Methods for first order ODEs:
---Trying classification methods ---
trying a quadrature
<-quadrature successful
```

$$y(x) = \frac{1}{4}x^4 - 5\cos(x) - 5x\sin(x) + \_C1 \tag{9.2}$$

Hence the general solution of the differential equation $\dfrac{dy}{dx} = x^3 - 5x\cos(x)$ is

$$y(x) = \frac{1}{4}x^4 - 5\cos(x) - 5x\sin(x) + \_C1.$$

(b)   We will now find the particular solution satisfying the condition $y(0) = 1$.

$initialvalues := y(0) = 1$

$$y(0) = 1 \tag{9.3}$$

$dsolve(\{ode, initialvalues\})$

```
Methods for first order ODEs:
---Trying classification methods ---
trying a quadrature
<-quadrature successful
```

$$y(x) = \frac{1}{4}x^4 - 5\cos(x) - 5x\sin(x) + 6 \tag{9.4}$$

Therefore the particular solution is $y(x) = \frac{1}{4}x^4 - 5\cos(x) - 5x\sin(x) + 6$. To graph this solution, we will use the *unapply* command to express the expression in **(9.4)** as a *Maple* function. We will label this function *y1*.

$y1 := unapply\left(rhs\left(\mathbf{(9.4)}\right), x\right)$

$$x \to \frac{1}{4}x^4 - 5\cos(x) - 5x\sin(x) + 6 \qquad \mathbf{(9.5)}$$

$plot\left(y1(x), x = -2..2\right)$

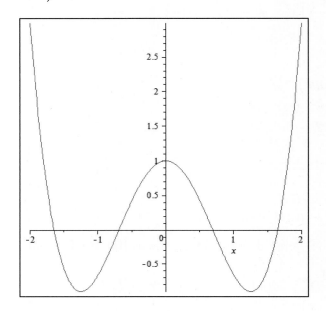

Definition: A differential equation is said to be *separable* if it can be written in the form $g(y)\,dy = f(x)\,dx$ where $f(x)$ is a function of the variable x, and $g(y)$ a function of the variable y.

Remark: To solve a separable differential equation, we can integrate both sides of the equation. The differential equation in example 1 is a separable differential equation.

Definition: A differential equation is said to be *linear* if it can be written in the form $\frac{dy}{dx} + p(x)y = q(x)$, where $p(x)$ and $q(x)$ are continuous functions on some intervals.

Example 2: Solve the differential equation $\dfrac{dy}{dx} = -\dfrac{(x-2)^3 y}{x}$.

Solution: Note that the differential equation is linear with $p(x) = \dfrac{(x-2)^3}{x}$ and $q(x) = 0$.

This equation will be solved using the *dsolve* command.

restart

$ode := \dfrac{d}{dx} y(x) = -\dfrac{(x-2)^3 \cdot y(x)}{x}$

$$\frac{d}{dx} y(x) = -\frac{(x-2)^3 \, y(x)}{x} \tag{9.6}$$

$infolevel[dsolve] := 3:$

dsolve(ode)

```
Methods for first order ODEs:
--- Trying classification methods ---
trying a quadrature
trying 1st order linear
<- 1st order linear successful
```

$$y(x) = \_C1\, x^8 e^{-\frac{1}{3} x \left(x^2 - 9x + 36 \right)} \tag{9.7}$$

Hence the general solution of the differential equation $\dfrac{dy}{dx} = -\dfrac{(x-2)^3 \, y}{x}$ is given by

$y(x) = \_C1\, x^8 e^{-\frac{1}{3} x \left(x^2 - 9x + 36 \right)}$.

Remarks:

1. One can use the *Maple* command *odeadvisor* in the package *with(DEtools)* to classify the differential equation according to standard textbooks. This information will be useful to select the method of solving the differential equation.

   with(DEtools):

   odeadvisor(ode)

   $$[\_separable] \tag{9.8}$$

   dsolve(ode, [separable])

   ```
   Classification methods on request
   Methods to be used are: [separable]
   ----------------------------------
   * Tackling ODE using method: separable
   --- Trying classification methods ---
   trying separable
   ```

```
<- separable successful
```

$$y(x) = \frac{x^8 e^{-\frac{1}{3}x^3 + 3x^2 - 12x}}{\_C1} \tag{9.9}$$

Hence the general solution of the differential equation $\dfrac{dy}{dx} = -\dfrac{(x-2)^3 y}{x}$ is given

by $y(x) = \dfrac{x^8 e^{-\frac{1}{3}x^3 + 3x^2 - 12x}}{\_C1}$. Notice that the only difference between this solution

and the one given in **(9.7)** is the way the constant is written.

2. We will now use *Maple* commands to show the steps (that display integrals) used

while solving the differential equation $\dfrac{dy}{dx} = -\dfrac{(x-2)^3 y}{x}$ as a separable equation.

restart

$ode := \dfrac{d}{dx} y(x) = -\dfrac{(x-2)^3 \cdot y(x)}{x}$

$$\frac{d}{dx} y(x) = -\frac{(x-2)^3 y(x)}{x} \tag{9.10}$$

dsolve (*ode*, [*separable*], *useInt*)

$$\int \frac{(x-2)^3}{x} \, dx + \int^{y(x)} \frac{1}{\_a} \, d\_a + \_C1 = 0 \tag{9.11}$$

value ((**9.11**))

$$\frac{1}{3} x^3 - 3x^2 + 12x - 8\ln(x) + \ln(y(x)) + \_C1 = 0 \tag{9.12}$$

Next we will solve equation **(9.12)** for $y(x)$ to obtain an explicit solution for the

differential equation $\dfrac{dy}{dx} = -\dfrac{(x-2)^3 y}{x}$.

isolate ((**9.12**), $y(x)$)

$$y(x) = e^{-\frac{1}{3}x^3 + 3x^2 - 12x - \_C1} x^8 \tag{9.13}$$

Notice that this solution is again similar to the ones found in **(9.7)** and **(9.9)**. Once again, the only difference is the way the constant is written.

3. Since the differential equation $\dfrac{dy}{dx} = -\dfrac{(x-2)^3 \, y}{x}$ is linear, we can use *Maple* commands to show the steps while solving the equation as a linear differential equation.

restart

$$ode := \frac{d}{dx} y(x) = -\frac{(x-2)^3 \cdot y(x)}{x}$$

$$\frac{d}{dx} y(x) = -\frac{(x-2)^3 \, y(x)}{x} \tag{9.14}$$

dsolve(ode,[linear],useInt)

$$y(x) = \_C1 \, e^{\int \left(-\frac{(x-2)^3}{x}\right) dx} \tag{9.15}$$

value((9.15))

$$y(x) = \_C1 \, e^{-\frac{1}{3}x^3 + 3x^2 - 12x + 8\ln(x)} \tag{9.16}$$

It is left to the student to check that this solution is equal to the solutions found earlier.

Example 3: Solve the differential equation $\dfrac{d}{dx}\left(y\ln(x)\right) = \dfrac{y}{x}$.

Solution: This equation is another example of a first order linear differential equation that is separable.

restart

$$dsolve\left(\frac{d}{dx}\left(y(x)\right) = \frac{y(x)}{x \cdot \ln(x)}, y(x)\right)$$

$$y(x) = \_C1 \ln(x) \tag{9.17}$$

The general solution of the differential equation is given by $y(x) = \_C1 \ln(x)$.

Remark: Now we will check whether $\dfrac{dy}{dx} = \dfrac{y}{x\ln(x)}$ is a separable or a linear differential equation.

restart

$$ode := \frac{d}{dx}\big(y(x)\big) = \frac{y(x)}{x \cdot \ln(x)}$$

$$\frac{d}{dx}y(x) = \frac{y(x)}{x\ln(x)} \qquad\qquad\qquad \textbf{(9.18)}$$

infolevel[*dsolve*] := 3 :

dsolve(*ode*)

```
Methods for first order ODEs:
--- Trying classification methods ---
trying a quadrature
trying 1st order linear
<- 1st order linear successful
```

$$y(x) = \_C1\ln(x) \qquad\qquad\qquad \textbf{(9.19)}$$

Hence $\dfrac{dy}{dx} = \dfrac{y}{x\ln(x)}$ is a linear differential equation. The following commands give

additional information for solving the differential equation using *Maple*.

with(*DEtools*) :

odeadvisor(*ode*)

$$[\_separable] \qquad\qquad\qquad \textbf{(9.20)}$$

dsolve(*ode*,[*separable*])

```
Classification methods on request
Methods to be used are: [separable]
----------------------------------
* Tackling ODE using method: separable
--- Trying classification methods ---
trying separable
<- separable successful
```

$$y(x) = \_C1\ln(x) \qquad\qquad\qquad \textbf{(9.21)}$$

Therefore, $\dfrac{dy}{dx} = \dfrac{y}{x\ln(x)}$ is a separable differential equation. We will now use *Maple*

commands to show the steps used while solving the differential equation $\dfrac{dy}{dx} = \dfrac{y}{x\ln(x)}$ as

a separable equation.

restart

$$ode := \frac{d}{dx}\big(y(x)\big) = \frac{y(x)}{x \cdot \ln(x)}$$

$$\frac{d}{dx}y(x) = \frac{y(x)}{x\ln(x)} \tag{9.22}$$

dsolve(ode,[separable],useInt)

$$\int \frac{1}{x\ln(x)}\,dx + \int^{y(x)}\frac{1}{\_a}\,d\_a + \_C1 = 0 \tag{9.23}$$

value((9.23))

$$\ln(\ln(x)) - \ln(y(x)) + \_C1 = 0 \tag{9.24}$$

isolate((9.24), y(x))

$$y(x) = e^{-\_C1}\ln(x) \tag{9.25}$$

Instead of solving the differential equation $\frac{dy}{dx}\ln(x) = \frac{y}{x}$ as a separable equation, we will use the fact that it is also a linear differential equation to solve it.

dsolve(ode,[linear],useInt)

$$y(x) = \_C1\,e^{\int \frac{1}{x\ln(x)}\,dx} \tag{9.26}$$

value((9.26))

$$y(x) = \_C1\ln(x) \tag{9.27}$$

Hence $y(x) = \_C1\ln(x)$ is a general solution of the differential equation $\frac{dy}{dx}\ln(x) = \frac{y}{x}$.

Example 4:

(a)   Solve the differential equation $\frac{dy}{dx} = x^2 + 3y$.

(b)   Find the particular solution that satisfies the initial condition $y(0) = 1$. Graph this solution on the interval $[0, 2]$.

(c)   Find the particular solutions for the initial condition $y(0) = c$, where c is a constant. Graph these solutions for $c = -2, -1, 0, 1, 2$.

Solution:

(a)   We will find the general solution of the differential equation $\frac{dy}{dx} = x^2 + 3y$.

   restart

   $infolevel[dsolve] := 3:$

   $ode := \frac{d}{dx}y(x) = x^2 + 3\cdot y(x)$

$$\frac{d}{dx} y(x) = x^2 + 3y(x) \qquad (9.28)$$

dsolve (*ode*)

```
Methods for first order ODEs:
--- Trying classification Methods ---
trying a quadrature
trying 1st order linear
<- 1st order linear successful
```

$$y(x) = -\frac{2}{27} - \frac{2}{9}x - \frac{1}{3}x^2 + e^{3x}\_C1 \qquad (9.29)$$

Hence the general solution is $y(x) = -\frac{2}{27} - \frac{2}{9}x - \frac{1}{3}x^2 + e^{3x}\_C1$.

(b)   We will now find the particular solution that satisfies the condition $y(0) = 1$.

initialvalues $:= y(0) = 1$

$$y(0) = 1 \qquad (9.30)$$

dsolve ({*ode, initialvalues*})

```
Methods for first order ODEs:
--- Trying classification methods ---
trying a quadrature
trying 1st order linear
<- 1st order linear successful
```

$$y(x) = -\frac{2}{27} - \frac{2}{9}x - \frac{1}{3}x^2 + \frac{29}{27}e^{3x} \qquad (9.31)$$

Hence the particular solution is $y(x) = -\frac{2}{27} - \frac{2}{9}x - \frac{1}{3}x^2 + \frac{29}{27}e^{3x}$. To   graph   the particular solution, we will define $y(x)$ as a *Maple* function.

$y1 := unapply\left(rhs(\textbf{9.31}), x\right)$

$$x \rightarrow -\frac{2}{27} - \frac{2}{9}x - \frac{1}{3}x^2 + \frac{29}{27}e^{3x} \qquad (9.32)$$

plot $\left(y1(x), x = 0..2\right)$

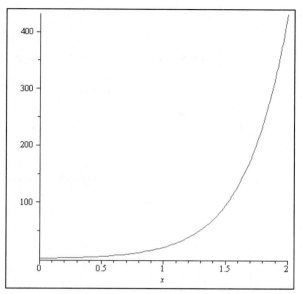

(c)   Now we will find the particular solution that satisfies $y(0) = c$ for any real number c.

restart
infolevel[*dsolve*] := 3 :

$ode := \dfrac{d}{dx} y(x) = x^2 + 3 \cdot y(x)$

$$\dfrac{d}{dx} y(x) = x^2 + 3y(x) \tag{9.33}$$

initialvalues := $y(0) = c$

$$y(0) = c \tag{9.34}$$

dsolve({*ode*, *initialvalues*})

```
Methods for first order ODEs:
--- Trying classification methods ---
trying a quadrature
trying 1st order linear
<- 1st order linear successful
```

$$y(x) = -\dfrac{2}{27} - \dfrac{2}{9}x - \dfrac{1}{3}x^2 + e^{3x}\left(c + \dfrac{2}{27}\right) \tag{9.35}$$

$y2 := unapply\left(rhs\left((9.35)\right), x, c\right)$

$$(x, c) \rightarrow -\dfrac{2}{27} - \dfrac{2}{9}x - \dfrac{1}{3}x^2 + e^{3x}\left(c + \dfrac{2}{27}\right) \tag{9.36}$$

$$plot\left(\left\{y2(x,c)\,\$\,c=-2..2\right\},x=0..2,thickness=2\right)$$

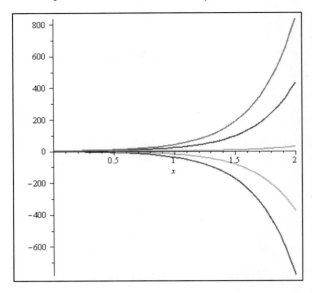

Higher Order Differential Equations

Example 5: Solve the second order differential equation $\dfrac{d^2y}{dx^2}-9y=0$ given the initial

conditions $y(0)=1$ and $\dfrac{dy}{dx}(0)=0$.

Solution: To find the particular solution of the differential equation that satisfies the given initial conditions we will use the *dsolve* command.

restart

infolevel[*dsolve*]:=3:

$$ode:=\dfrac{d^2}{dx^2}\,y(x)-9\cdot y(x)=0$$

$$\dfrac{d^2}{dx^2}\,y(x)-9\,y(x)=0 \tag{9.37}$$

initialvalues:=$y(0)=1,D(y)(0)=0$

$$y(0)=1,D(y)(0)=0 \tag{9.38}$$

dsolve({*ode,initialvalues*})

```
Methods for second order ODEs:
--- Trying classification methods ---
trying a quadrature
checking if the LODE has constant coefficients
<- constant coefficients successful
```

$$y(x) = \frac{1}{2}e^{3x} + \frac{1}{2}e^{-3x} \qquad\qquad \textbf{(9.39)}$$

Hence the solution of the second order differential equation $\dfrac{d^2y}{dx^2} - 9y = 0$ that satisfies

the given initial conditions is given by $y(x) = \dfrac{1}{2}e^{3x} + \dfrac{1}{2}e^{-3x}$.

Remark: To define the differential equation in example 5 in *Maple*, we could use the *diff* $(y(x), x, x)$ command.

ode := *diff* $(y(x), x, x) - 9 \cdot y(x) = 0$

$$\frac{d^2}{dx^2}y(x) - 9 \cdot y(x) = 0 \qquad\qquad \textbf{(9.40)}$$

Example 6: Solve the differential equation $\dfrac{d^2y}{dx^2} + 9y = 0$ given the initial conditions

$y(0) = 5$ and $\dfrac{dy}{dx}(0) = 2$. Graph the solution on the interval $[0, 2\pi]$.

Solution: We will first solve the differential equation with the given initial conditions.

restart
infolevel [*dsolve*] := 3 :

ode := $\dfrac{d^2}{dx^2}y(x) + 9 \cdot y(x) = 0$

$$\frac{d^2}{dx^2}y(x) + 9 \cdot y(x) = 0 \qquad\qquad \textbf{(9.41)}$$

initialvalues := $y(0) = 5, D(y)(0) = 2$

$$y(0) = 5, D(y)(0) = 2 \qquad\qquad \textbf{(9.42)}$$

dsolve $(\{ode, initialvalues\})$

```
Methods for second order ODEs:
--- Trying classification methods ---
trying a quadrature
checking if the LODE has constant coefficients
<- constant coefficients successful
```

$$y(x) = \frac{2}{3}\sin(3x) + 5\cos(3x) \qquad\qquad \textbf{(9.43)}$$

Hence the solution of the differential equation with the given initial conditions is given by $y(x) = \frac{2}{3}\sin(3x) + 5\cos(3x)$. We will now find the particular solution a *Maple* function and then we will graph the function.

$yl := unapply\left(rhs\left((\mathbf{9.43})\right), x\right)$

$$x \rightarrow \frac{2}{3}\sin(3x) + 5\cos(3x) \qquad\qquad (\mathbf{9.44})$$

$plot\left(yl(x), x = 0..2 \cdot \pi\right)$

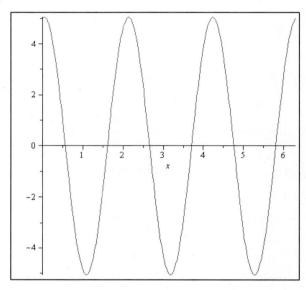

Example 7:

(a) Solve the differential equation $\dfrac{d^2 y}{dx^2} + y = e^x$.

(b) Find the particular solution that satisfies the initial conditions $y(0) = 0$ and $y'(0) = 1$ and plot the function.

Solution:

(a) We will first find the general solution.

$restart$

$infolevel\left[dsolve\right] := 3:$

$ode := \dfrac{d^2}{dx^2} y(x) + y(x) = e^x$

$$\frac{d^2}{dx^2} y(x) + y(x) = e^x \qquad (9.45)$$

dsolve(ode)

```
Methods for second order ODEs:
--- Trying classification methods ---
trying a quadrature
trying high order exact linear fully integrable
trying differential order: 2; linear nonhomogeneous
with symmetry [0,1]
trying a double symmetry of the form [xi=0, eta=F(x)]
-> Try solving first the homogeneous part of the ODE
   checking if the LODE has constant coefficients
   <- constant coefficients successful
   -> Determining now a particular solution to the non-
   homogeneous ODE
      trying a rational particular solution to g^(-
      1)*L*g = 1
      <- rational particular solution to g^(-1)*L*g = 1
      successful
<- solving first the homogeneous part of the ODE
successful
```

$$y(x) = \sin(x)\_C2 + \cos(x)\_C1 + \frac{1}{2}e^x \qquad (9.46)$$

Hence the general solution of the differential equation $\frac{d^2 y}{dx^2} + y = e^x$ is given by

$$y(x) = \sin(x)\_C2 + \cos(x)\_C1 + \frac{1}{2}e^x.$$

(b) We will now find the particular solution that satisfies the initial conditions $y(0) = 0$ and $y'(0) = 1$.

initialvalues $:= y(0) = 0, D(y)(0) = 1$

$$y(0) = 0, D(y)(0) = 1 \qquad (9.47)$$

dsolve({ode, initialvalues})

```
Methods for second order ODEs:
--- Trying classification methods ---
trying a quadrature
```

```
trying high order exact linear fully integrable
trying differential order: 2; linear nonhomogeneous
with symmetry [0,1]
trying a double symmetry of the form [xi=0, eta=F(x)]
-> Try solving first the homogeneous part of the ODE
   checking if the LODE has constant coefficients
   <- constant coefficients successful
   -> Determining now a particular solution to the non-
      homogeneous ODE
      trying a rational particular solution to g^(-
      1)*L*g = 1
      <- rational particular solution to g^(-1)*L*g = 1
      successful
<- solving first the homogeneous part of the ODE
successful
```

$$y(x) = \frac{1}{2}\sin(x) - \frac{1}{2}\cos(x) + \frac{1}{2}e^x \qquad (9.48)$$

Therefore the particular solution satisfying the given conditions is $y(x) = \frac{1}{2}\sin(x) - \frac{1}{2}\cos(x) + \frac{1}{2}e^x$. Next, we will define this solution as a *Maple* function and graph it.

$y1 := unapply\left(rhs\left((\mathbf{9.48})\right), x\right)$

$$x \to \frac{1}{2}\sin(x) - \frac{1}{2}\cos(x) + \frac{1}{2}e^x \qquad (9.49)$$

$plot\left(y1(x), x = -10..2\right)$

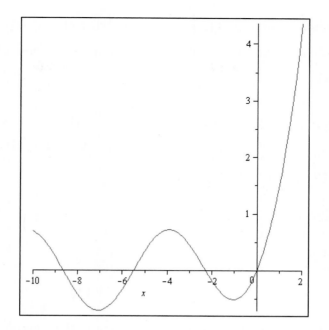

Example 8:

(a)   Solve the third-order differential equation $\dfrac{d^3 y}{dx^3} + \dfrac{dy}{dx} = 0$.

(b)   Find the particular solution that satisfies the initial conditions $y(0) = 0$, $y'(0) = 0$ and $y''(0) = 1$.

Solution:

(a)   We will first solve the differential equation.

   restart
   infolevel[*dsolve*] := 3:

   $ode := \dfrac{d^3}{dx^3} y(x) + \dfrac{d}{dx} y(x) = 0$

$$\dfrac{d^3}{dx^3} y(x) + \dfrac{d}{dx} y(x) = 0 \tag{9.50}$$

   dsolve (*ode*)

   ```
   Methods for third order ODEs:
   --- Trying classification methods ---
   trying a quadrature
   checking if the LODE has constant coefficients
   <- constant coefficients successful
   ```

$$y(x) = \_C1 + \_C2\sin(x) + \_C3\cos(x) \tag{9.51}$$

The general solution is $y(x) = \_C1 + \_C2\sin(x) + \_C3\cos(x)$. We will now find the particular solution that satisfies the initial conditions $y(0) = 0$, $y'(0) = 0$ and $y''(0) = 1$.

$$initialvalues := y(0) = 0, D(y)(0) = 0, D^{(2)}(y)(0) = 1$$

$$y(0) = 0, D(y)(0) = 0, D^{(2)}(y)(0) = 1 \qquad (9.52)$$

$dsolve(\{ode, initialvalues\})$

```
Methods for third order ODEs:
--- Trying classification methods ---
trying a quadrature
checking if the LODE has constant coefficients
<- constant coefficients successful
```

$$y(x) = 1 - \cos(x) \qquad (9.53)$$

Hence the particular solution satisfying the given initial conditions is $y(x) = 1 - \cos(x)$.

Bernoulli Differential Equations

Definition: A differential equation is *Bernoulli* if it can be written in the form $\dfrac{dy}{dx} + p(x)y = q(x)y^n$ where $p(x)$ and $q(x)$ are continuous functions on some interval, and *n* is a real number.

Example 9: Solve the differential equation $\dfrac{d}{dx}y = \dfrac{xy^2}{\sqrt{x^2 - 1}}$.

Solution: Note that the given differential equation is Bernoulli with $p(x) = 0$ and $q(x) = \dfrac{x}{\sqrt{x^2 - 1}}$.

restart
infolevel$[dsolve] := 3$:

$$ode := \frac{d}{dx}y(x) = \frac{x \cdot y(x)^2}{\sqrt{x^2 - 1}}$$

$$\frac{d}{dx}y(x) = \frac{x \cdot y(x)^2}{\sqrt{x^2 - 1}} \qquad (9.54)$$

dsolve(*ode*)

```
Method for first order ODEs:
--- Trying classification Methods ---
trying a quadrature
trying 1st order linear
trying Bernoulli
<- Bernoulli successful
```

$$y(x) = \frac{\sqrt{x^2-1}}{1-x^2+C1\sqrt{x^2-1}} \tag{9.55}$$

Hence the general solution of the Bernoulli differential equation is $y(x) = \dfrac{\sqrt{x^2-1}}{1-x^2+C1\sqrt{x^2-1}}$.

Example 10: Solve the differential equation $\dfrac{dy}{dx} = -\dfrac{\sqrt[3]{y}}{x^2}$.

Solution: The differential equation $\dfrac{dy}{dx} = -\dfrac{\sqrt[3]{y}}{x^2}$ is Bernoulli with $p(x)=0$, $q(x)=-\dfrac{1}{x^2}$ and $n=\dfrac{1}{3}$.

restart
infolevel[*dsolve*] := 3 :
$ode := \dfrac{d}{dx} y(x) = -\dfrac{\sqrt[3]{y(x)}}{x^2}$

$$\frac{d}{dx} y(x) = -\frac{y(x)^{\frac{1}{3}}}{x^2} \tag{9.56}$$

dsolve(*ode*)

```
Methods for first order ODEs:
--- Trying classification methods ---
trying a quadrature
trying 1st order linear
trying Bernoulli
<- Bernoulli successful
```

$$y(x)^{\frac{2}{3}} - \frac{2}{3x} - \_C1 = 0 \tag{9.57}$$

isolate((**9.57**), *y*(*x*))

$$y(x) = \left(\frac{2}{3x} + \_C1\right)^{\frac{3}{2}} \tag{9.58}$$

Hence the general solution of the differential equation $\dfrac{dy}{dx} = -\dfrac{\sqrt[3]{y}}{x^2}$ is

$$y(x) = \left(\frac{2}{3x} + \_C1 \right)^{\frac{3}{2}} .$$

Non Linear Differential Equations

Example 11: Solve the differential equation $\dfrac{d}{dx} y(x) = \dfrac{4 \cdot x^3 \cdot e^{y(x)}}{\sqrt{1-x^4}} .$

Solution: Note that the differential equation $ode := \dfrac{d}{dx} y(x) = \dfrac{4 \cdot x^3 \cdot e^{y(x)}}{\sqrt{1-x^4}}$ is non linear.

restart

infolevel$[dsolve] := 3:$

$ode := \dfrac{d}{dx} y(x) = \dfrac{4 \cdot x^3 \cdot e^{y(x)}}{\sqrt{1-x^4}}$

$$\frac{d}{dx} y(x) = \frac{4 \cdot x^3 \cdot e^{y(x)}}{\sqrt{1-x^4}} \qquad\qquad (9.59)$$

dsolve(ode)

```
Methods for first order ODEs:
--- Trying classification methods ---
trying a quadrature
trying 1st order linear
trying Bernoulli
trying separable
<- separable successful
```

$$y(x) = \ln\left(-\frac{1}{4\left(\displaystyle\int \frac{x^3}{\sqrt{1-x^4}} \, dx + \_C1 \right)} \right) \qquad\qquad (9.60)$$

value$((9.60))$

$$y(x) = \ln\left(-\frac{1}{4\left(\dfrac{1}{2} \dfrac{(x-1)(x+1)(x^2+1)}{\sqrt{1-x^4}} + \_C1 \right)} \right) \qquad\qquad (9.61)$$

Hence the general solution of the differential equation $\dfrac{d}{dx} y(x) = \dfrac{4 \cdot x^3 \cdot e^{y(x)}}{\sqrt{1-x^4}}$ is given by

$$y(x) = \ln\left(-\dfrac{1}{4\left(\dfrac{1}{2}\dfrac{(x-1)(x+1)(x^2+1)}{\sqrt{1-x^4}} + \_C1\right)}\right).$$

Example 12: Solve the differential equation $\dfrac{dy}{dx} = \dfrac{x}{y+\sin(y)}$.

Solution: Note that the differential equation $\dfrac{dy}{dx} = \dfrac{x}{y+\sin(y)}$ is not linear.

restart

$ode := \dfrac{d}{dx} y(x) = \dfrac{x}{y(x)+\sin(y(x))}$

$$\dfrac{d}{dx} y(x) = \dfrac{x}{y(x)+\sin(y(x))} \tag{9.62}$$

infolevel$[dsolve]:=3:$

dsolve(ode)

```
Methods for first order ODEs:
--- Trying classification methods ---
trying a quadrature
trying 1st order linear
trying Bernoulli
trying separable
<- separable successful
```

$$\dfrac{1}{2}x^2 - \dfrac{1}{2}y(x)^2 + \cos(y(x)) + \_C1 = 0 \tag{9.63}$$

isolate$((9.63), y(x))$

$$-\dfrac{1}{2}y(x)^2 + \cos(y(x)) = -\dfrac{1}{2}x^2 - \_C1 \tag{9.64}$$

Hence, in this case it was not possible to write the solution of the differential equation explicitly. This solution is given implicitly.

Direction Fields

The *direction field* of a differential equation consists of small arrows that are tangent to the graphs of the solution of the differential equation (called integral curves). The direction field is a graphical representation of the solution of the equation and helps analyze the behavior of the solutions. Indeed these arrows can serve as guides to sketch the integral curves of the differential equation.

Example 13:

(a)   Sketch the direction field of the differential equation $\dfrac{dy}{dx} = x^3 - 5x\cos(x)$.

(b)   Sketch the graphs of the solutions that satisfy the given initial conditions
$$y(0) = -2,\; y(0) = -1,\; y(0) = 0,\; y(0) = 1,\; y(0) = 2.$$

Solution:

(a)   To sketch the direction field for the differential equation, we need to activate the package *with(DEtools)* to enable the command *dfieldplot* .

restart
with(DEtools) :

$dfieldplot\left(\dfrac{d}{dx} y(x) = x^3 - 5 \cdot x \cdot \cos(x), y(x), x = -4..4, y = -5..5, title = 'direction\right.$

$field\ plot', color = x^3 - 5 \cdot x \cdot \cos(x), arrows = SLIM, dirfield = [10,10])$

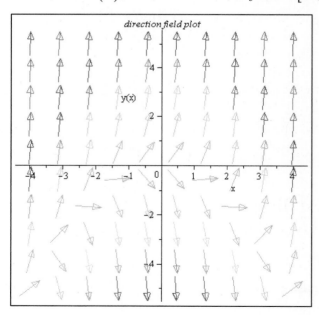

Note the different options included in the previous command (*title, color, arrow, dirfield*).

(b)   In order to sketch the direction field and the solution curves with *y*-intercepts -2, -1, 0, 1, and 2, we will use the command *DEplot*. Do not forget to activate the *with(DEtools)* package.

$$IC := \big[[0,c]\$c = -2..2\big]$$

$$\big[[0,-2],[0,-1],[0,0],[0,1],[0,2]\big] \tag{9.65}$$

$$DEplot\left(\frac{d}{dx}y(x) = x^3 - 5\cdot x\cdot\cos(x), [y(x)], x = -4..4, y = -5..5, IC,\right.$$

$$arrows = medium, color = [blue, red]\big)$$

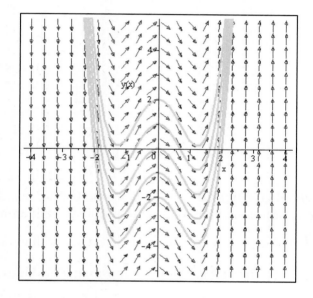

Observe that the arrows follow the curve and are tangent to the curve.

Example 14: Sketch the direction field and the integral curves of the differential equation $\dfrac{dy}{dx} = x^2 + 3y$ given the initial conditions

$$y(0) = -2,\ y(0) = -1,\ y(0) = 0,\ y(0) = 1,\ y(0) = 2.$$

Solution: We will activate the package *with(DEtools)* to enable the command *DEplot*.

restart
with(DEtools):
$$IC := \big[[0,c]\$c = -2..2\big]$$

$$\big[[0,-2],[0,-1],[0,0],[0,1],[0,2]\big] \tag{9.66}$$

$$DEplot\left(\frac{d}{dx}y(x)=x^2+3\cdot y(x),\left[y(x)\right],x=-4..4,y=-5..5,IC,arrows=medium,\right.$$

$$color=\left[blue,red\right])$$

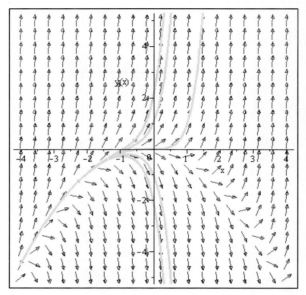

Euler's Method

Since many differential equations do not have explicit solutions, direction fields are used to approximate exact solutions of initial value problems. Given an initial value problem $\frac{dy}{dx}=f(x,y)$, $y(0)=y_0$, a step of size h, we define a sequence of points $y_n=y_{n-1}+hf(x_{n-1},y_{n-1})$, for $n\geq 1$, that are approximate values of the solution at the numbers x_0, $x_1=x_0+h$, $x_2=x_1+h$, ..., $x_n=x_{n-1}+h$. This is *Euler's method*.

Example 15: Given the initial value problem $\frac{dy}{dx}=e^x+y$, and $y(0)=1$.

(a)   Use Euler's method with step size 0.1 to construct an array of approximate values for the solution of the initial value problem. Find an estimate value for $y(1)$.

(b)   Solve the initial value problem. Compare the exact value to the approximate value found in (a). What is the error?

Solution:

(a)   We will first use Euler's method to construct an array where the third column is the sequence y_n defined above. We have $x_0=0$ and $h=0.1$.

> *restart*
> $A:=array(1..11,1..3):$

$A[1,1]:='n':A[1,2]:='x_n':A[1,3]:='y_n':A[2,1]:=1:A[2,2]:=0.1:A[2,3]:=2:$

for n **from** 3 **to** 11 **do** $A[n,1]:=n-1:A[n,2]:=0.1\cdot(n-1):A[n,3]:=A[n-1,3]+$

$0.1\cdot\left(e^{A[n-1,2]}+A[n-1,3]\right)$ **end do:**

print(A)

$$\begin{bmatrix} n & x_n & y_n \\ 1 & 0.1 & 1.2 \\ 2 & 0.2 & 1.430517092 \\ 3 & 0.3 & 1.695709077 \\ 4 & 0.4 & 2.000265866 \\ 5 & 0.5 & 2.349474923 \\ 6 & 0.6 & 2.749294542 \\ 7 & 0.7 & 3.206435876 \\ 8 & 0.8 & 3.728454735 \\ 9 & 0.9 & 4.323854301 \\ 10 & 1.0 & 5.002200042 \end{bmatrix}$$

(9.67)

Therefore an estimate for $y(1)$ is $y_{10}=5.002200042$.

(b)   We will now solve the initial value problem $\dfrac{dy}{dx}=e^x+y$, and $y(0)=1$. Note that the differential equation is linear, therefore the general solution can be written explicitly.

restart

infolevel$[dsolve]:=3:$

$ode:=\dfrac{d}{dx}y(x)=e^x+y(x)$

$$\frac{d}{dx}y(x)=e^x+y(x)$$

(9.68)

initialvalues $:=y(0)=1$

$$y(0)=1$$

(9.69)

dsolve$(\{ode,initialvalues\})$

```
Methods for first order ODEs:
--- Trying classification methods ---
trying a quadrature
trying 1st order linear
<- 1st order linear successful
```

$$y(x)=(x+1)e^x$$

(9.70)

Therefore the solution of the given initial value problem is $y(x) = (x+1)e^x$. Next, we will find the value of $y(1)$. To do so we will use the *subs* command.

$$evalf\left(subs\left(x=1, rhs\left(\left(\textbf{9.70}\right)\right)\right)\right)$$

$$5.436563656 \tag{9.71}$$

The exact value for $y(1)$ is 5.436563656, the approximate value found in (a) is $y_{10} = 5.002200042$. The error is $5.436563656 - 5.002200042 = 0.434363614$. Better approximations of $y(1)$ could be obtained if we consider smaller values of h.

Remark: The *Maple* command *dsolve$_{interactive}$* launches the *ODE Analyzer Assistant* which offers an interactive worksheet to solve differential equations and plot solution curves.

Exercises

1. Solve the differential equation $\dfrac{dy}{dx} = \dfrac{x}{e^x}$.

2. Solve the differential equation $\dfrac{dy}{dx} - 5x^2 y = 10x^2$.

3. Solve the differential equation $\dfrac{dy}{dx} = \dfrac{xy^2}{\sqrt{4x^2+1}}$.

4. Solve the differential equation $\dfrac{dy}{dx} = \dfrac{2}{3}(y^2 - 1)$.

5. Solve the differential equation $\dfrac{dy}{dx} = \ln y(y-1)$.

6. Find the particular solution of the differential equation $x\dfrac{dy}{dx} - y = x\ln(x)$ that satisfies $y(1) = 5$.

7. Sketch the direction field and the graph of the solution of the initial value problem $y' = x\cos(y)$ and $y(0) = 1$.

8. Find the particular solution of the differential equation $\dfrac{d^2 y}{dx^2} + 4y = 0$ that satisfies $y(0) = 1$ and $y'(0) = -2$.

9. Use Euler's method with step size 0.1 to construct an array of approximate construct an array of approximate values for the solution of the initial value problem $\dfrac{dy}{dx} = x - y$, $y(0) = 1$. Estimate the value of $y(1)$.

10. A tank of volume 150 liters is filled with a salty solution with a concentration of 0.6 kg per liter. To decrease the salt concentration, water is run into the solution. The quantity s of salt in the solution at any time satisfies the differential equation $\dfrac{ds}{dw} = -\dfrac{s}{150}$ where w is the amount of water which has run through the solution. How much water is needed to have a salt content of 60 kg?

11.   100 liters of brine solution is contained in a tank. The solution has a concentration of 0.5 kg of salt per liter. The resulting solution is kept uniform by stirring and is running out at the same rate it ran into the tank. The concentration of salt $s(t)$ at time t in the brine solution satisfies the differential equation $\dfrac{ds}{dt} = 3(0.5) - \dfrac{3}{100}s$. Find the time when the salt content is 70kg. We recall that the concentration of salt in the solution is the quantity of salt /volume of the solution.

12.   A college student needs a continuous cash flow of $20,000 per year for 5 years. He withdraws money from a fund earning interest at the annual rate of 6% compounded continuously. The amount of money A in the fund at any time t satisfies the differential equation $\dfrac{dA}{dt} = .06A - 20000$. How much money should be invested initially?

13.   A bank account earns interest at the annual rate of 8% compounded continuously. Money is deposited in a continuous cash flow at a rate of $1000 per year. How much money will be in the account after 10 years?

14.   An object of mass m on a vibrating spring with constant k is displaced s units from its equilibrium position. The function $s(t)$ satisfies the differential equation $m\dfrac{d^2s}{dt^2} = -k^2 s$.

   (a)   Solve the differential equation to find s.
   (b)   Given that $m = 3kg$ and $k = 243$, its initial position is $s = 0.4$ meters and its initial velocity is 0 m/s, find the position after 10 seconds.

15.   The population size of bacteria at time t (in hours), $P(t)$, satisfies the differential equation $\dfrac{dP}{dt} = 0.03P$. Given that the initial population is 100, what is the population after 2 hours?

16.   The population size $G(t)$ of gold fish in a pond satisfies the equation $G(t) = (B - D)\dfrac{dG}{dt}$ where B is the birth rate and D is the death rate. If we assume that the initial population of gold fish is 100, $B = 5\%$ and $G = 10\%$, how many gold fish will be left in the pond after 15 years?

17.   The size $X(t)$ of a tumor at time t satisfies the logistic equation $\dfrac{dX}{dt} = r\left(1 - \dfrac{X}{k}\right)X$. Solve the differential equation.

18.   The population size of a bacteria satisfies the equation $\dfrac{dP}{dt} = 2P(1000 - P)$. Given that the initial population is 100, what is the population after 10 years? Find the asymptotic population size $\lim_{t \to \infty} P(t)$.

19.   Below is a list of some commands used in this chapter. Describe the significance of each command. Can you find examples where each command is used?
        (a)   *DEplot*
        (b)   *dfieldplot*
        (c) $dsolve(ode)$
        (d) $dsolve(\{ode, initialvalues\})$
        (e) $dsolve(ode, [linear], useInt)$
        (f) $infolevel[dsolve] := n$
        (g) $initialvalues := y(0) = c$
        (h) $odeadvisor(ode)$
        (i) $with(DEtools)$

20.   Can you list new *Maple* commands that were used in this chapter but were not listed in exercise 19?

Chapter 10   Sequences and Series

The object of this chapter is to predict the convergence and divergence properties of sequences and series. Different tests will be employed and a greater appreciation of the usefulness of these tests will be stressed in the exploration problems. Approximation by Taylor's Polynomial will be made.

Sequences

Definition: A *sequence* is a function whose domain is a subset of the set of positive integers. We will use the notation $a(n)$ to denote the nth *term* (also called *general term*) of the sequence throughout this chapter and $\{a(n)\}$ to denote the sequence, for $n=1..\infty$.

Definition: We say that the infinite sequence $\{a(n)\}$ *converges* (or is convergent) if there is a real number L such that for every $\varepsilon > 0$ there is a positive integer N such that $|a(n)-L| < \varepsilon$ for all $n > N$. In this case we write $\lim\limits_{n\to\infty} a(n) = L$. If there is no such L, we say that the sequence *diverges* (or is divergent).

Example 1: Find the first five terms for the sequences whose the nth terms are:

(a) $a(n) = n^2$

(b) $b(n) = \dfrac{e^{3n}}{n}$

(c) $c(n) = \dfrac{(-1)^{n+1}}{n}$

Solution: We will use the command *seq* to find the desired terms.

(a)   We will enter the sequence $\{a(n)\}$ as a *Maple* function and then find the first five terms.

$$a := n \to n^2$$

$$n \to n^2 \tag{10.1}$$

$$seq\big(a(n), n=1..5\big)$$

$$1, 4, 9, 16, 25 \tag{10.2}$$

Hence the first five terms are 1, 4, 9, 16, and 25.

(b)   We will enter the sequence $\{b(n)\}$ as a *Maple* function and then find the first five terms.

$$b := n \to \frac{e^{3 \cdot n}}{n}$$

$$n \to \frac{e^{3n}}{n} \qquad (10.3)$$

$$seq\big(b(n), n = 1..5\big)$$

$$e^3, \frac{1}{2}e^6, \frac{1}{3}e^9, \frac{1}{4}e^{12}, \frac{1}{5}e^{15} \qquad (10.4)$$

Next we use the *evalf* command to find approximate values for these numbers.

$$evalf(\%)$$

$$20.08553692, 201.7143968, 2701.027976, 40688.69785, 6.538034744\ 10^5 \quad (10.5)$$

Recall that the % sign denotes the previous command. Therefore, the first five terms are $e^3, \frac{1}{2}e^6, \frac{1}{3}e^9, \frac{1}{4}e^{12}$, and $\frac{1}{5}e^{15}$ or as a floating point approximation 20.08553692, 201.7143968, 2701.027976, 40688.69785, $6.538034744\ 10^5$. Note that if we use the command $evalf\big((10.4)\big)$ we get the following error message.

$$evalf\big((10.4)\big)$$

```
Error, invalid input: evalf expects one or two
arguments but received 5
```

(c)   Once again we will enter the sequence $\{c(n)\}$ as a *Maple* function and then find the first five terms.

$$c := n \to \frac{(-1)^{n+1}}{n}$$

$$n \to \frac{(-1)^{n+1}}{n} \qquad (10.6)$$

$$seq\big(c(n), n = 1..5\big)$$

$$1, -\frac{1}{2}, \frac{1}{3}, -\frac{1}{4}, \frac{1}{5} \qquad (10.7)$$

Hence the first five terms are $1, -\frac{1}{2}, \frac{1}{3}, -\frac{1}{4}$, and $\frac{1}{5}$.

Remark: We can use the same command to find other terms of the sequence. For example the 18th through 21st terms of the sequence $\{n^2\}$ is shown.

$$a := n \rightarrow n^2$$

$$n \rightarrow n^2 \tag{10.8}$$

$$seq(a(n), n = 18..21)$$

$$324, 361, 400, 441 \tag{10.9}$$

Example 2: Use *Maple* commands to check if the sequences, whose nth terms are given below, converge or diverge.

(a) $\quad a(n) = \dfrac{2^{5n+1}}{n!}$.

(b) $\quad b(n) = 2^n$.

(c) $\quad c(n) = (-1)^n$.

Solution: We will use the *limit* command or $\lim\limits_{x \to a} f$ from the *Expression* palette to evaluate the limit, if it exists.

(a) We will define the sequence $\{a(n)\}$ as a *Maple* function and then find the limit.

$$a := n \rightarrow \dfrac{2^{5 \cdot n+1}}{n!}$$

$$n \rightarrow \dfrac{2^{5n+1}}{n!} \tag{10.10}$$

$$\lim\limits_{n \to \infty} a(n)$$

$$0 \tag{10.11}$$

The sequence $\{a(n)\}$ converges and its limit is 0.

(b) We will define the sequence $\{b(n)\}$ as a *Maple* function and then find the limit.

$$b := n \rightarrow 2^n$$

$$n \rightarrow 2^n \tag{10.12}$$

$$limit(b(n), n = \infty)$$

$$\infty \tag{10.13}$$

Since the limit is ∞, the sequence $\{b(n)\}$ is divergent.

(c) We will define the sequence $\{c(n)\}$ as a *Maple* function and then find the limit.

$$c := n \rightarrow (-1)^n$$

$$n \rightarrow (-1)^n \tag{10.14}$$

$$\lim_{n \to \infty} c(n)$$

$$-1..1 \tag{10.15}$$

Since there is no unique number that is the limit, the sequence $\{c(n)\}$ is divergent. Note that when n is even, the sequence $\{c(n)\}$ converges to 1, and when n is odd, the sequence converges to -1.

Example 3: Plot the first ten terms of the sequence with $a(n) = \dfrac{n}{2^n}$.

Solution: Let us first define the *Maple* sequence $a(n) = \dfrac{n}{2^n}$.

$$a := n \rightarrow \frac{n}{2^n}$$

$$n \rightarrow \frac{n}{2^n} \tag{10.16}$$

To plot a sequence (discrete points), we need to define points with coordinates $\left[n, a(n)\right]$, where $a(n)$ is the nth term of the sequence.

$$plot\left(\left[seq\left(\left[n, a(n)\right], n = 1..10\right)\right], style = point, color = blue, symbolsize = 15\right)$$

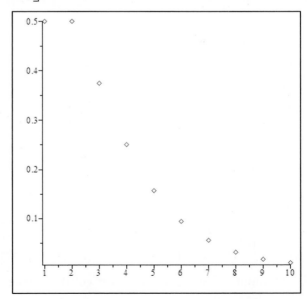

Example 4: Generate and plot the sequence with general term $a(n) = 1 - \dfrac{1}{n^2}$ for $n = 1..15$, then find its limit (if it exists).

Solution: Let us first define the general term $a(n) = 1 - \dfrac{1}{n^2}$ as a *Maple* function.

$a := n \to 1 - \dfrac{1}{n^2}$

$$n \to 1 - \frac{1}{n^2} \tag{10.17}$$

$seq(a(n), n = 1..15)$

$$0, \frac{3}{4}, \frac{8}{9}, \frac{15}{16}, \frac{24}{25}, \frac{35}{36}, \frac{48}{49}, \frac{63}{64}, \frac{80}{81}, \frac{99}{100}, \frac{120}{121}, \frac{143}{144}, \frac{168}{169}, \frac{195}{196}, \frac{224}{225} \tag{10.18}$$

$pointlist := \left[seq([n, a(n)], n = 1..15) \right]$

$$\left[[1, 0], \left[2, \frac{3}{4}\right], \left[3, \frac{8}{9}\right], \left[4, \frac{15}{16}\right], \left[5, \frac{24}{25}\right], \left[6, \frac{35}{36}\right], \left[7, \frac{48}{49}\right], \left[8, \frac{63}{64}\right], \left[9, \frac{80}{81}\right], \right. \tag{10.19}$$

$$\left. \left[10, \frac{99}{100}\right], \left[11, \frac{120}{121}\right], \left[12, \frac{143}{144}\right], \left[13, \frac{168}{169}\right], \left[14, \frac{195}{196}\right], \left[15, \frac{224}{225}\right] \right]$$

$plot(pointlist, 1..15, style = point, symbol = solidcircle, symbolsize = 15)$

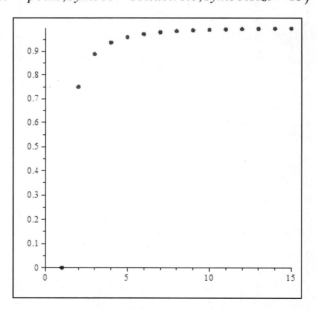

We notice that the elements of the sequence are getting closer to 1, as n gets larger. We will use *Maple* to find the limit.

$$\lim_{n\to\infty} a(n)$$

$$1 \qquad\qquad\qquad\qquad\qquad\qquad \textbf{(10.20)}$$

Hence $\lim\limits_{n\to\infty} a(n) = 1$.

Series

Definition: The kth-partial sum of the infinite series $\sum\limits_{n=1}^{\infty} a(n)$ is defined as

$$S_k = \sum_{n=1}^{k} a(n) = a(1) + a(2) + \ldots + a(k), \ k = 1..\infty.$$ The term $a(n)$ is called the nth term of the series.

Definition: We say that the series $\sum\limits_{n=1}^{\infty} a(n)$ *converges* if the sequence of partial sums $\{S_k\}$ converges. In this case we have $\sum\limits_{n=1}^{\infty} a(n) = \lim\limits_{k\to\infty} S_k$. If the sequence of partial sums diverges, we say that the series diverges.

Remark: Note that we have started the series at $n = 1$, we could also start the series at 0 or any other positive integer. Unless we need to specify the index, we will refer to an infinite series as $\sum a(n)$.

Theorem: If a series $\sum a(n)$ converges then $\lim\limits_{n\to\infty} a(n) = 0$.

Remark: The converse of this theorem is false.

Remark: If the series $\sum\limits_{n=m}^{\infty} a(n)$ converges for some positive integer $m \geq k$, then the series $\sum\limits_{n=k}^{\infty} a(n)$ also converges.

Properties of series

If $\sum\limits_{n=1}^{\infty} a(n)$ and $\sum\limits_{n=1}^{\infty} b(n)$ are convergent series, then

1. $c\sum\limits_{n=1}^{\infty} a(n)$ is convergent and $c\sum\limits_{n=1}^{\infty} a(n) = \sum\limits_{n=1}^{\infty} ca(n)$ for all real numbers c.

2. $\sum\limits_{n=1}^{\infty} a(n) + \sum\limits_{n=1}^{\infty} b(n)$ is convergent and $\sum\limits_{n=1}^{\infty} a(n) + \sum\limits_{n=1}^{\infty} b(n) = \sum\limits_{n=1}^{\infty} (a(n) + b(n))$

Example 5: Find the partial sums S_{10} and S_k of the series $\sum\limits_{n=1}^{\infty}\dfrac{(-1)^n}{2^n}$.

Solution: The *Maple* command *sum* will be used to evaluate the sum.

$$sum\left(\frac{(-1)^n}{2^n},n=1..10\right)$$

$$-\frac{342}{1024} \tag{10.21}$$

Now, let us find the partial sum S_k and simplify it.

$$sum\left(\frac{(-1)^n}{2^n},n=1..k\right)$$

$$-\frac{2}{3}\left(-\frac{1}{2}\right)^{k+1}-\frac{1}{3} \tag{10.22}$$

simplify$((\mathbf{10.22}))$

$$\frac{1}{3}\frac{(-1)^k}{2^k}-\frac{1}{3} \tag{10.23}$$

Remark: The *Maple* command *Sum* will return the inert form of the sum.

$$Sum\left(\frac{(-1)^n}{2^n},n=1..10\right)$$

$$\sum_{n=1}^{10}\frac{(-1)^n}{2^n} \tag{10.24}$$

Example 6: Determine whether the series $\sum\limits_{n=1}^{\infty}\dfrac{n^2-3n+1}{4n^2+5}$ converges.

Solution: We will find the limit of the nth *term* of the series $\sum\limits_{n=1}^{\infty}\dfrac{n^2-3n+1}{4n^2+5}$.

$$\lim_{n\to\infty}\frac{n^2-3\cdot n+1}{4\cdot n^2+5}$$

$$\frac{1}{4} \tag{10.25}$$

Since $\lim\limits_{n\to\infty}\dfrac{n^2-3n+1}{4n^2+5}\neq 0$, the series $\sum\limits_{n=1}^{\infty}\dfrac{n^2-3n+1}{4n^2+5}$ diverges. We will now use the *Maple* expression $\sum\limits_{i=k}^{n} f$ to verify the fact that the series diverges.

$$\sum_{n=1}^{\infty}\frac{n^2-3\cdot n+1}{4\cdot n^2+5}$$

$$\infty \hspace{8cm} \textbf{(10.26)}$$

This was expected as we know the series diverges.

The Limit Comparison Test: Given $\sum\limits_{n=1}^{\infty} a(n)$ and $\sum\limits_{n=1}^{\infty} b(n)$ two series with positive terms.

If $\lim\limits_{n\to\infty}\dfrac{a(n)}{b(n)}=L$ where $L>0$, then the series $\sum\limits_{n=1}^{\infty} a(n)$ and $\sum\limits_{n=1}^{\infty} b(n)$ are either both convergent or both divergent.

The p-series: The *p-series* $\sum\limits_{n=1}^{\infty}\dfrac{1}{n^p}$ converges if $p>1$, and diverges is $p\leq 1$.

Example 7: Determine whether the series $\sum\limits_{n=2}^{\infty}\dfrac{1}{n^3-1}$ converges or diverges.

Solution: We will use the limit comparison test with the *p-series* $\sum\limits_{n=2}^{\infty}\dfrac{1}{n^3}$ ($p=3$).

$$\lim_{n\to\infty}\frac{\dfrac{1}{n^3-1}}{\dfrac{1}{n^3}}$$

$$1 \hspace{8cm} \textbf{(10.27)}$$

Since the p-series $\sum\limits_{n=2}^{\infty}\dfrac{1}{n^3}$ converges, the series $\sum\limits_{n=2}^{\infty}\dfrac{1}{n^3-1}$ also converges.

Example 8: (Telescoping Sum)

(a)   Find $\sum\limits_{n=1}^{k}\dfrac{1}{n(n+1)}$.

(b)   Show that the series $\sum\limits_{n=1}^{\infty}\dfrac{1}{n(n+1)}$ converges and find its sum.

Solution:

(a)   We will use *Maple* to find the partial sum $\sum_{n=1}^{k} \frac{1}{n(n+1)}$.

$$\sum_{n=1}^{k} \frac{1}{n(n+1)}$$

$$-\frac{1}{k+1}+1 \tag{10.28}$$

Hence we have $\sum_{n=1}^{k} \frac{1}{n(n+1)} = 1 - \frac{1}{k+1}$.

(b)   We will now show that the series $\sum_{n=1}^{\infty} \frac{1}{n(n+1)}$ converges by checking if the sequence of partial sums obtained in part (a) converges.

$$\lim_{k \to \infty} \left(-\frac{1}{k+1} + 1 \right)$$

$$1 \tag{10.29}$$

Hence the series $\sum_{n=1}^{\infty} \frac{1}{n(n+1)}$ also converges and $\sum_{n=1}^{\infty} \frac{1}{n(n+1)} = 1$. We will verify this using *Maple* commands.

$$\sum_{n=1}^{\infty} \frac{1}{n(n+1)}$$

$$1 \tag{10.30}$$

Hence $\sum_{n=1}^{\infty} \frac{1}{n(n+1)} = 1$.

Example 9: (Telescoping Sum)

(a)   Find $\sum_{n=1}^{k} \ln\left(1 + \frac{1}{n}\right)$.

(b)   Show that the series $\sum_{n=1}^{k} \ln\left(1 + \frac{1}{n}\right)$ diverges.

Solution:

(a)   We will use *Maple* to find the partial sum $\sum_{n=1}^{k} \ln\left(1 + \frac{1}{n}\right)$.

$$\sum_{n=1}^{k} \ln\left(1 + \frac{1}{n}\right)$$

$$\sum_{n=1}^{k} \ln\left(\frac{n+1}{n}\right) \tag{10.31}$$

Notice that *Maple* returned the partial sum unevaluated. Let us now use the *simplify* command to find this partial sum.

simplify $\big((\mathbf{10.31})\big)$

$$\sum_{n=1}^{k} \ln\left(\frac{n+1}{n}\right) \tag{10.32}$$

$$\sum_{n=1}^{k} \ln\left(\frac{n+1}{n}\right)$$

$$\sum_{n=1}^{k} \ln\left(\frac{n+1}{n}\right) \tag{10.33}$$

Once again the partial sum is unevaluated. Let us now use the fact that $\ln\left(\dfrac{n+1}{n}\right)$

$= \ln(n+1) - \ln(n)$ to find the partial sum to see the result obtained by Maple and then simplify this result.

$$\sum_{n=1}^{k} \big(\ln(n+1) - \ln(n)\big)$$

$$\ln\big(\Gamma(k+2)\big) - \ln\big(\Gamma(k+1)\big) \tag{10.34}$$

$\xrightarrow{\text{simplify symbolic}}$

$$\ln(k+1) \tag{10.35}$$

Therefore, we have $\displaystyle\sum_{n=1}^{k} \ln\left(1+\frac{1}{n}\right) = \ln(k+1)$. Note that $\Gamma(x)$ is the gamma function when x is a positive integer and $\Gamma(x+1) = x!$.

(b)   We will now show that the series $\displaystyle\sum_{n=1}^{\infty} \frac{1}{n(n+1)}$ diverges by checking if the sequence of partial sums obtained in part (a) converges.

$$\lim_{k \to \infty} \big(\ln(k+1)\big)$$

$$\infty \tag{10.36}$$

Hence the series $\displaystyle\sum_{n=1}^{k} \ln\left(1+\frac{1}{n}\right)$ diverges.

The Integral Test: Let f be a positive and continuous function on the interval $[k,\infty]$. Let us assume that f is decreasing on the interval $[m,\infty)$ for $m \geq k$. We consider the sequence defined by $a(n) = f(n)$. If the improper integral $\int_k^\infty f(x)\,dx$ converges, so does the infinite series $\sum_{n=k}^{\infty} a(n)$, and if the improper integral diverges, so does the series.

Example 10: Determine whether the series $\sum_{n=1}^{\infty} ne^{-n^2}$ converges.

Solution: We will use the integral test to determine if the series converges or diverges. Let $f(x) = xe^{-x^2}$. Then $f(x)$ is positive and is continuous on the interval $[1,\infty)$. To determine whether $f(x)$ is decreasing, we will find the derivative $f'(x)$ and check whether $f'(x)$ is negative on the interval $[1,\infty)$.

$$f := x \rightarrow x \cdot e^{-x^2}$$

$$x \rightarrow xe^{-x^2} \tag{10.37}$$

$$\frac{d}{dx} f(x)$$

$$e^{-x^2} - 2x^2 e^{-x^2} \tag{10.38}$$

$$solve\left(\frac{d}{dx} f(x) < 0.0, \{x\}\right)$$

$$\{x < -0.7071067812\}, \{0.7071067812 < x\} \tag{10.39}$$

Hence $f(x)$ is decreasing for $x > 1$. This can be checked if we plot the graph of $f(x)$ on the interval $[1,\infty)$.

$$plot\left(f(x), x = 1..\infty\right)$$

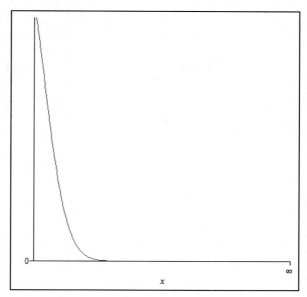

Next, we will find the value of the improper integral $\int_1^\infty f(x)\,dx$

$$\int_1^\infty f(x)\,dx$$

$$\frac{1}{2}e^{-1} \tag{10.40}$$

Hence the improper integral $\int_1^\infty f(x)\,dx$ converges. Therefore, the series $\displaystyle\sum_{n=1}^\infty ne^{-n^2}$ converges.

Remark: *Maple* does not evaluate the sum of the series $\displaystyle\sum_{n=0}^\infty n\cdot e^{-n^2}$. This is shown below.

$$\sum_{n=0}^\infty n\cdot e^{-n^2}$$

$$\sum_{n=0}^\infty n\,e^{-n^2} \tag{10.41}$$

Definition: An *alternating series* is a series whose terms alternate in signs.

The Alternating Series Test: Given an alternating series $\displaystyle\sum_{n=1}^\infty (-1)^n b(n)$, if $0 < b(n+1) \le b(n)$ for all n (decreasing sequence) and $\displaystyle\lim_{n\to\infty} b(n) = 0$, then the series $\displaystyle\sum_{n=1}^\infty (-1)^n b(n)$ is convergent.

Example 11: Find the sum of the series $\displaystyle\sum_{n=2}^{\infty}\frac{(-1)^n}{n-1}$ if it exists.

Solution: We will first use the alternating series test to show that the series $\displaystyle\sum_{n=2}^{\infty}\frac{(-1)^n}{n-1}$ converges and then we will find the sum of the series. Consider the sequence $\{b(n)\}$ with $b(n)=\dfrac{1}{n-1}$ and let $f(x)=\dfrac{1}{x-1}$. The function $f(x)$ is positive on the interval $[2,\infty)$. To show that $f(x)$ is decreasing, we will find the derivative $f'(x)$ and check that $f'(x)$ is negative on the interval $[2,\infty)$.

$$f:=x\to\frac{1}{x-1}$$

$$x\to\frac{1}{x-1} \tag{10.42}$$

$$\frac{d}{dx}f(x)$$

$$-\frac{1}{(x-1)^2} \tag{10.43}$$

Observe that $f'(x)$ is negative on the interval $[2,\infty)$ and hence the function $f(x)$ is decreasing on $[2,\infty)$. Let us use *Maple* commands to find $\lim\limits_{n\to\infty}b(n)=0$.

$$\lim_{n\to\infty}\frac{1}{n-1}$$

$$0 \tag{10.44}$$

Since all the conditions of the alternating series test are satisfied, the series $\displaystyle\sum_{n=2}^{\infty}\frac{(-1)^n}{n-1}$ converges. We will verify this using *Maple* commands and find $\displaystyle\sum_{n=2}^{\infty}\frac{(-1)^n}{n-1}$.

$$\sum_{n=2}^{\infty}\frac{(-1)^n}{n-1}$$

$$\ln(2) \tag{10.45}$$

Definition: We say that the series $\sum a(n)$ is *absolutely convergent* if the series $\sum|a(n)|$ converges.

Example 12: Show that the series $\sum_{n=1}^{\infty} \frac{(-1)^n}{n}$ converges but does not converge absolutely.

Solution: Using the alternating test as in the previous example, we can show that the series $\sum_{n=1}^{\infty} \frac{(-1)^n}{n}$ converges. Moreover, we can use *Maple* to find $\sum_{n=1}^{\infty} \frac{(-1)^n}{n}$.

$$\sum_{n=1}^{\infty} \frac{(-1)^n}{n}$$

$$-\ln(2) \tag{10.46}$$

Therefore, the series $\sum_{n=1}^{\infty} \frac{(-1)^n}{n}$ converges and its sum is $-\ln(2)$. On the other hand,

$\sum_{n=1}^{\infty} \left| \frac{(-1)^n}{n} \right| = \sum_{n=1}^{\infty} \frac{1}{n}$ is a *p*-series with *p* =1 (called the harmonic series), so it is divergent,

therefore the series $\sum_{n=1}^{\infty} \frac{(-1)^n}{n}$ does not converge absolutely. This could be checked using

Maple commands to find $\sum_{n=1}^{\infty} \left| \frac{(-1)^n}{n} \right|$

$$\sum_{n=1}^{\infty} \left| \frac{(-1)^n}{n} \right|$$

$$\infty \tag{10.47}$$

Remark: If a series is absolutely convergent, then it is convergent. The previous example shows that the converse statement is not true. If a series converges, but does not converge absolutely, we say that the series *converges conditionally.*

The Comparison Test: Given $\sum_{n=1}^{\infty} a(n)$ and $\sum_{n=1}^{\infty} b(n)$ two series with positive terms.

1. If $a(n) \leq b(n)$ for all n and $\sum_{n=1}^{\infty} b(n)$ converges, then $\sum_{n=1}^{\infty} a(n)$ converges.

2. If $a(n) \geq b(n)$ for all n and $\sum_{n=1}^{\infty} b(n)$ diverges, then $\sum_{n=1}^{\infty} a(n)$ diverges.

Example 13: Determine whether the series $\displaystyle\sum_{n=1}^{\infty} \frac{\sin(n)}{n^2}$ converges.

Solution: We will show that the series $\displaystyle\sum_{n=1}^{\infty} \frac{\sin(n)}{n^2}$ converge absolutely, therefore it converges. To do so we will use the comparison test. We have $\left|\dfrac{\sin(n)}{n^2}\right| \leq \dfrac{1}{n^2}$ for all n, and the series $\displaystyle\sum_{n=1}^{\infty} \frac{1}{n^2}$ converges since it a p-series with $p > 1$. In fact, we see that

$$\sum_{n=1}^{\infty} \frac{1}{n^2}$$

$$\frac{1}{6}\pi^2 \tag{10.48}$$

Therefore $\displaystyle\sum_{n=1}^{\infty} \frac{\sin(n)}{n^2}$ converges absolutely, hence it converges.

Example 14:

(a)   Determine whether the series $\displaystyle\sum_{n=1}^{\infty} \frac{(-1)^n \ln(n)}{n}$ converges or diverges.

(b)   Does the series converge absolutely?

Solution:

(a)   We will use the alternating series test to show that the series $\displaystyle\sum_{n=1}^{\infty} \frac{(-1)^n \ln(n)}{n}$ converges. Let $b(n) = \dfrac{\ln(n)}{n}$, for $n \geq 1$. Note that $b(n) > 0$ for $n \geq 1$. We will now check that the sequence $\{b(n)\}$ is decreasing.

$$f := x \to \frac{\ln(x)}{x}$$

$$x \to \frac{\ln(x)}{x} \tag{10.49}$$

$$\frac{\mathrm{d}}{\mathrm{d}x} f(x)$$

$$\frac{1}{x^2} - \frac{\ln(x)}{x^2} \tag{10.50}$$

$$solve\left(\frac{d}{dx}f(x)<0,\{x\}\right)$$

$$\{e<x\}\qquad\qquad\textbf{(10.51)}$$

Hence the function $f(x)$ is decreasing for $x \geq e$, which implies that the sequence $\{b(n)\}$ is decreasing for $n \geq 3$, as seen on the graph below.

$$plot\left(\left[\left[seq\left(\left[n,\frac{\ln(n)}{n}\right],n=3..100\right)\right],style=point,symbolsize=10\right)$$

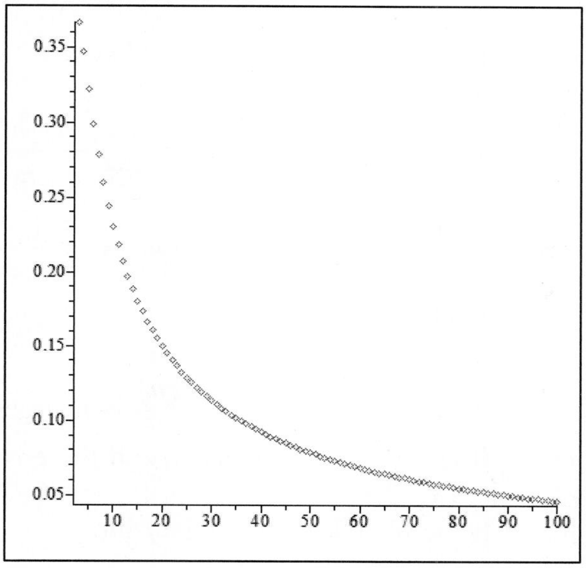

Let us now find $\displaystyle\lim_{n\to\infty}\frac{\ln(n)}{n}$.

$$\lim_{n\to\infty}\frac{\ln(n)}{n}$$

$$0\qquad\qquad\textbf{(10.52)}$$

As follows from the alternating series test, the series $\displaystyle\sum_{n=3}^{\infty}\frac{(-1)^n\ln(n)}{n}$ converges.

Hence the series $\displaystyle\sum_{n=1}^{\infty}\frac{(-1)^n\ln(n)}{n}$ also converges.

Chapter 10 Sequences and Series

(b)   To check if the series $\sum_{n=1}^{\infty} \dfrac{(-1)^n \ln(n)}{n}$ converge absolutely, we need to check if the

series $\sum_{n=1}^{\infty} \dfrac{\ln(n)}{n}$ converges. We will use the integral test with the function

$f(x) = \dfrac{\ln(x)}{x}$ defined in (a). Let us find the improper integral $\int_1^{\infty} f(x)\,dx$.

$$\int_1^{\infty} f(x)\,dx$$

$$\infty$$
                                                                                    (10.53)

Since the improper integral $\int_1^{\infty} f(x)\,dx$ diverges, the series $\sum_{n=1}^{\infty} \dfrac{\ln(n)}{n}$ also

diverges. This implies that the series $\sum_{n=1}^{\infty} \dfrac{(-1)^n \ln(n)}{n}$ does not converge

absolutely. But as seen in part (a) the series $\sum_{n=1}^{\infty} \dfrac{(-1)^n \ln(n)}{n}$ converges, so it

converges conditionally.

Geometric Series

Definition: A series $\sum a(n)$ is said to be *geometric* if there is a real number r such that $\dfrac{a(n+1)}{a(n)} = r$ for all n. The number r is called the *ratio*.

Theorem: A geometric series converges if the ration r satisfies $-1 < r < 1$.
In this case the sum of the series is $\sum a(n) = \dfrac{a(1)}{1-r}$.

Example 15: Determine whether the series $\sum_{n=1}^{\infty} \dfrac{(-2)^n}{3^{n-1}}$ is geometric. Does the series converge?

Solution: Define $a(n) = \dfrac{(-2)^n}{3^{n-1}}$ as a *Maple* function and find the ratio $r = \dfrac{a(n+1)}{a(n)}$.

$$a := n \to \dfrac{(-2)^n}{3^{n-1}}$$

$$n \to \frac{(-2)^n}{3^{n-1}} \tag{10.54}$$

$$\frac{a(n+1)}{a(n)}$$

$$\frac{(-2)^{n+1} 3^{n-1}}{3^n (-2)^n} \tag{10.55}$$

So $r = \dfrac{a(n+1)}{a(n)} = \dfrac{(-2)^{n+1} 3^{n-1}}{3^n (-2)^n}$. Now use the *Maple* command *simplify* to simplify the expression.

simplify $\big((\mathbf{10.54})\big)$

$$-\frac{2}{3} \tag{10.56}$$

Since the ratio $r = -\dfrac{2}{3}$ for all n, the series $\displaystyle\sum_{n=1}^{\infty} \dfrac{(-2)^n}{3^{n-1}}$ is geometric with ratio $r = -\dfrac{2}{3}$.

Given that $-1 < -\dfrac{2}{3} < 1$, the series $\displaystyle\sum_{n=1}^{\infty} \dfrac{(-2)^n}{3^{n-1}}$. Now let us find the sum of the series.

$$\sum_{n=1}^{\infty} a(n)$$

$$-\frac{6}{5} \tag{10.57}$$

Thus $\displaystyle\sum_{n=1}^{\infty} a(n) = -\dfrac{6}{5}$.

Remark: We will show that $\displaystyle\sum_{n=1}^{\infty} a(n) = -\dfrac{6}{5}$ using the formula of the sum of a geometric series.

$$r := -\frac{2}{3}$$

$$-\frac{2}{3} \tag{10.58}$$

$$\frac{a(1)}{1-r}$$

$$-\frac{6}{5} \tag{10.59}$$

The output in equation **(10.58)** is indeed the same number as in equation **(10.56)**.

Power Series

Definition: A series $\sum\limits_{n=0}^{\infty} a(n)(x-c)^n$ is called a *power series* about c, where x is a variable, $a(1)$, $a(2)$, .., $a(n)$, .. are the coefficients of the series.

Theorem: Given a power series $\sum\limits_{n=0}^{\infty} a(n)(x-c)^n$, then exactly one of the following statement is true:

1.   The series converges only at $x=c$. So the *radius of convergence* R is defined by $R=0$.
2.   The series converges for all values of x. So the *radius of convergence* R is defined by $R=\infty$.
3.   There is a positive number R, called the *radius of convergence*, so that the series converges for all values of x such that $|x-c|<R$ and diverges for $|x-c|>R$.

Theorem: *(Ratio Test)*

Given a power series $\sum\limits_{n=0}^{\infty} a(n)(x-c)^n$, let $C_n = a_n(x-c)^n$ then

1.   If $\lim\limits_{n\to\infty} \left| \dfrac{C(n+1)}{C(n)} \right| = 0$, then $R=\infty$.

2.   If $\lim\limits_{n\to\infty} \left| \dfrac{C(n+1)}{C(n)} \right| = \infty$, then $R=0$.

3.   If $\lim\limits_{n\to\infty} \left| \dfrac{C(n+1)}{C(n)} \right| = M|x-c|$, where M is a non-zero positive number, then

$R = \dfrac{1}{M}$.

Definition: The *interval of convergence* of a power series is the largest interval that contains all the values of x for which the series converges.

Example 16: Determine the radius and interval of convergence of the power series $\sum\limits_{n=1}^{\infty} \dfrac{2^{3n}\ln(n)}{n}(x-2)^n$.

Solution: We will enter the nth term, $C(n)$, of the power series as a *Maple* function of n and find and simplify, if necessary, the ratio $\left| \dfrac{C(n+1)}{C(n)} \right|$.

$$C := n \to \frac{2^{3 \cdot n} \cdot \ln(n)}{n} \cdot (x-2)^n$$

$$n \to \frac{2^{3n} \ln(n)(x-2)}{n} \qquad (10.60)$$

$$\left| \frac{C(n+1)}{C(n)} \right|$$

$$\left| \frac{2^{3n+3} \ln(n+1)(x-2)^{n+1} n}{(n+1) 2^{3n} \ln(n)(x-2)^n} \right| \qquad (10.61)$$

$simplify\big((\mathbf{10.61})\big)$

$$8 \left| \frac{\ln(n+1)(x-2) n}{(n+1)\ln(n)} \right| \qquad (10.62)$$

We will now evaluate the limit $\displaystyle\lim_{n\to\infty} \left| \dfrac{C(n+1)}{C(n)} \right|$.

$$\lim_{n\to\infty} 8 \cdot \left| \frac{n \cdot (x-2) \cdot \ln(n+1)}{(n+1) \cdot \ln(n)} \right|$$

$$8|x-2| \qquad (10.63)$$

Since $\displaystyle\lim_{n\to\infty} \left| \dfrac{C(n+1)}{C(n)} \right| = 8|x-2|$, the radius of convergence of the power series is $R = \dfrac{1}{8}$. We will now find the interval of convergence of the series. First we will solve the inequality $|x-2| < \dfrac{1}{8}$ using the *solve* command.

$$solve\left(|x-2| < \frac{1}{8}, \{x\} \right)$$

$$\left\{ \frac{15}{8} < x, x < \frac{17}{8} \right\} \qquad (10.64)$$

To determine what happens at the endpoints of the interval, we will begin by checking if the power series $\displaystyle\sum_{n=1}^{\infty}\frac{2^{3n}\ln(n)}{n}(x-2)^{n}$ converges at $x=\dfrac{15}{8}$.

$$subs\left(x=\frac{15}{8},\frac{2^{3n}\ln(n)}{n}(x-2)^{n}\right)$$

$$\frac{2^{3n}\ln(n)\left(-\dfrac{1}{8}\right)^{n}}{n} \tag{10.65}$$

$simplify\left(\left(\mathbf{10.65}\right)\right)$

$$\frac{(-1)^{n}\ln(n)}{n} \tag{10.66}$$

In example 14, we proved that the series $\displaystyle\sum_{n=1}^{\infty}\frac{(-1)^{n}\ln(n)}{n}$ converges, which means that $\dfrac{15}{8}$ is in the interval of convergence of the power series $\displaystyle\sum_{n=1}^{\infty}\frac{2^{3n}\ln(n)}{n}(x-2)^{n}$. Now let us see if the power series $\displaystyle\sum_{n=1}^{\infty}\frac{2^{3n}\ln(n)}{n}(x-2)^{n}$ converges at $x=\dfrac{17}{8}$.

$$subs\left(x=\frac{17}{8},\frac{2^{3n}\ln(n)}{n}(x-2)^{n}\right)$$

$$\frac{2^{3n}\ln(n)\left(\dfrac{1}{8}\right)^{n}}{n} \tag{10.67}$$

$simplify\left(\left(\mathbf{10.67}\right)\right)$

$$\frac{\ln(n)}{n} \tag{10.68}$$

Since the series $\displaystyle\sum_{n=1}^{\infty}\frac{\ln(n)}{n}$ diverges (see example 11), the number $\dfrac{17}{8}$ is not in the interval of convergence of the power series $\displaystyle\sum_{n=1}^{\infty}\frac{2^{3n}\ln(n)}{n}(x-2)^{n}$. Therefore, the interval of convergence of the series is $\left[\dfrac{15}{8},\dfrac{17}{8}\right)$.

Taylor and Maclaurin Series

Definition: If a function f is represented by a power series at c, in other terms $f(x) = \sum a(n)(x-c)^n$, for all x such that $|x-c| < R$ for $R > 0$, then the coefficients $a(n)$ are given by the formula

$$a(n) = \frac{f^{(n)}(c)}{n!}$$

and

$$f(x) = f(c) + f'(c)(x-c) + \frac{f''(c)}{2!}(x-c)^2 + \ldots + \frac{f^{(n)}(c)}{n!}(x-c)^n + \ldots$$

This series is called the _Taylor series of the function f at c_. In the special case where $c = 0$, then the series is called the _Maclaurin series of f_.

Definition: The _kth degree Taylor polynomial of f at c_ is defined by

$$T_k(x) = \sum_{i=0}^{k} \frac{f^{(i)}(c)}{i!}(x-c)^i.$$

Theorem: Let f be an infinitely differentiable function on an open interval I centered at c. Assume that there exists $M \geq 0$ such that $\left| f^{(i)}(x) \right| \leq M$, for all $i \geq 0$ and all x in I. Then f is represented by its Taylor series _at c_ on the interval I.

Example 17: Find the first 7 terms of the Maclaurin series of the function $x^3 e^x$.

Solution: We will use the _Maple_ command _TaylorApproximation_ in the $with(Student[Calculus1])$ package. In order to find the first 7 terms of the Maclaurin series of the function $x^3 e^x$ we will set the _order_ = 6. This command gives a 6th degree Taylor polynomial.

$with(Student[Calculus1])$:

$TaylorApproximation(x^3 \cdot e^x, order = 6)$

$$x^3 + x^4 + \frac{1}{2}x^5 + \frac{1}{6}x^6 \tag{10.69}$$

So the first 7 terms of the Maclaurin series of the function $x^3 e^x$ are $x^3 + x^4 + \frac{1}{2}x^5 + \frac{1}{6}x^6$.

This is the 6th degree Maclaurin polynomial of the function $x^3 e^x$.

Remarks:

1. If we want all the first seven Maclaurin expansions of the function $x^3 e^x$ we will set the *order* $= 0 .. 6$.

    $with\left(Student\left[Calculus1\right]\right):$

    $TaylorApproximation\left(x^3 \cdot e^x, order = 0 .. 6\right)$

    $$0, 0, 0, x^3, x^3 + x^4, x^3 + x^4 + \frac{1}{2}x^5, x^3 + x^4 + \frac{1}{2}x^5 + \frac{1}{6}x^6 \qquad \textbf{(10.70)}$$

2. We can use the *Maple* command *taylor* in the *with(MTM)* package to find the first 7 terms of the Maclaurin series of the function $x^3 e^x$.

    $with(MTM):$
    $taylor\left(x^3 e^x, 7\right)$

    $$x^3 + x^4 + \frac{1}{2}x^5 + \frac{1}{6}x^6 \qquad \textbf{(10.71)}$$

    Note that if we do not specify the number of terms in the expansion, the default number is 6. This means that the *taylor* command will compute the 5th degree Maclaurin polynomial.

    $taylor\left(x^3 e^x, x\right)$

    $$x^3 + x^4 + \frac{1}{2}x^5 \qquad \textbf{(10.72)}$$

3. If the package *with(MTM)* is not activated, the output of the *Maple* command *taylor* will be written differently.

    $restart$
    $taylor\left(x^3 e, x, 7\right)$

    $$x^3 + x^4 + \frac{1}{2}x^5 + \frac{1}{6}x^6 + O\left(x^7\right) \qquad \textbf{(10.73)}$$

    Again, if we do not specify the number of terms, we get the 5th degree Maclaurin polynomial.

    $taylor\left(x^3 e^x, x\right)$

    $$x^3 + x^4 + \frac{1}{2}x^5 + O\left(x^6\right) \qquad \textbf{(10.74)}$$

Example 18:

(a)   What are the first 3 terms of the Taylor series of the function $\dfrac{3}{1-5x}$ at -2?

(b)   Find the 3rd degree Taylor polynomial at -2.

Solution: We will activate the $with\left(Student[Calculus1]\right)$ package and use the $TaylorApproximation$ command.

(a)   Since we need to find the first 3 terms of the Taylor series of the function $\dfrac{3}{1-5x}$ at -2 we need to set $x=-2$ in the $TaylorApproximation$ command.

$with\left(Student[Calculus1]\right):$

$TaylorApproximation\left(\dfrac{3}{1-5x}, x=-2, order=2\right)$

$$\frac{993}{1331}+\frac{465}{1331}x+\frac{75}{1331}x^2 \qquad\qquad (10.75)$$

Thus the first 3 terms of the Taylor series of the function $\dfrac{3}{1-5x}$ at -2 are

$\dfrac{993}{1331}+\dfrac{465}{1331}x+\dfrac{75}{1331}x^2.$

(b)   Since we are finding the 3rd degree Taylor polynomial of the function $\dfrac{3}{1-5x}$ at -2 we need to set $x=-2$ and $order=3$ in the $TaylorApproximation$ command.

$with\left(Student[Calculus1]\right):$

$TaylorApproximation\left(\dfrac{3}{1-5x}, x=-2, order=3\right)$

$$\frac{13923}{14641}+\frac{9615}{14641}x+\frac{3075}{14641}x^2+\frac{375}{14641}x^3 \qquad\qquad (10.76)$$

Therefore, the first 3 terms of the Taylor series of the function $\dfrac{3}{1-5x}$ at 4 are

$\dfrac{13923}{14641}+\dfrac{9615}{14641}x+\dfrac{3075}{14641}x^2+\dfrac{375}{14641}x^3.$

Remark: We could use the *taylor* command to first 3 terms of the Taylor series of the function $\dfrac{3}{1-5x}$ at -2. This is seen as shown below.

restart

$$taylor\left(\frac{3}{1-5x}, x=-2, 4\right)$$

$$\frac{3}{11} + \frac{15}{121}(x+2) + \frac{75}{1331}(x+2)^2 + \frac{375}{14641}(x+2)^3 + O((x+2)^4) \qquad (10.77)$$

To convert the expression in equation **(10.73)** into a polynomial, we will use the *Maple* command *convert*

$$convert\left((\mathbf{10.73}), polynom\right)$$

$$\frac{63}{121} + \frac{15}{121}x + \frac{75}{1331}(x+2)^2 + \frac{375}{14641}(x+2)^3 \qquad (10.78)$$

simplify$\left((\mathbf{10.74})\right)$

$$\frac{13923}{14641} + \frac{9615}{14641}x + \frac{3075}{14641}x^2 + \frac{375}{14641}x^3 \qquad (10.79)$$

Example 19:
(a)   Find the first five Maclaurin polynomials of the function $x^2 + \cosh(x)$.
(b)   Graph the function and its Maclaurin polynomial approximations on the same window.

Solution:
(a)   We will use the *TaylorApproximation* command. Do not forget to activate the $with\left(Student\left[Calculus1\right]\right)$ package.

$$with\left(Student\left[Calculus1\right]\right):$$

$$TaylorApproximation\left(x^2 + \cosh(x), order = 4\right)$$

$$1, 1 + \frac{3}{2}x^2, 1 + \frac{3}{2}x^2, 1 + \frac{3}{2}x^2 + \frac{1}{24}x^4, 1 + \frac{3}{2}x^2 + \frac{1}{24}x^4 \qquad (10.80)$$

So the first five Maclaurin polynomials of the function $x^2 + \cosh(x)$ are

$$1, 1 + \frac{3}{2}x^2, 1 + \frac{3}{2}x^2, 1 + \frac{3}{2}x^2 + \frac{1}{24}x^4, 1 + \frac{3}{2}x^2 + \frac{1}{24}x^4$$

(b)   We will now graph the function $x^2 + \cosh(x)$ and its 5 Maclaurin polynomial approximations

$$TaylorApproximation\left(x^2+\cosh\left(x\right),order=1..5,output=plot,-4..4,\right.$$

$$view=\left[-4..4,0..40\right],functionoptions=\left[legend='x^2+\cosh\left(x\right)',\right.$$

$$\left.thickness=2\right],tayloroptions=\left[thickness=1\right]\right)$$

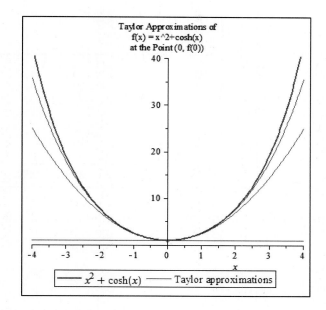

Looking at the graphs we could conjecture that the higher the degree of the Maclaurin polynomial, the better approximation it is to the function.

Example 20: Graph the function $\sin\left(x\right)$ and its 1st Taylor approximation at $\dfrac{\pi}{6}$ on the same window.

Solution: We will use the *TaylorApproximation* command here with the option $output=plot$. Do not forget to activate the $with\left(Student\left[Calculus1\right]\right)$ package.

$with\left(Student\left[Calculus1\right]\right):$

$$TaylorApproximation\left(\sin\left(x\right),x=\dfrac{\pi}{6},output=plot,-\dfrac{\pi}{4}..\dfrac{\pi}{2},view=\left[-\dfrac{\pi}{4}..\dfrac{\pi}{2},-1.5..1.5\right],\right.$$

$$\left.functionoptions=\left[legend='\sin\left(x\right)',thickness=2\right],tayloroptions=\left[thickness=1\right]\right)$$

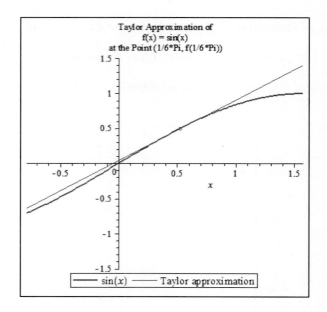

Notice that in this problem we are approximating the graph of $\sin x$ with the tangent line at $x = \dfrac{\pi}{6}$.

Remarks:

1.   One of the main applications of Taylor series is to approximate functions by polynomials, the *Maple* command *TaylorApproximation* in the package $with\left(Student\left[Calculus1\right]\right)$ launches a tutor interface where users can enter a function and choose the degree and center of the Taylor polynomial approximation. This interface also offers animation and plot options.

2.   Another important application of Taylor series is to find integrals of functions that we could not integrate using the tools introduced in chapter 7.

Example 21: Use the 8th degree Maclaurin polynomial of the function $f(x) = e^{-x^2}$ to approximate $\displaystyle\int_0^1 e^{-x^2}\,dx$

Solution: We will use the *TaylorApproximation* command to find the 8th degree Maclaurin polynomial of the function $f(x) = e^{-x^2}$. Do not forget to activate the $with\left(Student\left[Calculus1\right]\right)$ package.

$with\left(Student\left[Calculus1\right]\right):$

$TaylorApproximation\left(e^{-x^2}, order = 9\right)$

$$1 - x^2 + \frac{1}{2}x^4 - \frac{1}{6}x^6 + \frac{1}{24}x^8 \tag{10.81}$$

We will now convert **(10.77)** into a *Maple* function using the *unapply* command.

$f := unapply\left((\mathbf{10.77}), x\right)$

$$x \to 1 - x^2 + \frac{1}{2}x^4 - \frac{1}{6}x^6 + \frac{1}{24}x^8 \tag{10.82}$$

Let us find the definite integral of the polynomial function $\int_0^1 f(x)dx$

$\int_0^1 f(x)\,dx$

$$\frac{5651}{7560} \tag{10.83}$$

A floating-point approximation of the integral is given below.

$evalf\left((\mathbf{10.79})\right)$

$$0.7474867725 \tag{10.84}$$

Hence an approximation of $\int_0^1 e^{-x^2}dx$ is given by 0.7474867725. Notice that *Maple* does not provide a value for $\int_0^1 e^{-x^2}dx$ as seen below.

$\int_0^1 e^{-x^2}\,dx$

$$\frac{1}{2}\operatorname{erf}(1)\sqrt{\pi} \tag{10.85}$$

Exercises

1. List the first 7 terms of the sequence $\{n^3 e^{-n}\}$ and determine whether the sequence converges or diverges.

2. List the first 10 terms of the sequence $\left\{\dfrac{(2011n-2)!}{(2011n+2)!}\right\}$ and determine whether the sequence converges or diverges.

3. List the first 10 terms of the sequence $\left\{\left(2^n+3^n\right)^{\frac{1}{n}}\right\}$ and determine whether the sequence converges or diverges.

4. Determine whether each of the sequences whose general terms are given below has a limit as $c=-2$ and find its limiting value by hand. Mimic the steps done in example 4.

   (a) $a(n)=\dfrac{2n^3-5n+15}{3n^3+9n+1}$

   (b) $a(n)=\dfrac{12+(-1)^n n^3}{2n^2+3n+3}$

   (c) $a(n)=\dfrac{\ln\left(n^2\right)}{\sqrt[3]{2n}}$

   (d) $a(n)=2\cos(n\pi)$

   (e) $a(n)=\left(1-\dfrac{3}{5n}\right)^n$

4. Find the limit of the sequence $\left\{\dfrac{\cos\left(5n^7\right)}{1-3n}\right\}$ and determine whether the series

   $\displaystyle\sum_{n=1}^{\infty}\dfrac{\cos\left(5n^7\right)}{1-3n}$ converges.

5. Determine whether the series $\displaystyle\sum_{n=1}^{\infty}\dfrac{1}{n^4+1}$ converges or diverges. If it converges, find its sum.

6. Determine whether the series $\displaystyle\sum_{n=1}^{\infty}\dfrac{(-7)^n}{5^{3n-1}(n+2)}$ converges or diverges. If it

converges, find its sum.

7. Determine whether the series $\sum_{n=1}^{\infty} \frac{1}{n!}$ converges or diverges. If it converges, find its sum.

8. Determine whether the series $\sum_{n=1}^{\infty} \frac{1}{3^{n-1}}$ converges or diverges. If it converges, find its sum.

9. Determine whether the series $\sum_{n=1}^{\infty} \frac{3n^2 + n^4}{4 - 7n^4}$ converges or diverges. If it converges, find its sum.

10. Determine whether the series $\sum_{n=1}^{\infty} \frac{1}{(n+1)(n+2)}$ converges or diverges.

11. Determine whether the series $\sum_{n=1}^{\infty} \frac{1}{\arctan(n)}$ converges or diverges. If it converges, find its sum.

12. Determine whether the series $\sum_{n=2}^{\infty} \frac{(-1)^n}{n^3 \log(n)}$ converges absolutely.

13. Determine whether the series $\sum_{n=2}^{\infty} \frac{\cos(n)}{n^4 - 1}$ converges.

14. Determine whether the series $\sum_{n=1}^{\infty} \frac{(-1)^{n-1} \ln(n)}{n}$ is absolutely convergent, conditionally convergent or divergent.

15. Determine whether the series $\sum_{n=1}^{\infty} \sin\left(\frac{(-1)^n}{n}\right)$ converges.

16. Determine whether the series $\sum_{n=1}^{\infty} 3\left(-\frac{4}{5}\right)^{n-1}$ is geometric. Does the series converge?

17. Determine whether the series $\sum_{n=1}^{\infty} 3^{2n} 5^{1-n}$ is geometric. Does the series converge?

18. Determine the radius and interval of convergence of the power series $\displaystyle\sum_{n=1}^{\infty} \frac{(-1)^n (x-5)^n}{3n^2 + 1}$.

19. Determine the radius and interval of convergence of the power series $\displaystyle\sum_{n=1}^{\infty} \frac{(3x+3)^n}{n^n}$

20. Determine the radius and interval of convergence of the power series $\displaystyle\sum_{n=1}^{\infty} \frac{(x-1)^n}{(3n-1)!}$

21. Find the first 7 terms of the Maclaurin series of the function $\dfrac{\cos(x)}{1+x}$.

22. Find the first 7 terms of the Maclaurin series of the function $(1+x)^a$ where a is any real number.

23. Find the 8th degree Taylor polynomial of the function e^{-2e^x} at 3.

24. Graph the function $\cos(x)$ and its first 5 Maclaurin polynomial approximations on the same window.

25. Graph the function e^{-x^2} and its first 4 Maclaurin polynomial approximations on the same window.

26. Graph the function $\dfrac{1}{1-3x}$ and its first 4 Taylor polynomial approximations at -1 on the same window.

27. Use the 8th degree Maclaurin polynomial of the function $f(x) = \sin(x^2)$ to approximate $\displaystyle\int_0^1 \sin(x^2)\, dx$

28. Below is a list of some commands used in this chapter. Describe the significance of each command. Can you find examples where each command is used?

    (a) $convert(t, polynom)$

    (b) $plot\left(\left[seq\left([n, a(n)], n = k..m\right)\right], style = point, symbolsize = 15\right)$

    (c) $seq\left(a(n), n = k..m\right)$

(d) $Sum(a(n), n = k .. m)$

(e) $sum(a(n), n = k .. m)$

(f) $taylor(f(x), x = c, k)$

(g) $TaylorApproximation(f(x), order = 1..k)$

(h) $TaylorApproximation(f(x), order = 1..k, output = plot, a..b,$

$view = [a..b, c..d], functionoptions = [legend = 'f(x)', thickness = 4],$

$tayloroptions = [thickness = 1])$

(i) $taylor(f(x), x, k)$

(j) $with(MTM):$

29.   Can you list new *Maple* commands that were used in this chapter but were not listed in exercise 28?

Chapter 11   Parametric and Polar Curves

Students will be able to plot parametric and polar curves, find areas, and arc lengths. An appreciation of converting between coordinate systems will be examined. There will be a section on conics using *Maple* to plot polar coordinates.

Parametric Curves

Definition: Let f and g be continuous function on a closed interval $[a,b]$. The curve described by the equations $x = f(t)$ and $y = g(t)$, for $a \le t \le b$, is called a *parametric curve*. The equations $x = f(t)$ and $y = g(t)$ are called *parametric equations*.

Example 1: Sketch the curve described by $x(t) = 7(\cos t + t \sin t)$ and $y(t) = 7(\sin t - t \cos t)$, for $0 \le t \le 10\pi$.

Solution: We will use the *Maple* command *plot* to sketch the parametric curve.

$$plot\left(\left[7 \cdot (\cos(t) + t \cdot \sin(t)), 7 \cdot (\sin(t) - t \cdot \cos(t)), t = 0..10 \cdot \pi\right]\right)$$

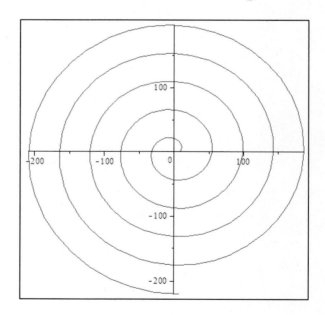

This curve is called the *involute of a circle* of radius 7.

Slope of the tangent line: The *slope of the tangent line* to the curve described by the parametric equations $x = f(t)$ and $y = g(t)$ at the point (x, y) is $\dfrac{dy}{dx} = \dfrac{\frac{dy}{dt}}{\frac{dx}{dt}}$ provided $\dfrac{dx}{dt} \neq 0$.

Example 2: Let C be the curve described by the parametric equations $x(t) = 5 \cdot \cos(t)$ and $y(t) = 3 \cdot \sin(t)$, for $0 \leq t \leq 2\pi$.

(a)   Sketch the curve C.

(b)   Find the slope of the tangent line to the curve C at $t = \dfrac{\pi}{3}$.

(c)   Sketch the curve and its tangent line at $t = \dfrac{\pi}{3}$ on the same axis.

(c)   Find points where the tangent line is horizontal, and points where the tangent line is vertical.

Solution:

(a)   We will use the plot command to sketch the graph of C.

$$x := t \rightarrow 5 \cdot \cos(t)$$

$$t \rightarrow 5 \cdot \cos(t) \qquad \textbf{(11.1)}$$

$$y := t \rightarrow 3 \cdot \sin(t)$$

$$t \rightarrow 3 \cdot \sin(t) \qquad \textbf{(11.2)}$$

$$plot\left(\left[x(t), y(t), t = 0..2 \cdot \pi\right], scaling = constrained\right)$$

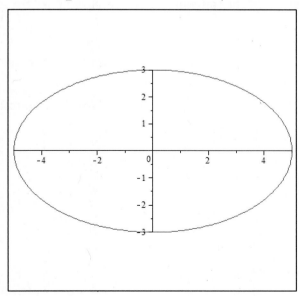

Observe that the curve C is an ellipse.

(b)   We will first find the slope of the tangent line

$$\dfrac{\dfrac{d}{dt}y(t)}{\dfrac{d}{dt}x(t)}$$

$$-\frac{3}{5}\frac{\cos(t)}{\sin(t)} \tag{11.3}$$

To find the slope at $t = \dfrac{\pi}{3}$, we will use the *subs* command to substitute $\dfrac{\pi}{3}$ for t in equation **(11.3)**.

$$m := subs\left(t = \frac{\pi}{3}, (\textbf{11.3})\right)$$

$$-\frac{3}{5}\frac{\cos\left(\dfrac{1}{3}\pi\right)}{\sin\left(\dfrac{1}{3}\pi\right)} \tag{11.4}$$

simplify $((\textbf{11.4}))$

$$-\frac{1}{5}\sqrt{3} \tag{11.5}$$

Hence the slope of the tangent line to the curve C at $t = \dfrac{\pi}{3}$ is $m = -\dfrac{1}{5}\sqrt{3}$.

(c)   The equation of the tangent line to the curve C at $t = \dfrac{\pi}{3}$ is described by $y = y\left(\dfrac{\pi}{3}\right) + m\left(x - x\left(\dfrac{\pi}{3}\right)\right)$. We will use *Maple* to simplify the right hand side of this equation.

$$y\left(\frac{\pi}{3}\right) + m\left(x - x\left(\frac{\pi}{3}\right)\right)$$

$$\frac{3}{2}\sqrt{3} - \frac{1}{5}\sqrt{3}\left(x - \frac{5}{2}\right) \tag{11.6}$$

L1 := *unapply* $((\textbf{11.6}), x)$

$$x \rightarrow \frac{3}{2}\sqrt{3} - \frac{1}{5}\sqrt{3}\left(x - \frac{5}{2}\right)$$ (11.7)

To graph the curve C and the tangent line $L1(x)$ on the same axis, we will activate the *with(plots)* package and use the *display* command, since the curve is a parametric curve and the tangent line is written using Cartesian coordinates.

with(plots):

$p1 \coloneqq plot\left(\left[x(t), y(t), t = 0..2 \cdot \pi\right], scaling = constrained\right):$

$$PLOT(...)$$ (11.8)

$p2 \coloneqq plot\left(L1(x), x = -6..13, y = -4..5\right):$

$$PLOT(...)$$ (11.9)

display(p1, p2)

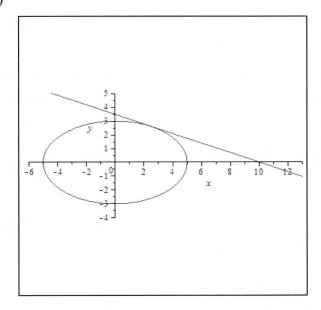

(d)   The tangent line to the curve C is horizontal if $\dfrac{dy}{dt} = 0$. We will solve this equation with *Maple* using the *Roots* command. Do not forget to activate the $with\left(Student\left[Calculus1\right]\right)$ package.

$with\left(Student\left[Calculus1\right]\right):$

$Roots\left(\dfrac{d}{dt} y(t) = 0, t = 0..2 \cdot \pi\right)$

$$\left[\frac{1}{2}\pi, \frac{3}{2}\pi\right]$$ (11.10)

We will now find points on the curve that correspond to $t = \dfrac{1}{2}\pi$ and $t = \dfrac{3}{2}\pi$.

$$x\left(\frac{\pi}{2}\right); x\left(\frac{3 \cdot \pi}{2}\right);$$

$$\begin{matrix} 0 \\ 0 \end{matrix}$$

$\quad$ **(11.11)**

$$y\left(\frac{\pi}{2}\right); y\left(\frac{3 \cdot \pi}{2}\right);$$

$$\begin{matrix} 3 \\ -3 \end{matrix}$$

$\quad$ **(11.12)**

Hence the tangent lines to the curve C at the points $(0,3)$ and $(0,-3)$ are horizontal.

The tangent line to the curve C is vertical if $\dfrac{dx}{dt} = 0$. We will solve this equation using the *Roots* command once again. We do not need to activate the $with\left(Student\left[Calculus1\right]\right)$ package as it has just been done.

$$Roots\left(\frac{d}{dt}x(t) = 0, t = 0 .. 2 \cdot \pi\right)$$

$$\left[0, \pi, 2\pi\right]$$

$\quad$ **(11.13)**

We will now find the find points on the curve that correspond to $t = 0$, $t = \pi$ and $t = 2\pi$.

$$x(0); x(\pi); x(2 \cdot \pi)$$

$$\begin{matrix} 5 \\ -5 \\ 5 \end{matrix}$$

$\quad$ **(11.14)**

$$y(0); y(\pi); y(2 \cdot \pi)$$

$$\begin{matrix} 0 \\ 0 \\ 0 \end{matrix}$$

$\quad$ **(11.15)**

Hence the tangent line to the curve C at the points $(5,0)$ and $(-5,0)$ are vertical.

Arc Length: Consider a parametric curve described by the equations $x = f(t)$ and $y = g(t)$, for $a \le t \le b$. We assume that $f'(t)$ and $g'(t)$ are continuous on the interval

$[a,b]$ and that the curve C does not intersect itself on the interval (a,b). Then the length

of C is $\displaystyle\int_a^b \sqrt{\left(\frac{dx}{dt}\right)^2 + \left(\frac{dy}{dt}\right)^2}\, dt$.

Example 3: Find the length of the curve C described by $x(t) = t^{\frac{3}{2}} - 1$ and $y(t) = 4t^2 + 2$ for $0 \le t \le 3$

Solution: We will enter the parametric equations as *Maple* functions.

$x := t \rightarrow t^{\frac{3}{2}} - 1$

$$t \rightarrow t^{3/2} - 1 \tag{11.16}$$

$y := t \rightarrow 4t^2 + 2$

$$t \rightarrow 4t^2 + 2 \tag{11.17}$$

We will use the *ArcLength* command to calculate the required arc length. As in the previous two examples we will need to activate the $with\left(Student\left[Calculus1\right]\right)$ package.

$with\left(Student\left[Calculus1\right]\right):$

$ArcLength\left(\left[x(t), y(t)\right], t = 0..3\right)$

$$\frac{81}{65536}\ln(2) + \frac{4635}{2048}\sqrt{259} - \frac{81}{65536}\ln\left(515 + 32\sqrt{259}\right) \tag{11.18}$$

$evalf\left((\textbf{11.18})\right)$

$$36.41527657 \tag{11.19}$$

Hence the arc length is 36.41527657.

Remarks:

1.   The integral used to make this calculation is seen below. We need to include the option *output = integral*.

$$ArcLength\left(\left[x(t), y(t)\right], t = 0..3, output = integral\right)$$

$$\int_0^3 \frac{1}{2}\sqrt{t(9 + 256t)}\, dt \tag{11.20}$$

When the option *output = plot* is used we see three graphs – the one in red is that of the function, the one in blue is the integrand, and the one in green is the graph of the integral as a function of *t*. This has been shown below the graph.

$$ArcLength\left(\left[\,x(t),y(t)\,\right],t=0\,..\,3,\,output=plot\right)$$

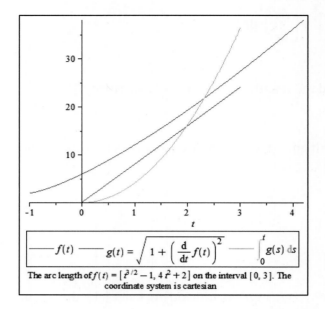

The arc length of $f(t) = [\,t^{3/2} - 1,\, 4\,t^2 + 2\,]$ on the interval $[\,0,\,3\,]$. The coordinate system is cartesian

2.   We could also use the definition of the arc length and *Maple* commands to find the arc length. But first we will need to check that the conditions of the arc length theorem are satisfied. Let us check that the derivatives $x'(t)$ and $y'(t)$ are continuous on the interval $[0,3]$.

$$x:=t \rightarrow t^{\frac{3}{2}} - 1$$

$$t \rightarrow t^{3/2} - 1 \qquad\qquad\qquad \textbf{(11.21)}$$

$$y:=t \rightarrow 4t^2 + 2$$

$$t \rightarrow 4t^2 + 2 \qquad\qquad\qquad \textbf{(11.22)}$$

$$\frac{d}{dt}x(t)$$

$$\frac{3}{2}\sqrt{t} \qquad\qquad\qquad \textbf{(11.23)}$$

$$\frac{d}{dt}y(t)$$

$$8\,t \qquad\qquad\qquad \textbf{(11.24)}$$

We will now sketch the curve C to check that it does not intersect itself on the interval $(0,3)$.

$$plot\left(\left[\,x(t),y(t),t=0\,..\,3\,\right]\right)$$

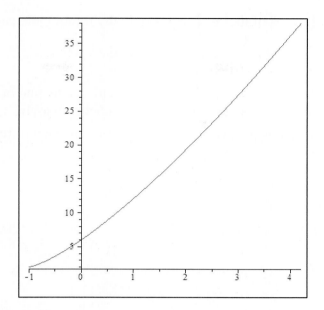

Based on the graph above, the curve C does not intersect itself on the interval $(0,3)$. We can now find the length of the parametric curve.

$$\int_0^3 \sqrt{\left(\frac{dx}{dt}\right)^2 + \left(\frac{dy}{dt}\right)^2}\, dt$$

$$\frac{81}{65536}\ln(3) + \frac{4635}{2048}\sqrt{259} - \frac{81}{65536}\ln\left(515 + 32\sqrt{259}\right) \qquad \textbf{(11.25)}$$

evalf $\left((\textbf{11.25})\right)$

$$36.41527657 \qquad \textbf{(11.26)}$$

Therefore the length of the curve is 36.41527657. Note that this number is the same as the one given in **(11.19)**.

Surface Area: Let C be a parametric curve described by the equations $x = f(t)$ and $y = g(t)$, for $a \le t \le b$. We assume that $f'(t)$ and $g'(t)$ are continuous on the interval $[a,b]$, and that the curve C does not intersect itself on the interval (a,b).

(a) If $g(t) \ge 0$ for $a \le t \le b$, then the area of the surface formed by rotating the curve C about the x-axis is $2\pi \int_a^b g(t) \sqrt{\left(\frac{dx}{dt}\right)^2 + \left(\frac{dy}{dt}\right)^2}\, dt$.

(b) If $f(t) \ge 0$ for $a \le t \le b$, then the area of the surface formed by rotating the curve C about the y-axis is $2\pi \int_a^b f(t) \sqrt{\left(\frac{dx}{dt}\right)^2 + \left(\frac{dy}{dt}\right)^2}\, dt$.

Example 4: Find the area of the surface S formed by rotating the curve C described by $x(t) = t^{\frac{3}{2}} - 1$ and $y(t) = 4t^2 + 2$, $0 \le t \le 3$, about the x-axis.

Solution: Notice that the parametric curve is the same as the one in the previous example. To use the surface area theorem, we need to check that $y(t) \ge 0$ for $0 \le t \le 3$. Since $y(t) = 4t^2 + 2$, it is always positive. We will now find the area of the surface S.

$$2 \cdot \pi \cdot \int_0^3 y(t) \cdot \sqrt{\left(\frac{dx}{dt}\right)^2 + \left(\frac{dy}{dt}\right)^2}\, dt$$

$$2\pi\left(-\frac{212500665}{8589934592}\ln(2) + \frac{42500133}{17179869184}\ln(3) + \frac{24215202999}{536870912}\sqrt{259} - \frac{42500133}{8589934592}\ln\left(\frac{1}{2} + \frac{1}{32}\sqrt{259}\right)\right)$$
(11.27)

evalf $((\mathbf{11.27}))$

$$4560.782362$$ (11.28)

Hence the area of the surface is 4560.782362.

Polar Curves

Definition of polar coordinates: Given a point O (called *the origin*) on a plane, and an array A (called *the polar axis*) with endpoint O, we define *the polar coordinate system* as the set of points $P = (r, \theta)$ where r is the distance from the point P to the origin O and θ is the angle (called *the polar angle* and measured anticlockwise) between the array A and the line segment of endpoints O and P.

Definition of polar curves: An equation $r = f(\theta)$ written in the polar coordinate system is called a *polar equation*. The graph of a polar equation is called a *polar curve*.

Example 5: Sketch the polar curves
(a) $r = 5\sin(4\theta)$, $0 \le \theta \le 2\pi$.
(b) $r = 6\cos(5\theta)$, $0 \le \theta \le 2\pi$.

Solution:
(a) To sketch the polar curve $r = 5\sin(4\theta)$, we will activate the *with*(*plots*) package.

$with(plots):$

$polarplot\left(\{5 \cdot \sin(4 \cdot \theta)\}, \theta = 0..2 \cdot \pi, scaling = constrained\right)$

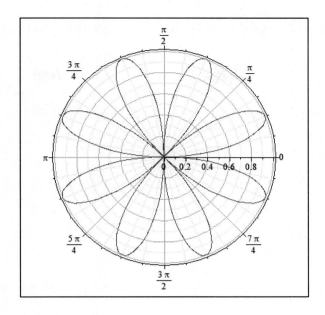

Notice that the "flower" sketched has 8 petals.

(b)   To sketch the polar curve $r = 6\cos(5\theta)$, we do not need to activate the *with*(*plots*) package, since this was done in part (a).

$$polarplot\left(\{6 \cdot \cos(5 \cdot \theta)\}, \theta = 0..2 \cdot \pi, scaling = constrained\right)$$

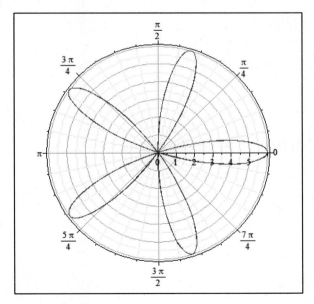

The "flower" sketched here has 5 petals.

Example 6: Sketch the polar curve $r = 20\sin\left(\dfrac{\theta}{11}\right)$, $0 \le \theta \le 11\pi$.

Solution: To sketch the polar curve, we will activate the $with(\,plots\,)$ package.

$with(\,plots\,)$:

$$polarplot\left(\left\{20\cdot\sin\left(\frac{\theta}{11}\right)\right\},\theta=0\,..\,11\cdot\pi,scaling=constrained\right)$$

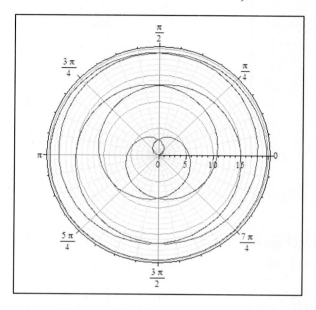

Example 7: Sketch the polar curve $r=e^{\frac{\theta}{7}}$, $0\le\theta\le10\pi$.

Solution: We will first activate the $with(\,plots\,)$ package.

$with(\,plots\,)$:

$$polarplot\left(\left\{e^{\frac{\theta}{7}}\right\},\theta=0\,..\,12\cdot\pi,scaling=constrained\right)$$

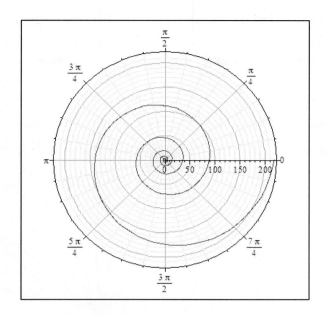

Example 8: Sketch the family of polar curves described by $r = \theta + n \sin(2\theta)$, $0 \le \theta \le 3\pi$, where n is an integer and $1 \le n \le 8$.

Solution: To sketch the polar curves, we will activate the $with(\,plots\,)$ package.

$with(\,plots\,)$:

$polarplot\left(\{\theta + n \cdot \sin(2 \cdot \theta)\$n = 1..8\}, \theta = 0..3 \cdot \pi, scaling = constrained, thickness = 2\right)$

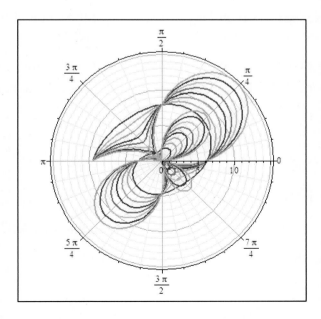

Remark: The *Maple* command *animate* in the *with(plots)* package offers an interactive worksheet to graph polar curves. As an exercise, use the following *Maple* command, and click on the graph below.

$with(plots):$

$animate(7-t\cdot\sin(t\cdot\theta),\theta=0..2\cdot\pi,t=-15..15,coords=\text{polar},color=blue,$

$\quad\quad thickness=2,numpoints=400,scaling=constrained,frames=35)$

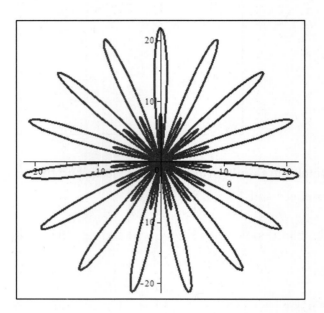

Slope and tangent lines to polar curves: To find the slope of the tangent line to the curve described by the polar equation $r=f(\theta)$ at the point (r,θ), we consider the parametric equations $x=r\cos\theta=f(\theta)\cos\theta$ and $y=r\sin\theta=f(\theta)\sin\theta$, and we use the formula for finding the slope that was given in the context of parametric equations.

We have $\dfrac{dy}{dx}=\dfrac{\dfrac{dy}{d\theta}}{\dfrac{dx}{d\theta}}$ provided $\dfrac{dx}{d\theta}\neq 0$

Example 9: Find the equation of the tangent line to the curve C described by $r=\dfrac{3}{4}+\sin\theta$, $0\leq\theta\leq 2\pi$, at the point corresponding to $\theta=\dfrac{\pi}{4}$. Graph the curve and the tangent line on the same axis.

Solution: We will define the parametric equations $x = r\cos\theta = \left(\dfrac{3}{4} + \sin(\theta)\right)\cos(\theta)$ and

$y = r\sin\theta = \left(\dfrac{3}{4} + \sin(\theta)\right)\sin(\theta)$ using *Maple*.

$x := \theta \to \left(\dfrac{3}{4} + \sin(\theta)\right) \cdot \cos(\theta)$

$$\theta \to \left(\dfrac{3}{4} + \sin(\theta)\right) \cdot \cos(\theta) \tag{11.29}$$

$y := \theta \to \left(\dfrac{3}{4} + \sin(\theta)\right)\sin(\theta)$

$$\theta \to \left(\dfrac{3}{4} + \sin(\theta)\right)\sin(\theta) \tag{11.30}$$

We will now find the slope of the tangent line $\dfrac{dy}{dx} = \dfrac{\frac{dy}{d\theta}}{\frac{dx}{d\theta}}$ at $\theta = \dfrac{\pi}{4}$

$\dfrac{\frac{d}{d\theta}\,y(\theta)}{\frac{d}{d\theta}\,x(\theta)}$

$$\dfrac{\cos(\theta)\sin(\theta) + \left(\dfrac{3}{4} + \sin(\theta)\right)\cos(\theta)}{\cos(\theta)^2 - \left(\dfrac{3}{4} + \sin(\theta)\right)\sin(\theta)} \tag{11.31}$$

simplify $\big((\mathbf{11.31})\big)$

$$\dfrac{\cos(\theta)\big(8\sin(\theta) + 3\big)}{-3\sin(\theta) + 7\cos(\theta)^2 - 4} \tag{11.32}$$

$m := subs\left(\theta = \dfrac{\pi}{4}, (\mathbf{11.32})\right)$

$$\dfrac{\cos\left(\dfrac{1}{4}\pi\right)\left(8\sin\left(\dfrac{1}{4}\pi\right) + 3\right)}{-3\sin\left(\dfrac{1}{4}\pi\right) + 7\cos\left(\dfrac{1}{4}\pi\right)^2 - 4} \tag{11.33}$$

simplify $\big((\mathbf{11.33})\big)$

$$-\dfrac{4}{3}\sqrt{2} - 1 \tag{11.34}$$

Hence the slope of the tangent line to the curve C at $\theta = \dfrac{\pi}{4}$ is $m = -\dfrac{4}{3}\sqrt{2} - 1$. We will now find the equation of the tangent line to the curve C at $\theta = \dfrac{\pi}{4}$ as we did in example 3.

$$y\left(\frac{\pi}{4}\right) + m\left(x - x\left(\frac{\pi}{4}\right)\right)$$

$$\frac{1}{2}\left(\frac{3}{4} + \frac{1}{2}\sqrt{2}\right)\sqrt{2} + \left(-\frac{4}{3}\sqrt{2} - 1\right)\left(x - \frac{1}{2}\left(\frac{3}{4} + \frac{1}{2}\sqrt{2}\right)\sqrt{2}\right) \tag{11.35}$$

$simplify\left((\mathbf{11.35})\right)$

$$\frac{17}{12}\sqrt{2} + 2 - \frac{4}{3}\sqrt{2}x - x \tag{11.36}$$

$L1 := unapply\left((\mathbf{11.36}), x\right)$

$$x \rightarrow \frac{17}{12}\sqrt{2} + 2 - \frac{4}{3}\sqrt{2}x - x \tag{11.37}$$

Hence the equation of the tangent line is $y = \dfrac{17}{12}\sqrt{2} + 2 - \dfrac{4}{3}\sqrt{2}x - x$. Next, we will graph the curve C and the tangent line $L1(x)$ on the same axis, we will activate the $with(plots)$ package and use the $display$ command.

$with(plots):$

$$p1 := polarplot\left(\left\{\frac{3}{4} + \sin(\theta)\right\}, \theta = 0 .. 2 \cdot \pi, scaling = constrained\right)$$

$$PLOT(...) \tag{11.38}$$

$p2 := plot\left(L1(x), x = -1..3, y = -5..5\right)$

$$PLOT(...) \tag{11.39}$$

$display(p1, p2)$

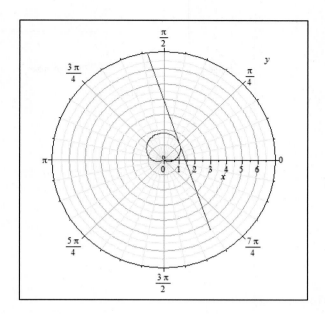

Areas in polar coordinates: The area enclosed by the polar curve $r = f(\theta)$ when $\theta_1 \le \theta \le \theta_2$ is $\dfrac{1}{2}\displaystyle\int_{\theta_1}^{\theta_2} r^2 d\theta$.

Example 10: Find the area of the region enclosed by the curve $r = 5\big(\sin(\theta)+1\big)$.

Solutions: We will begin by entering the curve as a *Maple* function.

$r := \theta \to 5 \cdot \big(\sin(\theta)+1\big)$

$$\theta \to 5\sin(\theta)+5 \qquad\qquad\qquad \textbf{(11.40)}$$

$with(plots):$

$polarplot\big(\{r(\theta)\}, \theta = 0..2\cdot\pi, scaling = constrained\big)$

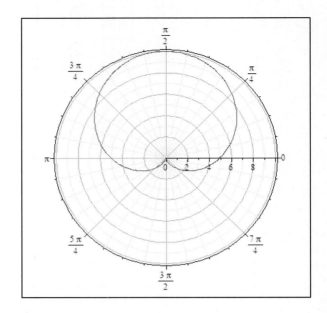

We will now find the area bounded by this curve.

$$\frac{1}{2}\int_0^\pi r(\theta)^2 d\theta$$

$$\frac{75}{2}\pi \qquad\qquad\qquad\qquad\qquad\text{(11.41)}$$

Therefore the area enclosed by this curve is $\dfrac{75}{2}\pi$.

Example 11: Find the area of the region enclosed by one petal of the curve $r = 5\sin(4\theta)$.

Solutions: We will begin by entering the curve as a *Maple* function.

$r := \theta \to 5 \cdot \sin(4 \cdot \theta)$

$$5\sin(4\theta) \qquad\qquad\qquad\qquad\qquad\text{(11.42)}$$

From example 5(a) it looks as though one petal is sketched when $0 \le \theta \le \dfrac{\pi}{4}$. We will verify this using the *polarplot* command. Do not forget to activate the *with(plots)* package.

$with(plots):$

$polarplot\left(\{r(\theta)\}, \theta = 0..\dfrac{\pi}{4}, scaling = constrained\right)$

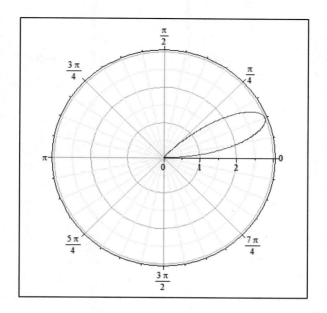

We will now find the area bounded by this curve.

$$\frac{1}{2} \cdot \int_0^{\frac{\pi}{4}} r(\theta)^2 d\theta$$

$$\frac{9}{16}\pi \qquad\qquad\qquad\qquad \textbf{(11.43)}$$

Hence the area enclosed by one petal is $\dfrac{9}{16}\pi$.

Example 12: Find the area of the region between the curves $r = -8\sin(\theta)$ and $r = 3 - 2\sin(\theta)$.

Solutions: We will begin by graphing the curves on the same axis.

$with(plots):$

$p1 := polarplot\left(\{-8 \cdot \sin(\theta)\}, \theta = 0..2 \cdot \pi, scaling = constrained, color = blue\right)$

$$PLOT(...) \qquad\qquad\qquad \textbf{(11.44)}$$

$p2 := polarplot\left(\{3 - 2 \cdot \sin(\theta)\}, \theta = 0..2 \cdot \pi, scaling = constrained\right)$

$$PLOT(...) \qquad\qquad\qquad \textbf{(11.45)}$$

$display(p1, p2)$

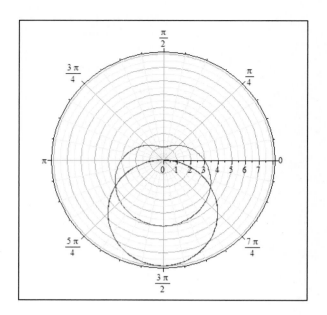

To find the area between the two curves, we will first find the intersection points. We will use the *Roots* command to solve the equation $-8\sin(\theta) = 3 - 2\sin(\theta)$. Do not forget to activate the $with\left(Student\left[Calculus1\right]\right)$ package.

$with\left(Student\left[Calculus1\right]\right):$
$Roots\left(-8 \cdot \sin(\theta) = 3 - 2 \cdot \sin(\theta), \theta = 0..2 \cdot \pi\right)$

$$\left[\frac{7}{6}\pi, \frac{11}{6}\pi\right]$$

(11.46)

We will now find the area between the two curves.

$$\frac{1}{2} \cdot \int_{\frac{7 \cdot \pi}{6}}^{\frac{11 \cdot \pi}{6}} \left(-8 \cdot \sin(\theta)\right)^2 d\theta - \frac{1}{2} \cdot \int_{\frac{7 \cdot \pi}{6}}^{\frac{11 \cdot \pi}{6}} \left(3 - 2 \cdot \sin(\theta)\right)^2 d\theta$$

$$\frac{3}{2}\sqrt{3} + 7\pi$$

(11.47)

Hence the area between the two curves is $\frac{3}{2}\sqrt{3} + 7\pi$.

Arc length: To length of the curve described by the polar equation $r = f(\theta)$, $\theta_1 \le \theta \le \theta_2$,

is $\int_{\theta_1}^{\theta_2} \sqrt{r^2 + \left(\frac{dr}{d\theta}\right)^2}\, d\theta$. To find the arc length of a polar curve with *Maple*, we will use the

ArcLength command in the $with\left(Student\left[Calculus1\right]\right)$ package.

Example 13: Find the length of the curve $r = 5\cos(4\theta)$, $0 \leq \theta \leq 2\pi$.

Solutions: We will begin by graphing the curves on the same axis.

$with\left(Student\left[Calculus1\right]\right):$

$ArcLength\left(5 \cdot \cos\left(4 \cdot \theta\right), \theta = 0 .. 2 \cdot \pi, coordinates = \text{polar}\right)$

$$80\text{EllipticE}\left(\frac{1}{4}\sqrt{15}\right) \tag{11.48}$$

$evalf\left(\left(\textbf{11.43}\right)\right)$

$$85.78421776 \tag{11.49}$$

Hence the arc length is 85.78421776.

Remarks: The integral used to make the calculation of the arc length is seen below. We need to include the option *output = integral*.

$ArcLength\left(5 \cdot \cos\left(4 \cdot \theta\right), \theta = 0 .. 2 \cdot \pi, coordinates = \text{polar}, output = integral\right)$

$$\int_0^{2\pi} 5\sqrt{16\sin\left(4\theta\right)^2 + \cos\left(4\theta\right)^2}\; d\theta \tag{11.50}$$

Chapter 11 Parametric and Polar Curves

Exercises

1.   Sketch the curve described by $x = 3 + 2t$ and $y = -1 + 4t$, for t in the interval $(-\infty, \infty)$.

2.   Sketch the curve described by $x = e^{\cos(t)}$ and $y = \sin(2t)$, for $0 \le t \le 2\pi$.

3.   Sketch the curve described by $x = 1 + 7\cos(t)$ and $y = -3 + 7\sin(t)$, for $0 \le t \le 2\pi$.

4.   Find the slope to the curve $x = \sin(t) - \sin(3t)$, $y = \cos(t) + \cos(3t)$, $0 \le t \le 2\pi$,

at $t = \dfrac{\pi}{3}$.

5.   Find the length of the curve C described by $x = 3\cos(t)$ and $y = 2\cos(2t)$ for

$0 \le t \le \dfrac{\pi}{2}$.

6.   Find the area of the surface S formed by rotating the curve C described by $x = 3\cos(t)$ and $y = 2\cos(2t)$, $0 \le t \le 3$, about the y-axis.

7.   Let C be the curve described by the parametric equations $x = \sin(t^3)$ and $y = \cos(t^3)$, for $0 \le t \le \dfrac{\pi}{2}$. Find the points where the tangent line is horizontal, and the points where the tangent line is vertical.

8.   Let C be the curve described by the parametric equations $x = \sin(t) - \sin(3t)$ and $y = \cos(t) + \cos(3t)$, for $0 \le t \le 2\pi$.
   (a)   Sketch the curve C.
   (b)   Find the slope of the tangent line to the curve C at $t = \dfrac{\pi}{3}$.
   (c)   Sketch the curve and its tangent at $t = \dfrac{\pi}{3}$ on the same axis.
   (c)   Find points where the tangent line is horizontal, and where the tangent line is vertical.

9.   Sketch the polar curve $r = \theta$, $0 \le \theta \le 5\pi$.

10.   Sketch the family of polar curves $r = n\cos\left(\dfrac{\theta}{3}\right)$, $0 \le \theta \le 10\pi$, where n is an integer and $1 \le n \le 15$.

11. Sketch the family of polar curves $r = 6 + n\cos(5\theta)$, $0 \le \theta \le 2\pi$, where n is an integer and $1 \le n \le 12$.

12. Find the equation of the tangent line to the curve C described by $r = \dfrac{3}{4} + \cos(\theta)$, $0 \le \theta \le 2\pi$ at the point corresponding to $\theta = \dfrac{\pi}{4}$. Graph the curve and the tangent line on the same axis.

13. Find the equation of the tangent line to the curve C described by $r = 3\sin(\theta)$, $0 \le \theta \le 2\pi$ at the point corresponding to $\theta = \dfrac{\pi}{3}$. Graph the curve and the tangent line on the same axis.

14. Find the equation of the tangent line to the curve C described by $r = 3\sin(2\theta)$, $0 \le \theta \le 2\pi$ at the point corresponding to $\theta = \dfrac{\pi}{4}$. Graph the curve and the tangent line on the same axis.

15. Find the area of the region enclosed by the curve $r = 3\cos^2(\theta)$.

16. Find the area of the region enclosed by the curve $r = \dfrac{3}{2}\sin(\theta)$.

17. Find the area of the region enclosed by one petal of the curve $r = 3\sin(2\theta)$.

18. Find the area of the region enclosed by one petal of the curve $r = 4\cos(3\theta)$.

19. Find the area of the region between the curves $r = 2 + \sin(\theta)$ and $r = 5\sin(\theta)$.

20. Find the area of the region between the curves $r = 3 + \sin(\theta)$ and $r = 4 - \sin^2(\theta)$.

21. Find the length of the curve $r = -2 + \sin(\theta)$, $0 \le \theta \le 2\pi$.

22. Find the length of the curve $r = 3 + \cos(2\theta)$, $0 \le \theta \le 2\pi$.

23.* Find the area of the region between the curves $r = 8\cos(\theta)$ and $r = -3 + 2\cos(\theta)$.

24. Below is a list of some commands used in this chapter. Describe the significance of each command. Can you find examples where each command is used?

    (a) $ArcLength\left(\left[x(t), y(t)\right], t = a .. b\right)$

    (b) $ArcLength\left(\left[x(t), y(t)\right], t = a .. b, output = integral\right)$

    (c) $ArcLength\left(\left[x(t), y(t)\right], t = a .. b, output = plot\right)$

    (d) $ArcLength\left(r(\theta), \theta = a .. b, coordinates = \text{polar}\right)$

    (e) $display\left(p1, p2\right)$

    (f) $p1 := plot\left(\left[x(t), y(t), t = a..b\right], scaling = constrained\right)$

    (g) $plot\left(\left[x(t), y(t), t = a..b\right], scaling = constrained\right)$

    (h) $polarplot\left(\{r(\theta)\}, \theta = 1..b, scaling = constrained\right)$

    (i) $Roots\left(f(t) = 0, t = a .. b\right)$

    (j) $with\left(plots\right):$

25. Can you list new *Maple* commands that were used in this chapter but were not listed in exercise 24?

Extensive problems will be examined in relation to the dot and cross products. Finding tangent vectors, equations of tangent planes and their plots will be done. Projections will also be done with the help of *Maple*.

Matrices

Definition: Let m and n be positive integers. An $m \times n$ *matrix* is a rectangular array of

the form $A = \begin{bmatrix} a_{11} & \cdots & a_{1j} & \cdots & a_{1n} \\ \vdots & & \vdots & & \vdots \\ a_{i1} & \cdots & a_{ij} & \cdots & a_{in} \\ \vdots & & \vdots & & \vdots \\ a_{m1} & \cdots & a_{mj} & \cdots & a_{mn} \end{bmatrix}$ with m rows and n columns. The element in the

i^{th} row and j^{th} column, namely a_{ij}, is the $(i, j)^{th}$ -element of the matrix. A *zero matrix* is a matrix where all the elements are zeros. A *square matrix* has the same number of rows and columns. A *row vector* is an $1 \times n$ matrix and a *column vector* is an $m \times 1$ matrix.

Remark: The $(i, j)^{th}$ -elements of our matrices are real numbers.

Definition: An $n \times n$ *diagonal matrix* is a square matrix where $a_{ij} = 0$ when $i \neq j$. The $n \times n$ identity matrix I_n is a diagonal matrix where $a_{ii} = 1$.

Definition: Let A and B be two $m \times n$ matrices with $(i, j)^{th}$ -elements a_{ij} and b_{ij}, respectively. Then the *sum* $A + B$ is an $m \times n$ matrix whose $(i, j)^{th}$ - element is $a_{ij} + b_{ij}$.

Definition: Let A be an $m \times n$ matrix with $(i, j)^{th}$ -element a_{ij} and let k be any real number. Then the *scalar multiplication* kA is an $m \times n$ matrix whose $(i, j)^{th}$ - element is ka_{ij}.

Definition: Let A be an $m \times n$ matrix and B be an $n \times p$ matrices with $(i, j)^{th}$ -elements a_{ij} and b_{ij}, respectively. Then the *product* AB is an $m \times p$ matrix whose $(i, j)^{th}$ - element c_{ij} is defined as $c_{ij} = a_{i1}b_{1j} + a_{i2}b_{2j} + ... + a_{ik}b_{kj}$.

Definition: Let A be an $m \times n$ matrix with $(i, j)^{th}$ -element a_{ij}. Then the *transpose* of A, denoted by A^T is an $n \times m$ matrix whose $(i, j)^{th}$ - element is a_{ji}.

Definition: Let A be an $n \times n$ square matrix. Then the *inverse* matrix, if it exists, is an $n \times n$ square, denoted by A^{-1}, such that $AA^{-1} = I_n = A^{-1}A$, where I_n is the $n \times n$ identity matrix.

In this chapter we will list properties of matrices that will be used in multi-variable calculus. One can enter matrices in *Maple* using the *Matrix Palette* on the left hand side of the screen or using the *matrix* command. We will use the *matrix* command to enter a matrix.

Example 1: Let $A = \begin{bmatrix} -3 & 2 & 1 \\ 0 & -7 & 11 \\ 9 & 6 & 5 \end{bmatrix}$ and $B = \begin{bmatrix} -13 & 1 & 1 \\ 3 & -1 & 6 \\ -6 & 2 & 1 \end{bmatrix}$. Use *Maple* commands to find:

(a) $-3A$
(b) $A + B$
(c) AB
(d) $A - 2B$

Solution: We will enter the matrices using the *Matrix* command.

$A := Matrix\left(\left[[-3,2,1],[0,-7,11],[9,6,5]\right]\right)$

$$\begin{bmatrix} -3 & 2 & 1 \\ 0 & -7 & 11 \\ 9 & 6 & 5 \end{bmatrix}$$
(12.1)

$B := Matrix\left(\left[[-13,1,1],[3,-1,6],[-6,2,1]\right]\right)$

$$\begin{bmatrix} -13 & 1 & 1 \\ 3 & -1 & 6 \\ -6 & 2 & 1 \end{bmatrix}$$
(12.2)

(a)   We will now find $-3A$.

$-3 \cdot A$

$$\begin{bmatrix} 9 & -6 & -3 \\ 0 & 21 & -33 \\ -27 & -18 & -15 \end{bmatrix}$$
(12.3)

Hence $-3A = \begin{bmatrix} 9 & -6 & -3 \\ 0 & 21 & -33 \\ -27 & -18 & -15 \end{bmatrix}$.

(b)   The sum of the two matrices is found below.

$A + B$

$$\begin{bmatrix} -16 & 3 & 8 \\ 3 & -8 & 17 \\ 1 & 8 & 6 \end{bmatrix}$$ **(12.4)**

Therefore, $A + B = \begin{bmatrix} -16 & 3 & 8 \\ 3 & -8 & 17 \\ 1 & 8 & 6 \end{bmatrix}$

(c)   The $evalm(A \,\&\cdot B)$ command is used to evaluate AB. Note that in this case we do not use the $\boxed{*}$ symbol to denote matrix multiplication.

$evalm(A \,\&\cdot B)$

$$\begin{bmatrix} 37 & -3 & -8 \\ -109 & 29 & -31 \\ -139 & 13 & 104 \end{bmatrix}$$ **(12.5)**

The product of the matrices A and B is $AB = \begin{bmatrix} 37 & -3 & -8 \\ -109 & 29 & -31 \\ -139 & 13 & 104 \end{bmatrix}$.

(d)   We will now find $A - 2B$.

$A - 2 \cdot B$

$$\begin{bmatrix} 23 & 0 & -13 \\ -6 & -5 & -1 \\ 25 & 2 & 3 \end{bmatrix}$$ **(12.6)**

Hence, we see that $A - 2 \cdot B = \begin{bmatrix} 23 & 0 & -13 \\ -6 & -5 & -1 \\ 25 & 2 & 3 \end{bmatrix}$.

Remarks:

1.   We could also use the $Matrix\left(m, n, \left[a_{11}, a_{12}, ..., a_{mn}\right]\right)$ command to generate an $m \times n$ matrix. In this case the dimensions of the matrix are entered first followed

by the elements in the matrix (entered row by row). This is shown below with the matrix A defined in example 1.

$$A := matrix\left(3, 3, [-3, 2, 1, 0, -7, 11, 9, 6, 5]\right)$$

$$\begin{bmatrix} -3 & 2 & 1 \\ 0 & -7 & 11 \\ 9 & 6 & 5 \end{bmatrix} \qquad \textbf{(12.7)}$$

Instead of using the *Matrix* command to generate matrices we could also use the *Matrix Palette* to insert the matrix A. Make sure that the number of rows and columns are 3 and then click on *Insert Matrix*. This is shown in the picture below.

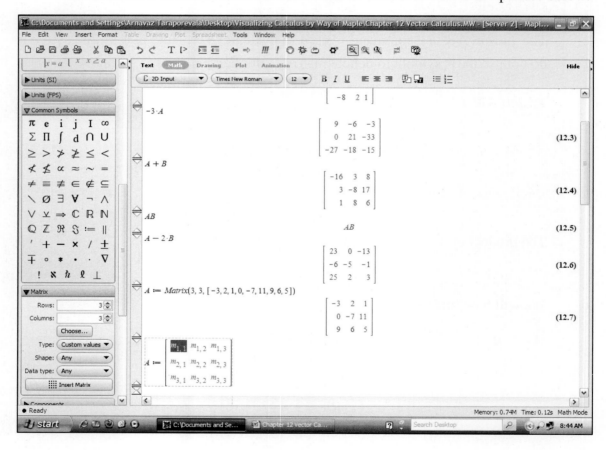

A matrix of the form $\begin{bmatrix} m_{1,1} & m_{1,2} & m_{1,3} \\ m_{2,1} & m_{2,2} & m_{2,3} \\ m_{3,1} & m_{3,2} & m_{3,3} \end{bmatrix}$ appears with $m_{1,1}$ that is highlighted.

Enter the number -3 and use use the $\boxed{Tab}$ key so that $m_{1,2}$ is highlighted. Enter the number 2 and use use the $\boxed{Tab}$ key so that $m_{1,3}$ is highlighted. Continue this

process till you have entered all the entries in the matrix A and then press the
$\boxed{Enter}$ key to get the following result.

$$A := \begin{bmatrix} -3 & 2 & 1 \\ 0 & -7 & 11 \\ 9 & 6 & 5 \end{bmatrix}$$

$$\begin{bmatrix} -3 & 2 & 1 \\ 0 & -7 & 11 \\ 9 & 6 & 5 \end{bmatrix} \qquad \textbf{(12.8)}$$

2. One can also use the *Matrix Palette* to insert an identity matrix. In order to insert a $4{\times}4$ identity matrix, make sure that the number of rows and columns are 4 and the *Type* is *Identity* and then click on *Insert Matrix*. This is shown in the picture below.

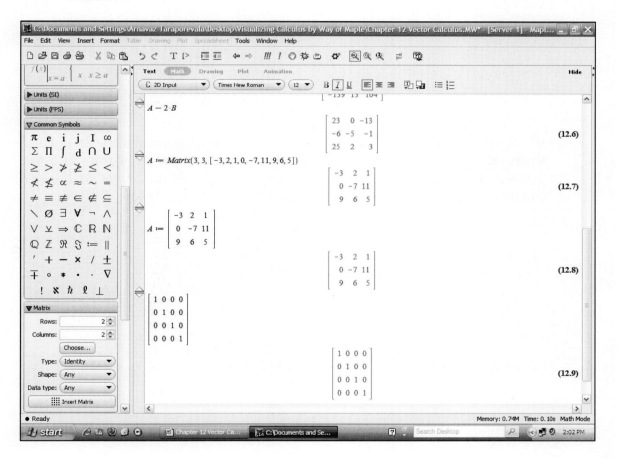

$$\begin{bmatrix} 1 & 0 & 0 & 0 \\ 0 & 1 & 0 & 0 \\ 0 & 0 & 1 & 0 \\ 0 & 0 & 0 & 1 \end{bmatrix}$$

$$\begin{bmatrix} 1 & 0 & 0 & 0 \\ 0 & 1 & 0 & 0 \\ 0 & 0 & 1 & 0 \\ 0 & 0 & 0 & 1 \end{bmatrix} \qquad (12.9)$$

3. The *diag* command can be used to enter a diagonal matrix. An example of a 5×5 diagonal matrix is below.

$$diag\,(1,-3,5,7,9)$$

$$\begin{bmatrix} 1 & 0 & 0 & 0 & 0 \\ 0 & -3 & 0 & 0 & 0 \\ 0 & 0 & 5 & 0 & 0 \\ 0 & 0 & 0 & 7 & 0 \\ 0 & 0 & 0 & 0 & 9 \end{bmatrix} \qquad (12.10)$$

We could also use the *Matrix Palette* to insert this diagonal matrix. In order to insert this 5×5 diagonal matrix, make sure that the number of rows is 5, the number of columns is 5, the *Type* is *Custom Values* and the *Shape* is *Diagonal* and then click on *Insert Matrix*. This is shown in the picture below.

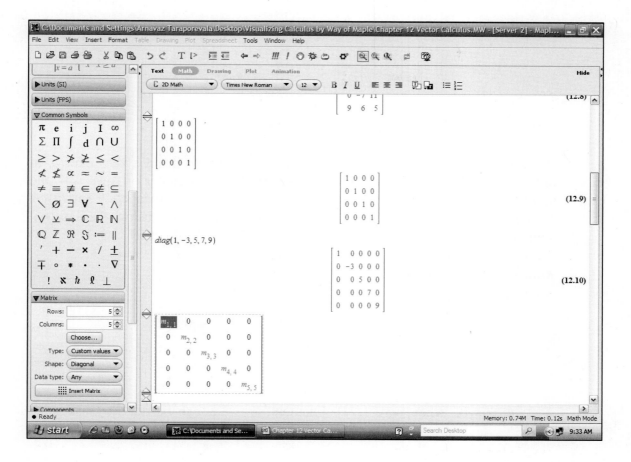

A matrix of the form $\begin{bmatrix} m_{1,1} & 0 & 0 & 0 & 0 \\ 0 & m_{2,2} & 0 & 0 & 0 \\ 0 & 0 & m_{3,3} & 0 & 0 \\ 0 & 0 & 0 & m_{4,4} & 0 \\ 0 & 0 & 0 & 0 & m_{5,5} \end{bmatrix}$ appears with $m_{1,1}$ that is

highlighted. Enter the number 1 and use the $\boxed{Tab}$ key so that $m_{2,2}$ is highlighted. Enter the number -3 and use the $\boxed{Tab}$ key so that $m_{3,3}$ is highlighted. Continue this process till you have entered all the entries in the diagonal matrix and then press the $\boxed{Enter}$ key to get the following result.

$$\begin{bmatrix} 1 & 0 & 0 & 0 & 0 \\ 0 & -3 & 0 & 0 & 0 \\ 0 & 0 & 5 & 0 & 0 \\ 0 & 0 & 0 & 7 & 0 \\ 0 & 0 & 0 & 0 & 9 \end{bmatrix}$$

$$\begin{bmatrix} 1 & 0 & 0 & 0 & 0 \\ 0 & -3 & 0 & 0 & 0 \\ 0 & 0 & 5 & 0 & 0 \\ 0 & 0 & 0 & 7 & 0 \\ 0 & 0 & 0 & 0 & 9 \end{bmatrix} \qquad (12.11)$$

4. By similar means one can also use the *Matrix Palette* to insert a zero matrix. In order to insert a 4×2 zero matrix, make sure that the number of rows is 4, the number of columns is 2 and the *Type* is *Zero-Filled* and then click on *Insert Matrix*. Another way of generating a zero matrix is by using the *Matrix* command.

$$Matrix\big(4,2,(i,j)\to0\big)$$

$$\begin{bmatrix} 0 & 0 \\ 0 & 0 \\ 0 & 0 \\ 0 & 0 \end{bmatrix} \qquad (12.12)$$

A 2×4 matrix, where any $(i,j)^{th}$-element, $a_{ij}=-6$ can be done by using the *Matrix* command.

$$Matrix\big(2,4,(i,j)\to-6\big)$$

$$\begin{bmatrix} -6 & -6 & -6 & -6 \\ -6 & -6 & -6 & -6 \end{bmatrix} \qquad (12.13)$$

We can use the *Matrix* command to construct a matrix where the $(i,j)^{th}$-element is $a_{ij}=\dfrac{2j}{1+j^2}$.

$$Matrix\left(2,3,(i,j)\to\dfrac{2\cdot i}{1+j^2}\right)$$

$$\begin{bmatrix} 1 & \dfrac{2}{5} & \dfrac{1}{5} \\ 2 & \dfrac{4}{5} & \dfrac{2}{5} \end{bmatrix} \qquad (12.14)$$

5. The *evalm( )* command is used to find the sum, product and scalar multiples of matrices in example 1.

$$A := Matrix\big(\big[[-3,2,1],[0,-7,11],[9,6,5]\big]\big)$$

$$\begin{bmatrix} -3 & 2 & 1 \\ 0 & -7 & 11 \\ 9 & 6 & 5 \end{bmatrix}$$

(12.15)

$$B := Matrix\big(\big[[-13,1,1],[3,-1,6],[-6,2,1]\big]\big)$$

$$\begin{bmatrix} -13 & 1 & 1 \\ 3 & -1 & 6 \\ -6 & 2 & 1 \end{bmatrix}$$

(12.16)

$$evalm(-3 \cdot A)$$

$$\begin{bmatrix} 9 & -6 & -3 \\ 0 & 21 & -33 \\ -27 & -18 & -15 \end{bmatrix}$$

(12.17)

$$evalm(A+B)$$

$$\begin{bmatrix} -16 & 3 & 8 \\ 3 & -8 & 17 \\ 1 & 8 & 6 \end{bmatrix}$$

(12.18)

$$evalm(A-2 \cdot B)$$

$$\begin{bmatrix} 23 & 0 & -13 \\ -6 & -5 & -1 \\ 25 & 2 & 3 \end{bmatrix}$$

(12.19)

The $A\,B$ command can be used to evaluate AB. However, the authors would advise the reader against using this command as was in the case of multiplying two real numbers in chapter 0.

$$A\,B$$

$$\begin{bmatrix} 37 & -3 & -8 \\ -109 & 29 & -31 \\ -139 & 13 & 104 \end{bmatrix}$$

(12.20)

Example 2: Let $A = \begin{bmatrix} -3 & 2 & 1 \\ 0 & -7 & 11 \\ 9 & 6 & 5 \end{bmatrix}$. Use *Maple* commands to find:

(a) A^4
(b)   the transpose of A
(c)   the inverse of A

Solution: We will enter the matrices using the *Matrix* command as we did in example 1.

$$A := Matrix\left(\left[[-3,2,1],[0,-7,11],[9,6,5]\right]\right)$$

$$\begin{bmatrix} -3 & 2 & 1 \\ 0 & -7 & 11 \\ 9 & 6 & 5 \end{bmatrix}$$
(12.21)

(a)   The matrix representing A^4 is calculated below.

A^4

$$\begin{bmatrix} -630 & -1718 & 3140 \\ 12771 & 11707 & -2354 \\ 2718 & 1038 & 10300 \end{bmatrix}$$
(12.22)

Therefore, $A^4 = \begin{bmatrix} -630 & -1718 & 3140 \\ 12771 & 11707 & -2354 \\ 2718 & 1038 & 10300 \end{bmatrix}$.

(b)   In order to get the transpose of the matrix A we will use the $A^{\%T}$ command.

$A^{\%T}$

$$\begin{bmatrix} -3 & 0 & 9 \\ 2 & -7 & 6 \\ 1 & 11 & 5 \end{bmatrix}$$
(12.23)

Hence the transpose of A is $A^T = \begin{bmatrix} -3 & 0 & 9 \\ 2 & -7 & 6 \\ 1 & 11 & 5 \end{bmatrix}$

(c)   The A^{-1} command is used to find the inverse of A.

A^{-1}

$$\begin{bmatrix} -\dfrac{101}{564} & -\dfrac{1}{141} & \dfrac{29}{564} \\ \dfrac{33}{188} & -\dfrac{2}{47} & \dfrac{11}{188} \\ \dfrac{21}{188} & \dfrac{3}{47} & \dfrac{7}{188} \end{bmatrix}$$
(12.24)

The inverse of A exists and $A^{-1} = \begin{bmatrix} -\dfrac{202}{564} & -\dfrac{1}{141} & \dfrac{29}{564} \\ \dfrac{33}{188} & -\dfrac{2}{47} & \dfrac{11}{188} \\ \dfrac{21}{188} & \dfrac{3}{47} & \dfrac{7}{188} \end{bmatrix}$.

Remarks: As in the case of the previous example we can use other commands to find the answers to example 2. This is shown below.

$evalm\left(A^4\right)$

$$\begin{bmatrix} -630 & -1718 & 3140 \\ 12771 & 11707 & -2354 \\ 2718 & 1038 & 10300 \end{bmatrix} \qquad \textbf{(12.25)}$$

To use an alternate way to find the transpose and inverse of A we first need to activate the $with\left(linalg\right)$ package.

$with\left(linalg\right):$

$transpose\left(A\right)$

$$\begin{bmatrix} -3 & 0 & 9 \\ 2 & -7 & 6 \\ 1 & 11 & 5 \end{bmatrix} \qquad \textbf{(12.26)}$$

$inverse\left(A\right)$

$$\begin{bmatrix} -\dfrac{101}{564} & -\dfrac{1}{141} & \dfrac{29}{564} \\ \dfrac{33}{188} & -\dfrac{2}{47} & \dfrac{11}{188} \\ \dfrac{21}{188} & \dfrac{3}{47} & \dfrac{7}{188} \end{bmatrix} \qquad \textbf{(12.27)}$$

Example 3: Let $A = \begin{bmatrix} -2 & 1 & 3 & -5 \\ 6 & -1 & 1 & 0 \\ 5 & 3 & -2 & 1 \end{bmatrix}$ and $B = \begin{bmatrix} -3 & -2 & 3 \\ -7 & 1 & 6 \end{bmatrix}$. Use *Maple* commands to find, if they exist:

(a)   *BA*

(b)   *AB*

(c) $A+B$

Solution: We will enter the matrices using the *Matrix* command.

$A := Matrix\big(\big[[-2,1,3,-5],[6,-1,1,0],[5,3,-2,1]\big]\big)$

$$\begin{bmatrix} -2 & 1 & 3 & -5 \\ 6 & -1 & 1 & 0 \\ 5 & 3 & -2 & 1 \end{bmatrix}$$ (12.28)

$B := Matrix\big(\big[[-3,-2,3],[-7,1,6]\big]\big)$

$$\begin{bmatrix} -3 & -2 & 3 \\ -7 & 1 & 6 \end{bmatrix}$$ (12.29)

(a)   The $evalm(B\,\&\cdot A)$ command is used to evaluate the matrix product BA. Since B is a 2×3 matrix and A is a 3×4 matrix we can find the matrix product BA. Note: the number of columns of A is the same as the number of rows of B

$evalm(B\,\&\cdot A)$

$$\begin{bmatrix} 37 & -3 & -8 \\ -109 & 29 & 31 \\ -139 & 13 & 104 \end{bmatrix}$$ (12.30)

The product of the matrices A and B is $BA = \begin{bmatrix} 37 & -3 & -8 \\ -109 & 29 & 31 \\ -139 & 13 & 104 \end{bmatrix}$.

(b)   Since A is a 3×4 matrix and B is a 2×3 matrix we cannot perform the operation AB, as the number of columns of A is different from the number of rows of B. This is verified below.

$evalm(A\,\&\cdot B)$
```
Error, (in linalg:-multiply), error in non matching
dimensions for vector/matrix product
```

(c)   Since A is a 3×4 matrix and B is a 2×3 matrix we cannot add the two matrices. This is verified below.

$A+B$
```
Error, in rtable/Sum, invalid arguments
```

Remarks: In example 3, since A is a 3×4 matrix and B is a 2×3 matrix we see that A^T is a 4×3 matrix and B^T is a 3×2 matrix. Hence, as the number of columns of A^T is the same as the number of rows of B^T, we can multiply A^T and B^T, in this order.

$$evalm\left(A^{\%T} \& \cdot B^{\%T}\right)$$

$$\begin{bmatrix} 9 & 50 \\ 8 & 10 \\ -17 & -32 \\ 18 & 41 \end{bmatrix}$$

(12.30)

Vectors

Definition: A _vector_ in $\mathbb{R}^n$ is an array of the form $v = \langle v_1, v_2, ..., v_n \rangle$, where $v_1, v_2, ..., v_n$ are called the components of the vector.

Remark: A vector is denoted using the _Bold Font_. From now on we will only consider three dimensional vectors, namely, vectors in $\mathbb{R}^3$, but this can be extended to higher dimensions.

Remark: Consider two points $P = (x_1, y_1, z_1)$ and $Q = (x_2, y_2, z_2)$. The vector with initial point (x_1, y_1, z_1) and terminal point (x_2, y_2, z_2) is $\overrightarrow{PQ} = p = \langle x_2 - x_1, y_2 - y_1, z_2 - z_1 \rangle$.

Definition: Let $v = \langle v_1, v_2, v_3 \rangle$ be a vector in $\mathbb{R}^3$. Then the _magnitude_ of v, denoted by $\|v\|$, is defined as the number $\|v\| = \sqrt{v_1^2 + v_2^2 + v_3^2}$.

Definition: Let $v = \langle v_1, v_2, v_3 \rangle$ be a three dimensional vector and let k be any real number. Then the _scalar multiplication_ kv is the vector in $\mathbb{R}^3$ defined as $kv = \langle kv_1, kv_2, kv_3 \rangle$.

Definition: Let $v = \langle v_1, v_2, v_3 \rangle$ and $w = \langle w_1, w_2, w_3 \rangle$ be two vectors in $\mathbb{R}^3$. Their _dot product_, denoted by $v \cdot w$, is defined as the number $v \cdot w = v_1 w_1 + v_2 w_2 + v_3 w_3$

Definition: Let $v = \langle v_1, v_2, v_3 \rangle$ and $w = \langle w_1, w_2, w_3 \rangle$ be two vectors in $\mathbb{R}^3$. If their dot product $v \cdot w = 0$, then v and w are said to be _normal or orthogonal to each another._

Definition: Let $v = \langle v_1, v_2, v_3 \rangle$ and $w = \langle w_1, w_2, w_3 \rangle$ be two vectors in $\mathbb{R}^3$. Their _cross product_, denoted by $v \times w$, is defined as the vector $v \times w = \langle v_2 w_3 - v_3 w_2, v_1 w_3 - v_3 w_1, v_1 w_2 - v_3 w_2 \rangle$.

Remarks:

1.   Let $v = \langle v_1, v_2, v_3 \rangle$ and $w = \langle w_1, w_2, w_3 \rangle$ be two vectors in $\mathbb{R}^3$. Their cross product $v \times w$ is normal to both v and w and the plane containing the vectors v and w.

2.   Let $v = \langle v_1, v_2, v_3 \rangle$ and $w = \langle w_1, w_2, w_3 \rangle$ be two vectors in $\mathbb{R}^3$. Then $v \times w = 0$ if and only if v and w are parallel.

Definition: Let $v = \langle v_1, v_2, v_3 \rangle$ and $w = \langle w_1, w_2, w_3 \rangle$ be two vectors. The *horizontal projection* of the vector v onto w (or the *projection of the vector v onto a line in the direction of the vector w*) is given by the formula $(v \cdot w) \dfrac{w}{\|w\|^2}$.

Remark: If w is a *unit vector* then this formula becomes $(v \cdot w) w$. Note that *Maple* calls the horizontal projection of the vector v onto w the *projection.*

Definition: Let $v = \langle v_1, v_2, v_3 \rangle$ and $w = \langle w_1, w_2, w_3 \rangle$ be two vectors. The *vertical projection* of the vector v onto w is given by the formula $w - (v \cdot w) \dfrac{w}{\|w\|^2}$.

Remark: *Maple* calls this the *orthogonal complement.*

Definition: A vector v that is perpendicular to a plane is called a *normal vector.*

Remark: The equation of a plane through a point $P = (x_0, y_0, z_0)$ with a normal vector $v = \langle v_1, v_2, v_3 \rangle$ is given by $v_1(x - x_0) + v_2(y - y_0) + v_3(z - z_0) = 0$. Another way of finding the equation of this plane is to use $v \cdot p = 0$ where $p = \langle x - x_0, y - y_0, z - z_0 \rangle$.

Example 4: Let $v = \langle -3, 2, 4 \rangle$ and $w = \langle 1, 0, -6 \rangle$. Use *Maple* commands to find:
(a)   the magnitude of the vector v
(b) $v \cdot w$
(c) $v \times w$
(d)   the angle between v and w, expressed in radians and degrees
(e)   the horizontal and vertical projections of the vector v onto w
(f)   the plot showing the projection of v onto w .

Solution: We will first enter the vectors v and w using *Maple* commands. Note that while entering in the vectors in *Maple* we do not use the *Bold Font.*

$v := \langle -3, 2, 4 \rangle$

$$\begin{bmatrix} -3 \\ 2 \\ 4 \end{bmatrix} \qquad \textbf{(12.32)}$$

$$w := \langle 1, 0, -6 \rangle$$

$$\begin{bmatrix} 1 \\ 0 \\ -6 \end{bmatrix} \qquad \textbf{(12.33)}$$

Note that *Maple* enters the vector as a column vector.

(a)   The *Norm*$(v, 2)$ command is used to find the magnitude of the vector v. The *with*(*LinearAlgebra*) package must be activated.

with(*LinearAlgebra*):
Norm$(v, 2)$

$$\sqrt{29} \qquad \textbf{(12.34)}$$

The magnitude of the vector v is $\sqrt{29}$, i.e. $\|v\| = \sqrt{29}$.

(b)   The *DotProduct* command will be used to find $v \cdot w$. We do not need to activate the *with*(*LinearAlgebra*) package as this has been done in part (a) of this example.

DotProduct(v, w)

$$-27 \qquad \textbf{(12.35)}$$

Hence, $v \cdot w = -27$.

(c)   In order to find the cross product of the two vectors we use the *CrossProduct* command. In this case again we do not need to activate the *with*(*LinearAlgebra*) package.

CrossProduct(v, w)

$$\begin{bmatrix} -12 \\ -14 \\ -2 \end{bmatrix} \qquad \textbf{(12.36)}$$

Therefore, $v \times w = \langle -12, -14, -2 \rangle$.

(d)   The *VectorAngle* command will be used to find the angle between the two vectors *v* and *w*. Once again we do not need to activate the *with*(*LinearAlgebra*) package.

VectorAngle(*v*, *w*)

$$\pi - \arccos\left(\frac{27}{1073}\sqrt{29}\sqrt{37}\right)$$ **(12.37)**

evalf((**12.37**))

$$2.539689271$$ **(12.38)**

evalf(*convert*(**12.38**), *degrees*)

$$145.5134765 \; degrees$$ **(12.39)**

Therefore, the angle between *v* and *w* is 2.539689271 radians = 145.5134765°.

(e)   We will now find the horizontal and vertical projections of the vector *v* onto *w*

$$DotProduct(v, w) \cdot \frac{w}{Norm(w, 2)^2}$$

$$\begin{bmatrix} -\dfrac{27}{37} \\ 0 \\ \dfrac{162}{37} \end{bmatrix}$$ **(12.40)**

$$w - DotProduct(v, w) \cdot \frac{w}{Norm(w, 2)^2}$$

$$\begin{bmatrix} \dfrac{64}{37} \\ 0 \\ -\dfrac{384}{37} \end{bmatrix}$$ **(12.41)**

Therefore the horizontal and vertical projections of the vector *v* onto *w* are $\left\langle -\dfrac{27}{37}, 0, \dfrac{162}{37} \right\rangle$ and $\left\langle \dfrac{64}{37}, 0, -\dfrac{384}{37} \right\rangle$ respectively.

(f)   The *ProjectionPlot* command will be used to find the projection of *v* onto *w*. We will activate the *with*(*Student*[*LinearAlgebra*]) package and set the *infolevel*[*Student*[*LinearAlgebra*]] equal to 1.

with(*Student*[*LinearAlgebra*]):

infolevel[*Student*[*LinearAlgebra*]]:=1:

ProjectionPlot(*v*,*w*)

```
Vector: <-3, 2, 4>
Projection: <-.7297, -0., 4.378>
Orthogonal complement: <-2.270, 2., -.3784>
Norm of orthogonal complement: 3.049
```

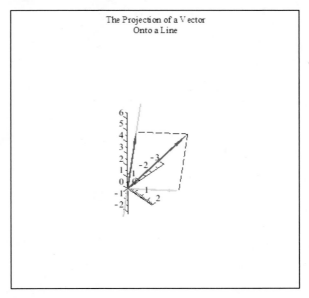

Right click on the plot and click on Scaling Constrained to get the plot below.

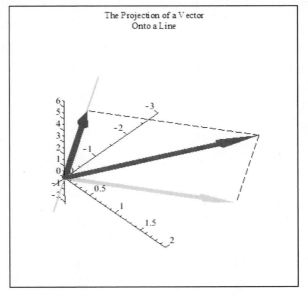

Remarks:

1. The $with(Student[LinearAlgebra])$ package can be used to find $v \& x \ w$ (the cross product) and $v.w$ (the dot product). The command to find the angle between two vectors is the same command as in the $with(LinearAlgebra)$ package.

$$v := \langle -3, 2, 4 \rangle$$

$$\begin{bmatrix} -3 \\ 2 \\ 4 \end{bmatrix} \qquad \qquad \textbf{(12.42)}$$

$$w := \langle 1, 0, -6 \rangle$$

$$\begin{bmatrix} 1 \\ 0 \\ -6 \end{bmatrix} \qquad \qquad \textbf{(12.43)}$$

$$v \& x \ w$$

$$\begin{bmatrix} -12 \\ -14 \\ -2 \end{bmatrix} \qquad \qquad \textbf{(12.44)}$$

$$v.w$$

$$-27 \qquad \qquad \textbf{(12.45)}$$

2. We can use the *crossprod*, *dotprod* and *angle* command to find the dot product of the two vectors v and w and the angle between the two vectors v and w using the $with(linalg)$ package. With this package the result of the cross product command comes in the form of a $1 \times m$ matrix.

$$v := \langle -3, 2, 4 \rangle$$

$$\begin{bmatrix} -3 \\ 2 \\ 4 \end{bmatrix} \qquad \qquad \textbf{(12.46)}$$

$$w := \langle 1, 0, -6 \rangle$$

$$\begin{bmatrix} 1 \\ 0 \\ -6 \end{bmatrix} \qquad \qquad \textbf{(12.47)}$$

$with(linalg):$

$crossprod(v, w)$

$$\begin{bmatrix} -12 & -14 & -2 \end{bmatrix} \tag{12.48}$$

$dotprod\,(v,w)$

$$-27 \tag{12.49}$$

$angle\,(v,w)$

$$\pi - \arccos\left(\frac{27}{1073}\sqrt{29}\sqrt{37}\right) \tag{12.50}$$

Example 5: Graph the plane $x + y + z = 1$ in the first octant in $\mathbb{R}^3$.

Solution: We will graph this plane using the *plot3d* command.

$plot3d\,(1 - x - y, x = 0\,..\,1, y = 0\,..\,1, axes = normal)$

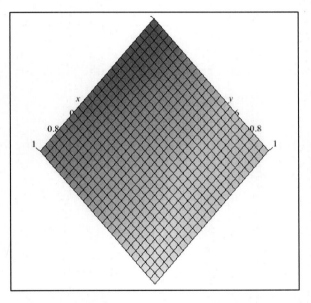

In order to get a better view we will change the *view* and *orientation* options in this *plot3d* command.

$plot3d\,(1 - x - y, x = 0\,..\,1, y = 0\,..\,1, axes = normal,$
$$orientation = [25, 55], view = [0\,..\,1.2, 0\,..\,1.2, 0\,..\,1.2])$$

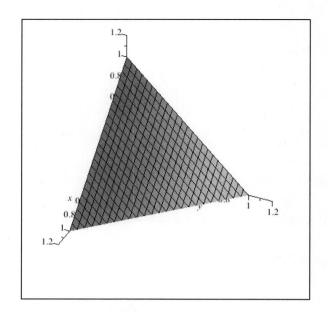

Example 6: Find the equation of a plane through the point $p = \left(-3, 1, -\sqrt{2}\right)$ with a normal vector $v = \langle -4, 2, 7 \rangle$. Graph this plane.

Solution: We will first enter the point $p = \left(-3, 1, -\sqrt{2}\right)$ and the vector $v = \langle -4, 2, 7 \rangle$ using *Maple* commands.

$p := \left(-3, 1, -\sqrt{2}\right)$

$$-3, 1, -\sqrt{2} \qquad\qquad \textbf{(12.51)}$$

$v = \langle -4, 2, 7 \rangle$

$$\begin{bmatrix} -4 \\ 2 \\ 7 \end{bmatrix} \qquad\qquad \textbf{(12.52)}$$

Hence the equation of the plane is obtained by simplifying $-4\left(x - (-3)\right) + 2\left(y - 1\right) + 7\left(z - \left(-\sqrt{2}\right)\right) = 0$

$simplify\left(-4 \cdot \left(x - (-3)\right) + 2 \cdot \left(y - 1\right) + 7 \cdot \left(z - \left(-\sqrt{2}\right)\right) = 0\right)$

$$-4\,x - 14 + 2\,y + 7\,z + 7\sqrt{2} = 0 \qquad\qquad \textbf{(12.53)}$$

Hence the equation of the plane is $4\,x - 2\,y - 7\,z = -14 + 7\sqrt{2}$. Before graphing this plane we will use this equation to express z as a function of x and y.

$$solve\left((\mathbf{12.53}),z\right)$$

$$2+\frac{4}{7}x-\frac{2}{7}y-\sqrt{2} \qquad\qquad (\mathbf{12.54})$$

Hence $z=\frac{4}{7}x-\frac{2}{7}y+2-\sqrt{2}$. We will now graph this plane.

$$plot3d\left(2+\frac{4}{7}x-\frac{2}{7}y-\sqrt{2},x=-6..6,y=-6..6,axes=normal\right)$$

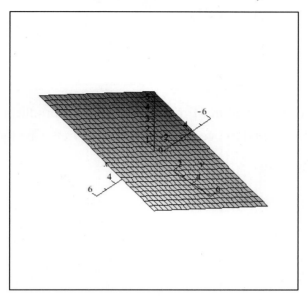

Remarks:

1. We can modify this command to get a better view of the plane by changing the *view* and *orientation* options.

$$plot3d\left(2+\frac{4}{7}x-\frac{2}{7}y-\sqrt{2},x=0..6,y=0..6,axes=normal,\right.$$

$$\left.orientation=\left[45,45\right],view=\left[-3..7,-3..7,-3..7\right]\right)$$

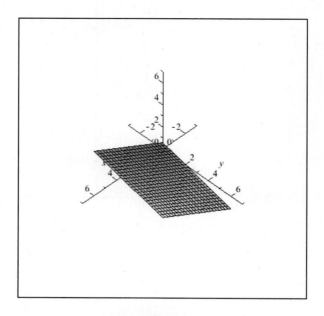

2.   We can find the equation of the plane using dot products too. Consider the point $p = \left(-3, 1, -\sqrt{2}\right)$ and the vector $v = \langle -4, 2, 7 \rangle$. Let q be the point (x, y, z). Define the vector $w = \overrightarrow{pq}$.

$v = \langle -4, 2, 7 \rangle$

$$\begin{bmatrix} -4 \\ 2 \\ 7 \end{bmatrix}$$ (12.55)

$p := \left(-3, 1, -\sqrt{2}\right)$

$$-3, 1, -\sqrt{2}$$ (12.56)

$q := (x, y, z)$

$$x, y, z$$ (12.57)

$w = \langle q[1] - p[1], q[2] - p[2], q[3] - p[3] \rangle$

$$\begin{bmatrix} x + 3 \\ y - 1 \\ z + \sqrt{2} \end{bmatrix}$$ (12.58)

We will find the dot product of the vectors v and w. Do not forget to activate the $with(LinearAlgebra)$ package.

$with(LinearAlgebra)$:

$DotProduct(v, w)$

$$-4\,x-14+2\,y+7\,z+7\sqrt{2} \qquad\qquad \textbf{(12.59)}$$

Therefore, $v\cdot w = -4\,x-14+2\,y+7\,z+7\sqrt{2}$. Hence the equation of the plane is obtained by simplifying solving $v\cdot w = 0$ to get $-4\,x-14+2\,y+7\,z+7\sqrt{2}=0$. Hence the equation of the plane is $4\,x-2\,y-7\,z=-14+7\sqrt{2}$.

Example 7: Find the equation of a plane through the points $p=(1,2,3)$, $q=(-1,1,3)$, and $r=(4,-5,6)$. Graph this plane.

Solution: We will first enter the points $p=(2,-3,1)$, $q=(-1,1,3)$, and $r=(4,-5,6)$.

$$p := (1,2,3)$$

$$1,2,3 \qquad\qquad \textbf{(12.60)}$$

$$q = (-1,1,3)$$

$$-1,1,3 \qquad\qquad \textbf{(12.61)}$$

$$r = (4,-5,6)$$

$$4,-5,6 \qquad\qquad \textbf{(12.62)}$$

Let us now find the vectors $v = \overrightarrow{pq}$ and $w = \overrightarrow{pr}$.

$$v = \langle q[1]-p[1], q[2]-p[2], q[3]-p[3]\rangle$$

$$\begin{bmatrix} -2 \\ -1 \\ 0 \end{bmatrix} \qquad\qquad \textbf{(12.63)}$$

$$w = \langle r[1]-p[1], r[2]-p[2], r[3]-p[3]\rangle$$

$$\begin{bmatrix} 3 \\ -7 \\ 3 \end{bmatrix} \qquad\qquad \textbf{(12.64)}$$

We will find the cross product of the vectors v and w. Do not forget to activate the $with(LinearAlgebra)$ package.

$with(LinearAlgebra):$

$CrossProduct(v,w)$

$$\begin{bmatrix} -3 \\ 6 \\ 17 \end{bmatrix} \qquad\qquad (12.65)$$

Therefore, $v \times w = \langle -3, 6, 17 \rangle$. Hence the equation of the plane is obtained by simplifying

$$-3(x-1) + 6(y-2) + 17(z-3) = 0$$

$$-3 \cdot (x-1) + 6 \cdot (y-2) + 17 \cdot (z-3) = 0$$
$$-3\,x - 60 + 6\,y + 17\,z = 0 \qquad\qquad (12.66)$$

Hence the equation of the plane is $-3\,x + 6\,y + 17\,z = 60$. Before graphing this plane we will use this equation to express z as a function of x and y.

$solve\big((12.66), z\big)$

$$\frac{60}{17} + \frac{3}{17}\,x - \frac{6}{17}\,y \qquad\qquad (12.67)$$

Hence $z = \dfrac{60}{17} + \dfrac{3}{17}\,x - \dfrac{6}{17}\,y$. We will now graph this plane.

$$plot3d\left(\frac{60}{17} + \frac{3}{17}\,x - \frac{6}{17}\,y, x = -4..4, y = -4..4, axes = normal \right)$$

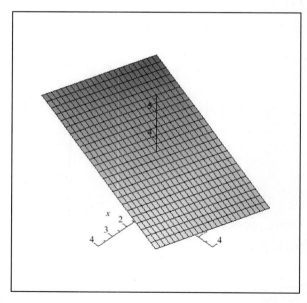

Another view of this plane is given below.

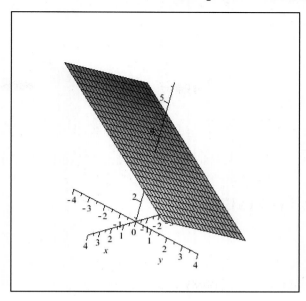

Exercises

1 – 10 Let $A = \begin{bmatrix} 5 & -2 & 0 \\ -3 & 1 & 9 \\ 0 & -2 & 7 \end{bmatrix}$, $B = \begin{bmatrix} -13 & 1 & 1 \\ 3 & -1 & 6 \\ -6 & 2 & 1 \end{bmatrix}$, and $C = \begin{bmatrix} 7 & 2 & 1 & 6 \\ -2 & -3 & 0 & 9 \\ 2 & 1 & 3 & -5 \end{bmatrix}$ and

$B = \begin{bmatrix} -3 & -2 & 3 \\ -7 & 1 & 6 \end{bmatrix}$. Use *Maple* commands to find, if they exist:

1. $14C$

2. $A + B$

3. AB

4. $B - 5A$

5. B^3

6. $A + C$

7. CB

8.   the transpose of A

9.   the inverse of B

10. AC

11 – 18 Use *Maple* commands to find:
(a)   the magnitude of the vector v
(b) $v \cdot w$
(c) $v \times w$
(d)   the angle between v and w, expressed in radians and degrees
(e)   the plot showing the projection of v onto w.

11. $v = \langle -3, 2, 5 \rangle$ and $w = \langle 6, -4, 10 \rangle$

12. $v = \langle -5, 2, 7 \rangle$ and $w = \langle 11, -19, \sqrt{2} \rangle$

13. $v = \langle -3, 2 \rangle$ and $w = \langle 1, -8 \rangle$

14. $v = \langle -1, -1, -2 \rangle$ and $w = \langle 6, \sqrt{6}, -\sqrt{2} \rangle$

15. $v = \langle 0, -1, \sqrt{3} \rangle$ and $w = \langle e, \pi, -\pi \rangle$

16. $v = \langle -11, 99 \rangle$ and $w = \langle 8, 3 \rangle$

17. $v = \langle a, a, a^2 \rangle$ and $w = \langle -a, 2a, a^5 \rangle$, where a is a non-zero real number.

18. $v = \langle 1, 1, t \rangle$ and $w = \langle -3, -3, t^4 \rangle$, t is any real number.

19 Find and graph the equation of a plane that passes through the three points $(0, -3, 11)$, $(-\sqrt{5}, 1, 3)$, and $(6, -5, 6)$.

20. Find and graph the equation of the plane that contains the point $(6, -5, 6)$ and is parallel to the plane $5x - 3y + z = \sqrt{2011}$.

21. Find and graph the equation of the plane that contains the point $(6, -5, 6)$ and is perpendicular to the plane $5x - 3y + z = \sqrt{2011}$.

22. Are the planes $5x - 3y + z = \sqrt{2011}$ and $2x + 5y - z = 9$ perpendicular? Use Maple to verify your answer and graph the two planes to illustrate your claim.

23. Show that the planes $5x - 3y + 10z = 6$ and $5x + 5y - z = 9$ are perpendicular? Find an equation of a plane that is perpendicular to both of these planes and passes through their point of intersection. Use Maple to verify your answer and graph the planes to illustrate your claim.

24. Below is a list of some *Maple* commands used in this chapter. Describe the significance of each command. Can you find examples where each command is used?

   (a) $CrossProduct(v, w)$

   (b) $dotprod(v, w)$

   (c) $evalm(A \& \cdot B)$

   (d) $inverse(A)$

   (e) $Matrix(m, n, [a_{11}, a_{12}, ..., a_{mn}])$

   (f) $Norm(v, 2)$

   (g) $VectorAngle(v, w)$

(h)   *with*(*LinearAlgebra*)

25.   Can you list new *Maple* commands that were used in this chapter but were not listed in exercise 24?

Chapter 13   Limits and Continuity of Multivariate Functions

We will examine the limits and check the continuity of multivariate functions by way of various examples.

Definitions

Definition of a limit: Let f be a real valued function of two variables. Let (c,d) be in $\mathbb{R}^2$ and L be a real number. We say that the *limit of* $f(x,y)$ *as* (x,y) *approaches* (c,d) is L, denoted by $\lim\limits_{(x,y)\to(c,d)} f(x,y) = L$, if for every $\varepsilon > 0$ there is a corresponding $\delta > 0$ such that if $0 < \sqrt{(x-c)^2 + (y-d)^2} < \delta$ then $|f(x,y) - L| < \varepsilon$. (Note that (x,y) must be in the domain of the function f.)

Remarks:

1.   From the above definition, the point (c,d) must be "close" to the domain of the function.

2.   We know from chapter 2, that the limit of a real valued function of one variable exists if its left-hand limit and right-hand limit exist and are equal. However, in the case of real valued functions of two variables, the point (x,y) can "approach" the point (c,d) along different paths. Let f be a real valued function of two variables. If f approaches two different limits as (x,y) approaches (c,d) along two different paths, then $\lim\limits_{(x,y)\to(c,d)} f(x,y)$ does not exist.

Definition: Let f be a real valued function of two variables. Let (c,d) be in the domain of f. We say that f *is continuous at* (c,d) if $\lim\limits_{(x,y)\to(c,d)} f(x,y)$ exists and $\lim\limits_{(x,y)\to(c,d)} f(x,y) = f(c,d)$. The function f *is a continuous function* if f is continuous at every point (c,d) in the domain of f.

Definition: Let f be a real valued function of two variables. Let (c,d) be a point in $\mathbb{R}^2$ (it may or may not be in the domain of f). We say that f *has a removable discontinuity at* (c,d) if $\lim\limits_{(x,y)\to(c,d)} f(x,y)$ exists and $\lim\limits_{(x,y)\to(c,d)} f(x,y) \neq f(c,d)$.

Remarks: If f has a removable discontinuity at (c,d), we can define a new function g as $g(x,y) = f(x,y)$ for every $(x,y) \neq (c,d)$ and $g(c,d) = \lim\limits_{(x,y)\to(c,d)} f(x,y)$. Then g is a continuous function.

Contour Curves

Definition: Let f be a real valued function of two variables. Let C be a constant such that $f(x,y)=C$ has a solution. We refer to $f(x,y)=C$ as a *contour curve* or a *level curve*.

Remarks: Contours give us insight into the continuity of functions in $\mathbb{R}^3$. The contours are equivalent to sketching the graph of $f(x,y)$ for different heights C. The following examples illustrate the important properties of contours.

Example 1: Let $f(x,y)=e^{-(x^2+y^2)}$. Construct a contour plot of twenty contours in two dimensions for the surface.

Solution: We will begin by entering f as a *Maple* function. In order to construct the contour plot we will need to activate the $with(plots)$ package.

$$f := (x,y) \rightarrow e^{-(x^2+y^2)}$$

$$(x,y) \rightarrow e^{-x^2-y^2} \tag{13.1}$$

In order to construct the contour plot we will need to activate the $with(plots)$ package. If the number of contours is not specified, then the default number of contours is eight.

$with(plots):$

$contourplot(f(x,y), x=-0.5..0.5, y=-0.5..0.5, contours=20)$

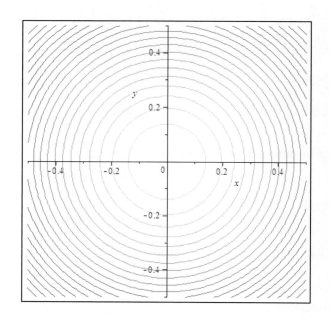

Example 2: Let $f(x, y) = e^{-\left(x^2 + y^2\right)}$. Construct a surface plot with its contours lifted at different heights.

Solution: We will begin by entering f as a *Maple* function.

$f := (x, y) \rightarrow e^{-\left(x^2 + y^2\right)}$

$$(x, y) \rightarrow e^{-x^2 - y^2} \qquad\qquad (13.2)$$

We will now construct a three-dimensional contour plot on the surface of the curve. Do not forget to activate the $with(plots)$ package.

$with(plots):$

$s1 := contourplot3d\left(f(x, y), x = -0.5 .. 0.5, y = -0.5 .. 0.5, contours = 20, axes = box,\right.$
$\qquad\left. color = blue, thickness = 2\right)$

$$PLOT3D(\ldots) \qquad\qquad (13.3)$$

$s2 := plot3d\left(f(x, y), x = -0.5 .. 0.5, y = -0.5 .. 0.5, axes = box, color = grey, transparency = 0.1\right)$

$$PLOT3D(\ldots) \qquad\qquad (13.4)$$

$display(s1, s2)$

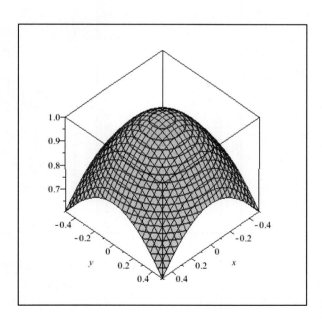

This illustration shows how the collection of contours in example 1 of $f(x, y) = e^{-\left(x^2 + y^2\right)}$ are lifted from the page of the text to appropriate heights to generate a three dimensional surface.

Example 3: Let $f(x, y) = e^{-50\left(x^2 + y^2\right)}$. Construct a contour plot of twenty contours in two dimensions for the surface. Construct a surface plot with its contours lifted at different heights.

Solution: We will begin by entering f as a *Maple* function. In order to construct the contour plot we will need to activate the *with(plots)* package.

$f := (x, y) \rightarrow e^{-50 \cdot \left(x^2 + y^2\right)}$

$$(x, y) \rightarrow e^{-50 x^2 - 50 y^2} \qquad \qquad \textbf{(13.5)}$$

In order to construct the contour plot we will need to activate the *with(plots)* package.

with(plots):

contourplot$\left(f(x, y), x = -0.5 .. 0.5, y = -0.5 .. 0.5, contours = 20, grid = [125, 125]\right)$

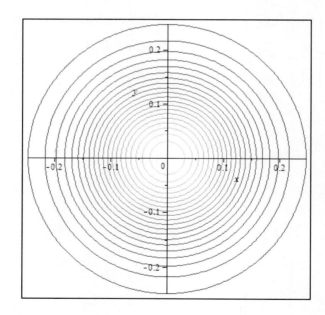

Observe that these contour lines are closer than those in example 1. In fact, if you compare the contour plot for the function with that in example 1, over the same interval, the closeness of the contours is apparent. We will now construct a three-dimensional contour plot on the surface of the curve.

$s1 :=$ *contourplot3d* $\left(f(x, y), x = -0.5 .. 0.5, y = -0.5 .. 0.5, contours = 20, axes = box,\right.$

$\left. color = blue, thickness = 2\right)$

$$PLOT3D(...) \qquad \qquad \textbf{(13.6)}$$

$$s2 := plot3d\left(f\left(x, y\right), x = -0.5 .. 0.5, y = -0.5 .. 0.5, axes = box, color = grey, transparency = 0.1\right)$$

$$PLOT3D\left(...\right) \tag{13.7}$$

$display\left(s1, s2\right)$

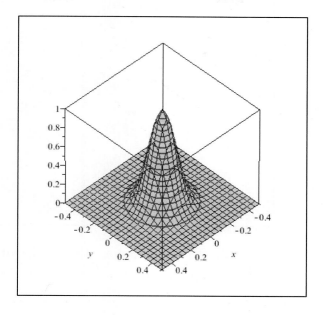

This illustration shows how the collection of contours of $f\left(x, y\right) = e^{-50\left(x^2 + y^2\right)}$ are lifted from the page of the text to appropriate heights to generate a three dimensional surface.

Example 4: Let $f\left(x, y\right) = \dfrac{\left|xy\right|}{x^2 + y^2}$. Construct a contour plot of twenty contours in two dimensions for the surface. Construct a surface plot with its contours lifted at different heights.

Solution: We will begin by entering f as a *Maple* function. In order to construct the contour plot we will need to activate the $with\left(plots\right)$ package.

$$f := \left(x, y\right) \rightarrow \frac{\left|x \cdot y\right|}{x^2 + y^2}$$

$$\left(x, y\right) \rightarrow \frac{\left|x\, y\right|}{x^2 + y^2} \tag{13.8}$$

In order to construct the contour plot we will need to activate the $with\left(plots\right)$ package.

$with\left(plots\right):$

$contourplot\left(f\left(x, y\right), x = -0.5 .. 0.5, y = -0.5 .. 0.5, contours = 20, grid = \left[150, 150\right]\right)$

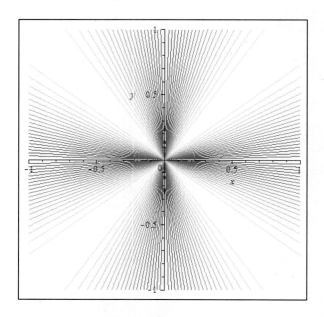

We will now construct a three-dimensional contour plot on the surface of the curve. We do not need to activate the $with(plots)$ package.

$s1 := contourplot3d\left(f(x, y), x = -0.5 .. 0.5, y = -0.5 .. 0.5, contours = 20, axes = box,\right.$
$\left. color = blue, thickness = 2, grid = [150, 150]\right)$

$$PLOT3D(...)\tag{13.9}$$

$s2 := plot3d\left(f(x, y), x = -0.5 .. 0.5, y = -0.5 .. 0.5, axes = box, color = grey, transparency\right.$
$\left. = 0.1, grid = [150, 150]\right)$

$$PLOT3D(...)\tag{13.10}$$

$display(s1, s2)$

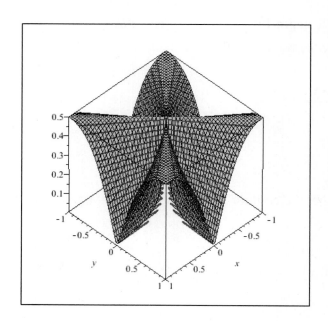

Observe that the surface is discontinuous at the point(s) where the contours intersect. Why?

Remarks:

1.   Functions with removable discontinuities will not have intersecting distinct contours.

2.   It is important to note that if distinct contours intersect then the limit at that point does not exist. Suppose we consider two distinct contours $f(x,y) = C_1$ and $f(x,y) = C_2$, where $C_1 \neq C_2$. If these contours intersect at some point (c,d) it means that $f(x,y)$ approaches two different values at (c,d) along different curves. This implies that $\lim_{(x,y) \to (c,d)} f(x,y)$ does not exist.

Limits and Continuity

Example 5: Let $f(x,y) = \dfrac{xy}{x^2 + 2xy + y^2 + 7}$. Find $\lim_{(x,y) \to (-7,3)} f(x,y)$, if it exists. Is f continuous at $(x,y) = (-7,3)$? Give reasons for your answer. Is f a continuous function? Graph this function.

Solution: We will begin by entering f as a *Maple* function.

$$f := (x,y) \to \frac{x \cdot y}{x^2 + 2 \cdot x \cdot y + y^2 + 7}$$

$$(x,y) \to \frac{xy}{x^2 + 2xy + y^2 + 7} \qquad \textbf{(13.11)}$$

Since $x^2 + 2xy + y^2 = (x+y)^2$, we will complete the square for the denominator of f to find the domain of f. Do not forget to activate the $with\left(Student[Precalculus]\right)$ package.

$with\left(Student[Precalculus]\right)$

$CompleteSquare\left(denom\left(f(x,y)\right)\right)$

$$(x+y)^2 + 7 \qquad \textbf{(13.12)}$$

Hence f is a rational function where its denominator is positive and therefore is never zero. Therefore, the domain of f is $\mathbb{R}^2$. We will now find $\lim\limits_{(x,y)\to(-7,3)} f(x,y)$. Since we are no longer considering functions of one variable, we cannot use $\lim\limits_{x\to a} f$ from the *Expression Palette* on the left side of the screen. Instead we will use the *limit* command.

$limit\left(f(x,y), \{x=-7, y=3\}\right)$

$$-\frac{21}{23}$$ (13.13)

Hence $\lim\limits_{(x,y)\to(-7,3)} f(x,y) = -\dfrac{21}{23}$. Furthermore, as

$f(-7,3)$

$$-\frac{21}{23}$$ (13.14)

we see that $\lim\limits_{(x,y)\to(-7,3)} f(x,y) = -\dfrac{21}{23} = f(-7,3)$. Therefore, f is continuous at $(-7,3)$. Since f is a rational function and its denominator is never zero, f is a continuous function.

$plot3d\left(f(x,y), x=-3..3, y=-3..3, axes=box\right)$

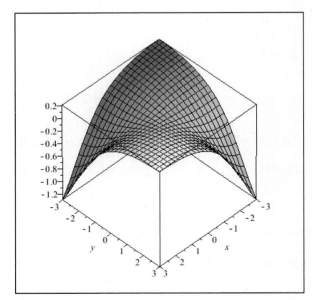

Example 6: Graph the function $f(x, y) = \dfrac{xy}{x^2 + xy + y^2}$. Find $\lim\limits_{(x,y) \to (0,0)} f(x, y)$, if it exists.

Use a contour plot to verify your answer. Use *Maple* commands to justify your answer.

Solution: We will begin by entering f as a *Maple* function and graph it.

$$f := (x, y) \to \frac{x \cdot y}{x^2 + x \cdot y + y^2}$$

$$(x, y) \to \frac{x\,y}{x^2 + x\,y + y^2} \qquad\qquad (13.15)$$

$$plot3d\left(f(x, y), x = -0.1..0.1, y = -0.1..0.1, axes = box\right)$$

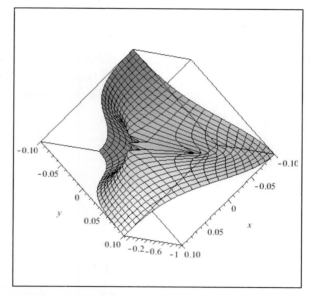

The plot is obtained by clicking on the graph and rotating it. We will now construct a contour plot of f. Do not forget to activate the *with(plots)* package.

$$with(plots):$$

$$contourplot\left(f(x, y), x = -3..3, y = -3..3, contours = 20, grid = [125,125]\right)$$

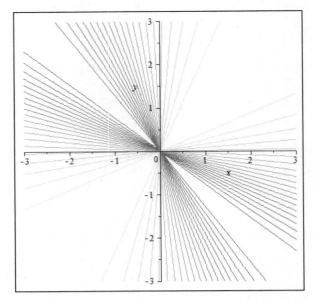

Based on the contour plot above, we see that distinct contours intersect at the origin. We can speculate that the limit at the origin does not exist (why?). We will now use *Maple* commands to verify our answer by approaching the origin along different paths.

$limit\left(f\left(x,y\right),\{x=y\}\right)$

$$\frac{1}{3}$$ **(13.16)**

$limit\left(f\left(x,y\right),\{x=-y\}\right)$

-1 **(13.17)**

Hence, if we approach $\left(0,0\right)$ along the path $x=y$ we get a limit of $\dfrac{1}{3}$ and if we approach $\left(0,0\right)$ along the path $x=-y$ we get a limit of -1. Therefore, $\displaystyle\lim_{(x,y)\to(0,0)} f\left(x,y\right)$ does not exist.

Remark: Note that as $\displaystyle\lim_{(x,y)\to(0,0)} f\left(x,y\right)$ does not exist, f is not continuous at $\left(0,0\right)$. In fact $\left(0,0\right)$ is not in the domain of f. We will verify this using *Maple* commands.

$f:=\left(x,y\right)\to \dfrac{x\cdot y}{x^2+x\cdot y+y^2}$

$$\left(x,y\right)\to \frac{x\,y}{x^2+x\,y+y^2}$$ **(13.18)**

$f\left(0,0\right)$

Error, (in f) numeric exception: division by zero

Let us find the value of $\lim\limits_{(x,y)\to(0,0)} f(x,y)$ using the *limit* command from example 1.

$limit\left(f(x,y),\{x=0,y=0\}\right)$

$$limit\left(\frac{x\,y}{x^2+x\,y+y^2},\{x=0,y=0\}\right) \qquad\qquad \textbf{(13.19)}$$

If the *Maple* output comes this way, we cannot conclude that the limit does not exist. It may or may not exist. We will see this in the next example too.

Example 7: Let $f(x,y)=\dfrac{x^2|y|}{x^2+y^2}$. Graph the function f. Find $\lim\limits_{(x,y)\to(0,0)} f(x,y)$, if it exists. Is f continuous at $(0,0)$? Give reasons for your answer.

Solution: We will begin by entering f as a *Maple* function and evaluate the function at $(0,0)$.

$f:=(x,y)\to\dfrac{x^2\cdot|y|}{x^2+y^2}$

$$(x,y)\to\frac{x^2|y|}{x^2+y^2} \qquad\qquad \textbf{(13.20)}$$

$f(0,0)$
Error, (in f) numeric exception: division by zero

Observe that $(0,0)$ is not in the domain of f. Hence f is not continuous at $(0,0)$. We will now plot this function.

$plot3d\left(f(x,y),x=-3..3,y=-3..3,axes=box\right)$

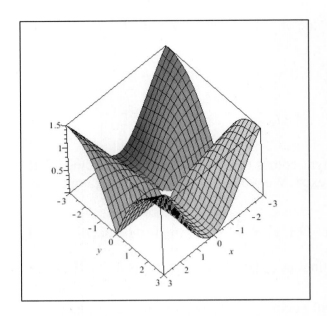

We will now check and see if $\lim\limits_{(x,y)\to(0,0)} f(x,y)$ exists using contour plots. Do not forget to activate the *with(plots)* package.

with(plots):

contourplot $\left(f(x,y), x=-3..3, y=-3..3, contours=25, grid=[75,75] \right)$

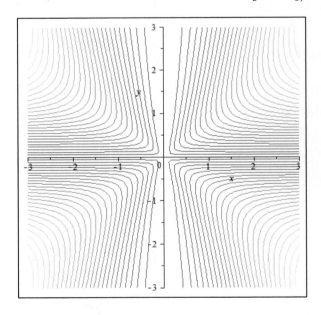

Since the distinct contours do not intersect in the above contour plot, it looks as though the limit exists. We will verify this using the Squeeze Theorem. Observe that

$f(x,y) = \dfrac{x^2|y|}{x^2 + y^2} \geq 0$. This is verified in the plot below which is sketched in a small rectangle around $(0,0)$ on the xy-plane.

$$plot3d\left(\left[\,f(x,y),0\right], x = -0.01..0.01,\, y = -0.01..0.01,\, axes = box\right)$$

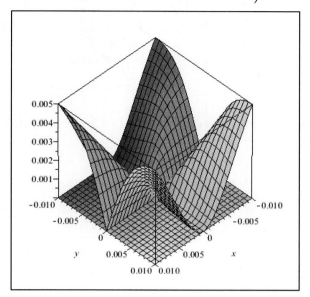

Furthermore, $x^2 \leq x^2 + y^2$ so that $x^2|y| \leq (x^2 + y^2)|y|$, we see that

$f(x,y) = \dfrac{x^2|y|}{x^2 + y^2} \leq \dfrac{(x^2 + y^2)|y|}{x^2 + y^2} = |y|$. This is verified in the plot below.

$$plot3d\left(\left[\,f(x,y),|y|\right], x = -0.01..0.01,\, y = -0.01..0.01,\, axes = box\right)$$

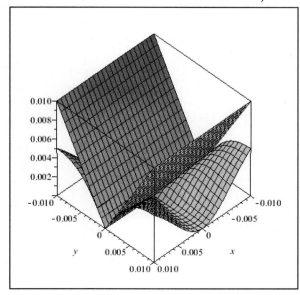

Hence $0 \le f(x, y) \le 2|y|$. Furthermore, as

$$limit\left(|y|, y = 0\right)$$

$$0 \qquad\qquad (13.21)$$

we see that $\lim_{(x,y)\to(0,0)} |y| = 0$. Furthermore, $\lim_{(x,y)\to(0,0)} 0 = 0$. Therefore, by the Squeeze Theorem (see chapter 2), $\lim_{(x,y)\to(0,0)} f(x, y)$ exists and $\lim_{(x,y)\to(0,0)} f(x, y) = 0$.

Remarks:

1. Note that, as in the case of example 2, if we use the *limit* command we do not get any output.

$$f := (x, y) \to \frac{x^2 \cdot |y|}{x^2 + y^2}$$

$$(x, y) \to \frac{x^2 |y|}{x^2 + y^2} \qquad\qquad (13.22)$$

$$limit\left(f(x, y), \{x = 0, y = 0\}\right)$$

$$limit\left(\frac{x^2 |y|}{x^2 + y^2}, \{x = 0, y = 0\}\right) \qquad\qquad (13.23)$$

2. One can compute the limits using a polar transformation too. Let $x = r\cos(\theta)$ and $y = r\sin(\theta)$. Then $\lim_{(x,y)\to(0,0)} f(x, y) = \lim_{r\to 0, \theta \to \text{any angle}} f\left(r\cos(\theta), r\sin(\theta)\right)$. Note that $x^2 + y^2 = r^2 \cos^2(\theta) + r^2 \sin^2(\theta) = r^2$. Then

$$\lim_{(x,y)\to(0,0)} f(x, y) = \lim_{(x,y)\to(0,0)} \frac{x^2 \cdot |y|}{x^2 + y^2} = \lim_{r\to 0, \theta \to \text{any angle}} \frac{r^2 \cos^2(\theta)|r\sin(\theta)|}{r^2}$$

$$= \lim_{r\to 0, \theta \to \text{any angle}} \left[\cos^2(\theta)|r\sin(\theta)|\right] = 0$$

Hence $\lim_{(x,y)\to(0,0)} f(x, y) = 0$.

Example 8: Graph the function $f(x, y) = \dfrac{\sin(x^2 + y^2)}{x^2 + y^2}$. Find $\lim_{(x,y)\to(0,0)} f(x, y)$, if it exists. Use a contour plot to verify your answer. Can you use *Maple* commands to justify your answer? Why? Is f continuous at $(0,0)$? Can you define a continuous function g such that $f(x, y) = g(x, y)$ at every point except $(0,0)$?

Solution: We will begin by entering f as a *Maple* function and graph it. The plot is obtained by clicking on the graph and rotating it.

$$f := (x, y) \rightarrow \frac{\sin\left(x^2 + y^2\right)}{x^2 + y^2}$$

$$(x, y) \rightarrow \frac{\sin\left(x^2 + y^2\right)}{x^2 + y^2} \qquad\qquad \textbf{(13.24)}$$

$plot3d\left(f\left(x, y\right), x = -3\,..\,3, y = -3\,..\,3, axes = box\right)$

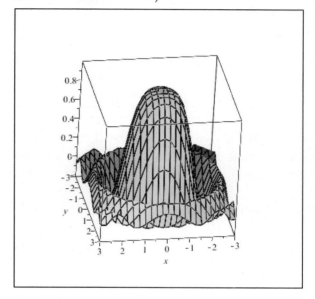

Based on the above plot, it is reasonable to assume that $\lim\limits_{(x,y)\to(0,0)} f\left(x, y\right)$ exits. We will now construct a contour plot of f. Do not forget to activate the $with\left(plots\right)$ package.

$with\left(plots\right):$

$contourplot\left(f\left(x, y\right), x = -3\,..\,3, y = -3\,..\,3, contours = 10, grid = \left[75, 75\right]\right)$

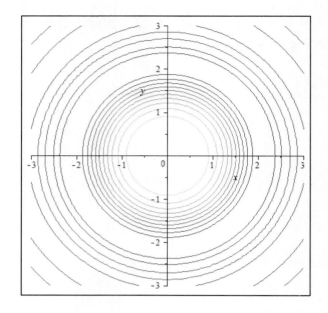

Based on the contour plot above, we see that distinct contours do not intersect at the origin. We can speculate that the limit at the origin exists (why?). We will now use *Maple* commands to verify our answer by approaching the origin along different paths.

$limit\left(f\left(x,y\right),\left\{x=0,y=0\right\}\right)$

$$limit\left(\frac{\sin\left(x^2+y^2\right)}{x^2+y^2},\left\{x=0,y=0\right\}\right)$$ (13.25)

This command does not give us an output. However, if we choose to find the limit in the following way we see that the value is the same.

$limit\left(limit\left(f\left(x,y\right),\left\{x=y\right\}\right),\left\{y=0\right\}\right)$

$$1$$ (13.26)

$limit\left(limit\left(f\left(x,y\right),\left\{x=-3\cdot y\right\}\right),\left\{y=0\right\}\right)$

$$1$$ (13.27)

Hence, if we approach $\left(0,0\right)$ along the path $x=y$ we get a limit of 1 and if we approach $\left(0,0\right)$ along the path $x=-3y$ we get a limit of 1. However, this does not constitute a proof that the limit is 1. We could compute the limits using a polar transformation just as we did in the remark before this example to establish that the value of this limit is indeed 1. Let $x=r\cos\left(\theta\right)$ and $y=r\sin\left(\theta\right)$. Then $\displaystyle\lim_{\left(x,y\right)\to\left(0,0\right)}f\left(x,y\right)=\lim_{r\to0,\theta\to\text{any angle}}f\left(r\cos\left(\theta\right),r\sin\left(\theta\right)\right)$ and $x^2+y^2=$ $r^2\cos^2\left(\theta\right)+r^2\sin^2\left(\theta\right)=r^2$. Furthermore,

$$\lim_{(x,y)\to(0,0)} f(x,y) \quad = \lim_{(x,y)\to(0,0)} \frac{\sin\left(x^2+y^2\right)}{x^2+y^2} = \lim_{r\to0,\theta\to\text{any angle}} \frac{\sin\left(r^2\right)}{r^2} = 1.$$

Hence $\lim_{(x,y)\to(0,0)} f(x,y)=1$. Since $(0,0)$ is not in the domain of f, it is not continuous at $(0,0)$.

We will now define a function g as

$$g(x,y) = \begin{cases} \dfrac{\sin\left(x^2+y^2\right)}{x^2+y^2} & \text{for } (x,y)\neq(0,0) \\ 1 & \text{for } (x,y)=(0,0) \end{cases}$$

Then g is a continuous function (why?).

Exercises

1 – 8   Sketch a contour plot of each of the following functions in two dimensions for the surface. Construct a surface plot with its contours lifted at different heights. Are these functions continuous? Carefully explain your reasoning in each case.

1. $f(x, y) = \dfrac{xy \sin(xy)}{x^2 + 2xy + y^2 - 11}$

2. $f(x, y) = \tan(xy)$

3. $f(x, y) = (\sin(x) + \cos(y)) e^{xy}$

4. $f(x, y) = x^{y^x}$

5. $f(x, y) = \dfrac{x^2 y}{-11 + \sqrt{x^2 + 2xy + y^2}}$

6. $f(x, y) = \tan^{-1}\left(\dfrac{y}{x}\right)$

7. $f(x, y) \dfrac{x^5 y^2 - xy^4}{x + 3y - 2}$

8. $f(x, y) = \dfrac{x^4 + y^4}{\sin(x^2 - y^2)}$

9 – 19 Graph each of the functions. Find $\lim\limits_{(x,y)\to(0,0)} f(x, y)$, if it exists. Use a contour plot to verify your answer.

9. $f(x, y) = \dfrac{x^3 y + \sqrt{y}}{x^2 + y^2}$

10. $f(x, y) = \dfrac{\tan(x^2 + y^2)}{x^2 + y^2}$

11. $f(x, y) = \dfrac{|xy|}{x^2 + y^2}$

12.　　$f(x,y) = \dfrac{e^{x^2+y^2} - 1}{x^2 + y}$

13.　　$f(x,y) = \dfrac{\sin\left(5x^2 y^2\right)}{6x^2 y^2}$

14.　　$f(x,y) = \dfrac{\ln\left(5x^2 + 5y^2\right)}{x^2 + y^2}$

15.　　$f(x,y) = \dfrac{x^2 y\, e^{xy} - y^3 \sin(x+y)}{-3 + \sqrt{x^2 + y^2 + 9}}$

16.　　$f(x,y) = x^{y^x}$

17.　　$f(x,y) = x^y y^x$

18.　　$f(x,y) = \sqrt{\dfrac{\tan\left(x^2 + y^2\right)}{x^2 + y^2}}$

19.　　$f(x,y) = \sqrt{2\sin^2(xy) - \cos 2(xy) - 1}$

20.　　Below is a list of some *Maple* commands used in this chapter. Describe the significance of each command. Can you find examples where each command is used?

　　　　(a)　　*CompleteSquare(expr)*

　　　　(b)　　$limit\left(f(x,y), \{x = c, y = d\}\right)$

　　　　(c)　　$plot3d\left(f(x,y), x = a\,..\,b, y = c\,..\,d, axes = box\right)$

　　　　(d)　　*with(plots)*

　　　　(e)　　*with(Student[Precalculus])*

21.　　Can you list new *Maple* commands that were used in this chapter but were not listed in exercise 20?

Chapter 14 Partial Derivatives

We will calculate partial derivatives, generate tangent planes, find local extrema, plot gradient functions and examine, by both graphical and theoretical means the concept of the gradient.

Partial Derivatives

Definition: Let f be a real valued function of two variables such that $\lim_{h \to 0} \left(\dfrac{f(x+h, y) - f(x, y)}{h} \right)$ exists. Then *the partial derivative of $f(x, y)$ with respect to*

x, denoted by $f_x(x, y)$, is defined as $f_x(x, y) = \lim_{h \to 0} \left(\dfrac{f(x+h, y) - f(x, y)}{h} \right)$.

Definition: Let f be a real valued function of two variables such that $\lim_{k \to 0} \left(\dfrac{f(x, y+k) - f(x, y)}{k} \right)$ exists. Then *the partial derivative of $f(x, y)$ with respect to*

y, denoted by $f_y(x, y)$, is defined as $f_y(x, y) = \lim_{k \to 0} \left(\dfrac{f(x, y+k) - f(x, y)}{k} \right)$.

Remark: The alternate notations for the partial derivative of f with respect to x are $\dfrac{\partial f}{\partial x}$ or $\dfrac{\partial}{\partial x} f(x, y)$ or $D_x f$ and the alternate notations for the partial derivative of f with respect to y are $\dfrac{\partial f}{\partial y}$ or $\dfrac{\partial}{\partial y} f(x, y)$ or $D_y f$. These partial derivatives are also known as the first partial derivatives.

Definition: Let f be a real valued function of two variables. Then *the second partial derivatives of $f(x, y)$* (provided they exist) are defined below:

$$f_{xx} = \frac{\partial}{\partial x} \left(\frac{\partial f}{\partial x} \right) = \frac{\partial^2 f}{\partial x^2}$$

$$f_{xy} = \frac{\partial}{\partial y} \left(\frac{\partial f}{\partial x} \right) = \frac{\partial^2 f}{\partial y \partial x}$$

$$f_{yx} = \frac{\partial}{\partial x} \left(\frac{\partial f}{\partial y} \right) = \frac{\partial^2 f}{\partial x \partial y}$$

$$f_{yy} = \frac{\partial}{\partial y} \left(\frac{\partial f}{\partial y} \right) = \frac{\partial^2 f}{\partial y^2}$$

Example 1: Graph the function $f(x, y) = e^{\cos(x)} \sin(3x^2 y)$. Find the first partial derivatives of f and graph them.

Solution: We will enter f as a *Maple* function in two variables and graph it. Please do not use the letter e from the keyboard to type the exponential function.

$$f := (x, y) \rightarrow e^{\cos(x)} \cdot \sin(3 \cdot x^2 \cdot y)$$

$$(x, y) \rightarrow e^{\cos(x)} \sin(3x^2 y) \qquad \textbf{(14.1)}$$

$$plot3d\left(f(x, y), x = -1..1, y = -1..1, axes = box\right)$$

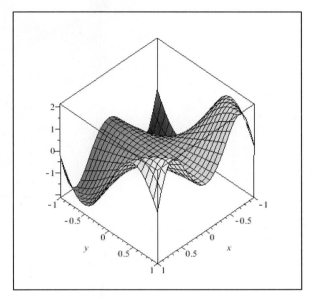

In order to evaluate the partial derivative with respect to x, click on $\dfrac{\partial}{\partial x} f$ from the *Expression palette*, insert x, then use the $\boxed{\text{Tab}}$ key and insert $f(x, y)$. Repeat the same procedure to find the partial derivative with respect to y.

$$\frac{\partial}{\partial x} f(x, y)$$

$$-\sin(x) e^{\cos(x)} \sin(3x^2 y) + 6 e^{\cos(x)} \cos(3x^2 y) x y \qquad \textbf{(14.2)}$$

$$\frac{\partial}{\partial y} f(x, y)$$

$$3 e^{\cos(x)} \cos(3x^2 y) x^2 \qquad \textbf{(14.3)}$$

Therefore, we see that $f_x(x, y) = -\sin(x)\, e^{\cos(x)} \sin(3x^2\, y) + 6\, e^{\cos(x)} \cos(3x^2\, y)\, x\, y$ and $f_y(x, y) = 3\, e^{\cos(x)} \cos(3x^2\, y)\, x^2$. We will now graph these partial derivatives.

$$plot3d\left(\frac{\partial}{\partial x} f(x, y), x = -1..1, y = -2..2, axes = box\right)$$

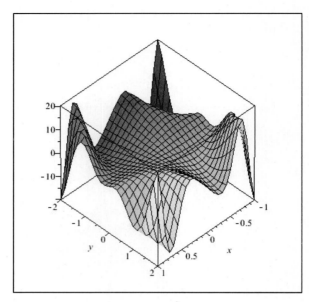

$$plot3d\left(\frac{\partial}{\partial y} f(x, y), x = -1..1, y = -2..2, axes = box\right)$$

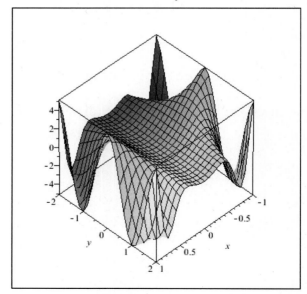

Example 2: Let $f(x, y) = -4e^{-x} \sin(2xy)$. Find the all the second partial derivatives of f and graph them.

Solution: We will enter f as a *Maple* function in two variables and find the second partial derivatives along with their plots.

$f := (x, y) \rightarrow -4 \cdot e^{-x} \cdot \sin(2 \cdot x \cdot y)$

$$(x, y) \rightarrow -4 e^{-x} \sin(2 x y) \tag{14.4}$$

$\dfrac{\partial}{\partial x}\left(\dfrac{\partial}{\partial x} f(x, y)\right)$

$$-4 e^{-x} \sin(2 x y) + 16 e^{-x} \cos(2 x y) y + 16 e^{-x} \sin(2 x y) y^2 \tag{14.5}$$

Hence $f_{xx} = -4 e^{-x} \sin(2 x y) + 16 e^{-x} \cos(2 x y) y + 16 e^{-x} \sin(2 x y) y^2$. A plot of this partial derivative is shown below.

$plot3d\left(\dfrac{\partial}{\partial x}\left(\dfrac{\partial}{\partial x} f(x, y)\right), x = -1..1.5, y = -1..1.5, axes = box\right)$

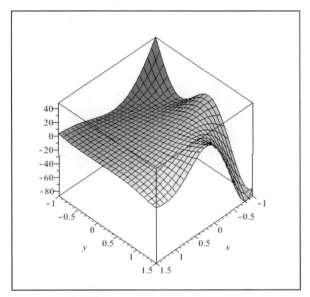

$\dfrac{\partial}{\partial y}\left(\dfrac{\partial}{\partial x} f(x, y)\right)$

$$8 e^{-x} \cos(2 x y) x + 16 e^{-x} \sin(2 x y) x y - 8 e^{-x} \cos(2 x y) \tag{14.6}$$

$\dfrac{\partial}{\partial x}\left(\dfrac{\partial}{\partial y} f(x, y)\right)$

$$8 e^{-x} \cos(2 x y) x + 16 e^{-x} \sin(2 x y) x y - 8 e^{-x} \cos(2 x y) \tag{14.7}$$

Therefore, $f_{xy} = 8 e^{-x} \cos(2 x y) x + 16 e^{-x} \sin(2 x y) x y - 8 e^{-x} \cos(2 x y) = f_{yx}$. A plot of these partial derivatives is shown.

$$plot3d\left(\frac{\partial}{\partial y}\left(\frac{\partial}{\partial x}f\left(x,y\right)\right),x=-1..1,y=-1..1,axes=box\right)$$

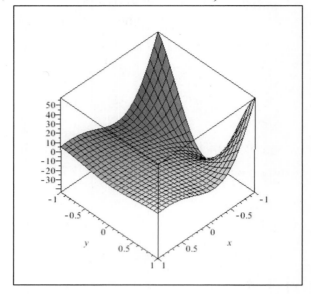

$$\frac{\partial}{\partial y}\left(\frac{\partial}{\partial y}f\left(x,y\right)\right)$$

$$16\,e^{-x}\sin\left(2\,x\,y\right)x^2 \tag{14.8}$$

Hence $f_{yy} = 16\,e^{-x}\sin\left(2\,x\,y\right)x^2$. A plot of this partial derivative is shown.

$$plot3d\left(\frac{\partial}{\partial x}\left(\frac{\partial}{\partial x}f\left(x,y\right)\right),x=-1..1,y=-1..1,axes=box\right)$$

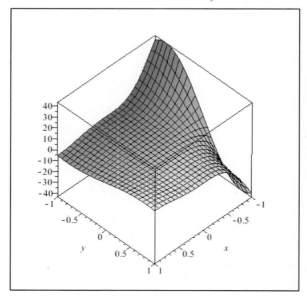

Remarks: We can find the partial derivatives of any function of two variables using other techniques, two of which are given below.

1. In order for the output of the partial derivatives to be *Maple* functions, we will use the $D[i]$ or the $D[i, j]$ command where the values of i or j are 1 when we are finding the partial derivative with respect to x and the values of i or j are 2 when we are finding the partial derivative with respect to y.

$$f := (x, y) \rightarrow -4 \cdot e^{-x} \cdot \sin(2 \cdot x \cdot y)$$

$$(x, y) \rightarrow -4 e^{-x} \sin(2 x y) \qquad \textbf{(14.9)}$$

$D[1](f)$

$$(x, y) \rightarrow 4 e^{-x} \sin(2 x y) - 8 e^{-x} \cos(2 x y) y \qquad \textbf{(14.10)}$$

$D[2](f)$

$$(x, y) \rightarrow -8 e^{-x} \cos(2 x y) x \qquad \textbf{(14.11)}$$

$D[1,1](f)$

$$(x, y) \rightarrow -4 e^{-x} \sin(2 x y) + 16 e^{-x} \cos(2 x y) y + 16 e^{-x} \sin(2 x y) y^2 \textbf{(14.12)}$$

$D[1,2](f)$

$$(x, y) \rightarrow 8 e^{-x} \cos(2 x y) x + 16 e^{-x} \sin(2 x y) x y - 8 e^{-x} \cos(2 x y) \qquad \textbf{(14.13)}$$

$D[2,1](f)$

$$(x, y) \rightarrow 8 e^{-x} \cos(2 x y) x + 16 e^{-x} \sin(2 x y) x y - 8 e^{-x} \cos(2 x y) \qquad \textbf{(14.14)}$$

$D[2,2](f)$

$$(x, y) \rightarrow 16 e^{-x} \sin(2 x y) x^2 \qquad \textbf{(14.15)}$$

2. We can also use the *diff* command to find partial derivatives. The examples below are the results for the partial derivatives f_x, f_{xy}, and f_{yy}, respectively.

$diff(f(x, y), x)$

$$4 e^{-x} \sin(2 x y) - 8 e^{-x} \cos(2 x y) y \qquad \textbf{(14.16)}$$

$diff(f(x, y), x, y)$

$$8 e^{-x} \cos(2 x y) x + 16 e^{-x} \sin(2 x y) x y - 8 e^{-x} \cos(2 x y) \qquad \textbf{(14.17)}$$

$diff(f(x, y), y, y)$

$$16 e^{-x} \sin(2 x y) x^2 \qquad \textbf{(14.18)}$$

Theorem: Let f be a real valued function of two variables such that the second partial derivatives f_{xy} and f_{yx} are continuous in an open set G. Then $f_{xy}(x,y) = f_{yx}(x,y)$ for every (x,y) in G.

Example 3: Let $f(x,y) = \begin{cases} \dfrac{xy^3 - x^3 y}{y^2 + x^2} & \text{if} \quad (x,y) \neq (0,0) \\ 0 & \text{if} \quad (x,y) = (0,0) \end{cases}$. Then $f_{xy}(0,0) \neq f_{yx}(0,0)$.

Solution: We will enter f as a piecewise *Maple* function and graph it.

$$f := (x,y) \rightarrow piecewise\left(x \neq 0 \quad \textbf{and} \quad y \neq 0, \frac{x \cdot y^3 - x^3 \cdot y}{y^2 + x^2}, x = 0 \quad \textbf{and} \quad y = 0, 0 \right)$$

$$(x,y) \rightarrow piecewise\left(x \neq 0 \quad \textbf{and} \quad y \neq 0, \frac{x y^3 - x^3 y}{y^2 + x^2}, x = 0 \quad \textbf{and} \quad y = 0, 0 \right) \qquad \textbf{(4.19)}$$

$f(x,y)$

$$\frac{x y^3 - x^3 y}{y^2 + x^2} \qquad \textbf{(14.20)}$$

$f(0,0)$

$$0 \qquad \textbf{(14.21)}$$

Note that the only output when $f(x,y)$ was activated in equation **(14.20)** was $\frac{x y^3 - x^3 y}{y^2 + x^2}$. However, when the next command, namely $f(0,0)$, was activated, the answer of 0 was obtained which agrees with $f(0,0) = 0$. We need to take this into account when we plot the function below, by specifying the point $(0,0,0)$.

$$plot3d\left(\left[f(x,y), (0,0,0) \right], x = -1..1, y = -1..1, axes = box, grid = [80,80] \right)$$

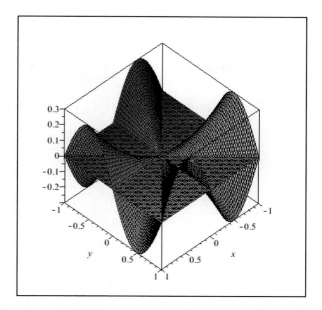

We will now use the definition of the first partial derivatives to find $f_x(0, y)$ and $f_y(x, 0)$.

$$fx := (x, y) \rightarrow \lim_{h \to 0} \frac{f(x+h, y) - f(x, y)}{h}$$

$$(x, y) \rightarrow \lim_{h \to 0} \frac{f(x+h, y) - f(x, y)}{h} \tag{14.22}$$

$$fx(0, y)$$

$$y \tag{14.23}$$

$$fy := (x, y) \rightarrow \lim_{k \to 0} \frac{f(x, y+k) - f(x, y)}{k}$$

$$(x, y) \rightarrow \lim_{k \to 0} \frac{f(x, y+k) - f(x, y)}{k} \tag{14.24}$$

$$fy(x, 0)$$

$$-x \tag{14.25}$$

Therefore, we see that $fx(0, y) = y$ and $fy(x, 0) = -x$. We will now proceed to find the second partial derivatives $f_{xy}(0, 0)$ and $f_{yx}(0, 0)$, in that order.

$$\lim_{k \to 0} \frac{fx(0, k) - fx(0, 0)}{k}$$

$$1 \tag{14.26}$$

$$\lim_{h \to 0} \frac{fy(h, 0) - fy(0, 0)}{h}$$

$$-1 \qquad\qquad\qquad\qquad\qquad \textbf{(14.27)}$$

Therefore, we see that $f_{xy}(0,0)=1$ and $f_{yx}(0,0)=-1$ so that $f_{xy}(0,0) \neq f_{yx}(0,0)$. Does this violate the theorem that states that $f_{xy}(x,y)=f_{yx}(x,y)$? We will answer this question shortly. We will now find $f_{xy}(x,y)$ and $f_{yx}(x,y)$ when $(x,y) \neq (0,0)$.

$$\frac{\partial}{\partial y}\left(\frac{\partial}{\partial x}f(x,y)\right)$$

$$\frac{3y^2-3x^2}{x^2+y^2}-\frac{2\left(3xy^2-x^3\right)x}{\left(x^2+y^2\right)^2}-\frac{2\left(y^3-3x^2y\right)y}{\left(x^2+y^2\right)^2}+\frac{8\left(xy^3-x^3y\right)yx}{\left(x^2+y^2\right)^3} \qquad \textbf{(14.28)}$$

$simplify\left((\textbf{14.28})\right)$

$$-\frac{9y^2x^4-9y^4x^2-y^6+x^6}{\left(x^2+y^2\right)^3} \qquad\qquad \textbf{(14.29)}$$

$$\frac{\partial}{\partial x}\left(\frac{\partial}{\partial y}f(x,y)\right)$$

$$\frac{3y^2-3x^2}{x^2+y^2}-\frac{2\left(3xy^2-x^3\right)x}{\left(x^2+y^2\right)^2}-\frac{2\left(y^3-3x^2y\right)y}{\left(x^2+y^2\right)^2}+\frac{8\left(xy^3-x^3y\right)yx}{\left(x^2+y^2\right)^3} \qquad \textbf{(14.30)}$$

$simplify\left((\textbf{14.30})\right)$

$$-\frac{9y^2x^4-9y^4x^2-y^6+x^6}{\left(x^2+y^2\right)^3} \qquad\qquad \textbf{(14.31)}$$

Note that in this case we see that $f_{xy}(x,y)=f_{yx}(x,y)$. We will give a *Maple* contour graph below. Do not forget to activate the *with(plots)* package.

$with(plots):$

$$contourplot\left(-\frac{9y^2x^4-9y^4x^2-y^6+x^6}{\left(x^2+y^2\right)^3},x=-1..1,y=-1..1,contours=9,grid=[200,200]\right)$$

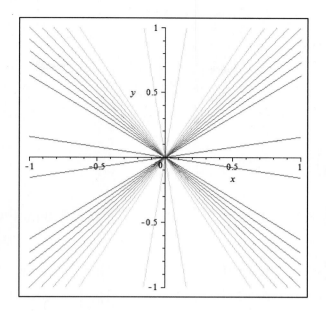

Recall from chapter 13 that the point at which distinct contours intersect is the point where the limit does not exist. Hence from the contour plot above $\lim_{(x,y)\to(0,0)} f_{xy}(x,y)$ does not exist. This indicates that the partial derivatives $f_{xy}(x,y)$ and $f_{yx}(x,y)$ are not continuous at $(0,0)$. Therefore, the hypothesis for the theorem before this example was not satisfied at $(0,0)$.

Tangent Planes

In chapter 3, we found the equation of the tangent line to a curve $y = f(x)$ when $x = x_1$. We will now find the equation of a tangent plane in the case of a surface $z = f(x,y)$. We can think of the tangent plane to a surface as a plane that consists of all tangent lines to the surface at that point. There will be infinitely many tangent lines.

Definition: Let f be a real valued function of two variables such that its partial derivatives f_x and f_y are continuous. Then the equation of the *tangent plane* to the surface $z = f(x,y)$ when $z_1 = (x_1, y_1)$ is $z - z_1 = f_x(x_1, y_1)(x - x_1) + f_y(x_1, y_1)(y - y_1)$.

Example 4: Let $f(x,y) = x^3 y + 2xy$. Find the equation of the tangent plane to the surface $z = f(x,y)$ when $(x,y) = (1,2)$. Plot the surface f along with its tangent plane at the point $(1,2)$.

Solution: We will enter f as a *Maple* function and find the first partial derivatives of this function using the $D[1](f)$ and $D[2](f)$ commands.

$$f := (x, y) \rightarrow x^3 \cdot y + 2 \cdot x \cdot y$$

$$(x, y) \rightarrow x^3 y + 2 x y \qquad \text{(14.32)}$$

$$D[1](f)$$

$$(x, y) \rightarrow 3 x^2 y + 2 y \qquad \text{(14.33)}$$

$$D[2](f)$$

$$(x, y) \rightarrow x^3 + 2 x \qquad \text{(14.34)}$$

Note that the first derivatives are continuous. We will now find the equation of the tangent plane.

$$L := (x, y) \rightarrow f(1,2) + D[1](f)(1,2) \cdot (x-1) + D[2](f)(1,2) \cdot (y-2)$$

$$(x, y) \rightarrow f(1,2) + D_1(f)(1,2)(x-1) + D_2(f)(1,2)(y-2) \qquad \text{(14.35)}$$

$$L(x, y)$$

$$-10 + 10 x + 3 y \qquad \text{(14.36)}$$

Therefore, the equation of the tangent plane is $z = 10 x + 3 y - 10$. We will now graph the function along with its tangent plane at the point $(x, y) = (1, 2)$.

$$plot3d\left(\left[f(x, y), L(x, y)\right], x = -3.3 .. 3.1, y = -3.3 .. 2.3, axes = box, color = \left[yellow, red\right]\right)$$

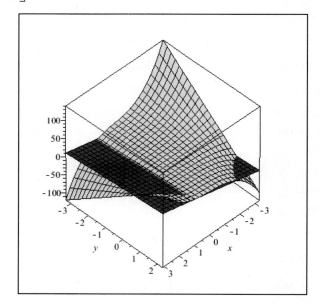

Linearization

Let f be a real valued function of two variables such that its partial derivatives f_x and f_y are continuous. Then the equation of the tangent plane to the surface $z = f(x, y)$ at the point (x_1, y_1, z_1) is $z - z_1 = f_x(x - x_1) + f_y(y - y_1)$. If the value of (x, y) is in a small enough rectangle around the point (x_1, y_1), then the graph of the function $f(x, y)$ looks like the graph of the tangent plane. We can say that the value of $f(x, y)$ is approximately $f(x, y) \approx f(x_1, y_1) + f_x(x, y)(x - x_1) + f_y(x, y)(y - y_1)$. This process is called *the linear approximation* or *the tangent plane approximation* of the function f at (x_1, y_1). In this case, the relative error is $\left| \dfrac{\text{Estimated Value-Exact Value}}{\text{Exact Value}} \right|$

Example 5: Find the linearization at $\left(0, \dfrac{1}{2}\right)$ of $f(x, y) = e^{-x} \sin(\pi y)$. Use your answer to approximate $f(-0.001, 0.483)$. Find the relative error of your answer.

Solution: We will enter f as a *Maple* function and find the first partial derivatives of this function using the $D[1](f)$ and $D[2](f)$ commands.

$f := (x, y) \rightarrow e^{-x} \cdot \sin(\pi \cdot y)$

$$(x, y) \rightarrow e^{-x} \sin(\pi y) \tag{14.37}$$

$D[1](f)$

$$(x, y) \rightarrow -e^{-x} \sin(\pi y) \tag{14.38}$$

$D[2](f)$

$$(x, y) \rightarrow e^{-x} \cos(\pi y) \pi \tag{14.39}$$

Note that the first derivatives are continuous. We will now find the equation of the tangent plane.

$$L := (x, y) \rightarrow f\left(0, \frac{1}{2}\right) + D[1](f)\left(0, \frac{1}{2}\right) \cdot (x - 0) + D[2](f)\left(0, \frac{1}{2}\right) \cdot \left(y - \frac{1}{2}\right)$$

$$(x, y) \rightarrow f\left(0, \frac{1}{2}\right) + D_1(f)\left(0, \frac{1}{2}\right) x + D_2(f)\left(0, \frac{1}{2}\right)\left(y - \frac{1}{2}\right) \tag{14.40}$$

$L(x, y)$

$$1 - x \tag{14.41}$$

Therefore, the linearization of f at $\left(0, \dfrac{1}{2}\right)$ is $L(x, y) = 1 - x$. We will now use this linearization to estimate $f(-0.001, 0.483)$ and find the relative error.

$L(-0.001, 0.483)$

$$1.001 \tag{14.42}$$

$$\left| \frac{f(-0.001, 0.483) - L(-0.001, 0.483)}{f(-0.001, 0.483)} \right|$$

$$0.001427354517 \tag{14.43}$$

Hence an estimate of $f(-0.001, 0.483)$ is 1.001. The relative error is 0.001427354517. A graph of the function is given below. Note that the graph below is obtained after clicking on the graph and rotating it.

$$plot3d\left(\left[f(x, y), L(x, y) \right], x = -1 .. 2, y = -2 .. 2, axes = box \right)$$

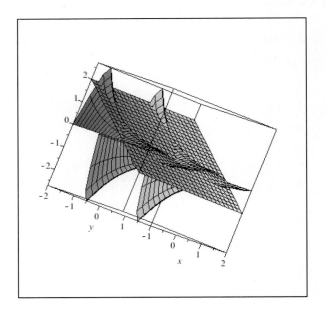

Local Maxima, Local Minima, and Saddle Points

Definition: Let f be a real valued function of two variables. A point (x_1, y_1) is said to be a *critical point* if the equations $f_x(x_1, y_1) = 0$ and $f_y(x_1, y_1) = 0$ are satisfied.

Definition: Let f be a real valued function of two variables such that the second partial derivatives f_{xy} and f_{yx} are continuous in an open set G. The *discriminant* is defined as the function $D(x, y) = f_{xx}(x, y) f_{yy}(x, y) - f_{xy}^{\,2}(x, y)$.

Definition: Let f be a real valued function of two variables. The function f has a *local maximum* at (x_1, y_1) if $f(x_1, y_1) \geq f(x, y)$ for every point (x, y) in a small enough area around (x_1, y_1).

Definition: Let f be a real valued function of two variables. The function f has a *local minimum* at (x_1, y_1) if $f(x_1, y_1) \leq f(x, y)$ for every point (x, y) in a small enough area around (x_1, y_1).

The Second Partial Derivatives Test: Let f be a real valued function of two variables such that its second partial derivatives f_{xx}, f_{xy} and f_{yy} are continuous in an open set around the point (x_1, y_1). Suppose that (x_1, y_1) is a critical point of the function f. Then

(a) If $D(x_1, y_1) > 0$ and $f_{xx}(x_1, y_1) > 0$ then (x_1, y_1) is a local minimum.

(b) If $D(x_1, y_1) > 0$ and $f_{xx}(x_1, y_1) < 0$ then (x_1, y_1) is a local maximum.

(c) If $D(x_1, y_1) < 0$ then (x_1, y_1) is called a *saddle point*.

Remark: If the above conditions are not satisfied then the point (x_1, y_1) is not a local extremum or a saddle point.

Example 6: Let $f(x, y) = x^6 + 2y^4 - 4x^2 - y^2 - 5x^2 y^2$. Find the local extrema and saddle points of this function. Verify your answer by graphing f using the *plot3d* command and constructing a contour plot.

Solution: We will enter f as a *Maple* function and find the first and second partial derivatives. We enter the coefficient -5 of $x^2 y^2$ as -5.0 to get floating-decimal answers.

$f := (x, y) \rightarrow x^6 + 2 \cdot y^4 - 4 \cdot x^2 - y^2 - 5.0 \cdot x^2 \cdot y^2$

$$(x, y) \rightarrow x^6 + 2y^4 - 4x^2 - y^2 - 5.0 x^2 y^2 \qquad \textbf{(14.44)}$$

$D[1](f)$

$$(x, y) \rightarrow 6x^5 - 8x - 10.0xy^2 \qquad \textbf{(14.45)}$$

$D[2](f)$

$$(x, y) \rightarrow 8y^3 - 2y - 10.0x^2 y \qquad \textbf{(14.46)}$$

$D[1,1](f)$

$$(x, y) \rightarrow 30x^4 - 8 - 10.0y^2 \qquad \textbf{(14.47)}$$

$D[1,2](f)$

$$(x, y) \rightarrow -20.0\, x\, y \qquad\qquad (14.48)$$

$$D[2,2](f)$$

$$(x, y) \rightarrow 24\, y^2 - 2 - 10.0\, x^2 \qquad\qquad (14.49)$$

Hence $f_x = 6\, x^5 - 8\, x - 10.0\, x\, y^2$, $f_y = 8\, y^3 - 2\, y - 10.0\, x^2\, y$, $f_{xx} = 30\, x^4 - 8 - 10.0\, y^2$, $f_{xy} = -20.0\, x\, y^2$, and $f_{yy} = 24\, y^2 - 2 - 10.0\, x^2$.

We will first find the critical points, which are called cp .

$$cp := \left[solve \big(\{ D[1](f)(x, y) = 0, D[2](f)(x, y) = 0 \}, \{x, y\} \big) \right]$$
$$[\{x = 0., y = 0.\}, \{x = 1.074569932, y = 0.\}, \{x = 1.074569932I, y = 0.\}, \qquad (14.50)$$
$$\{x = -1.074569932, y = 0.\}, \{x = -1.074569932\,I, y = 0.\}, \{x = 0., y = 0.5000000000\},$$
$$\{x = 0., y = -0.5000000000\}, \{x = 1.650888583, y = 1.912273880\},$$
$$\{x = -1.650888583, y = 1.912273880\}, \{x = 0.8013112874I, y = 0.7433873311I\},$$
$$\{x = -0.8013112874\,I, y = 0.7433873311I\}, \{x = 1.650888583, y = -1.912273880\},$$
$$\{x = -1.650888583, y = -1.912273880\}, \{x = 0.8013112874\,I, y = -0.7433873311I\},$$
$$\big(30\, x^4 - 8 - 10.0\, y^2 \big) \big(24\, y^2 - 2 - 10.0\, x^2 \big) - 400.00\, x^2\, y^2$$

Since the ciritcal points are not complex numbers, we will remove the solutions that have $I = \sqrt{-1}$. We will use the *remove* command to obtain the real-valued critical points, which will be denoted by cpr .

$$cpr := remove(has, cp, I)$$
$$[\{x = 0., y = 0.\}, \{x = 1.074569932, y = 0.\}, \{x = -1.074569932, y = 0.\}, \qquad (14.51)$$
$$\{x = 0., y = 0.5000000000\}, \{x = 0., y = -0.5000000000\},$$
$$\{x = 1.650888583, y = 1.912273880\}, \{x = -1.650888583, y = 1.912273880\},$$
$$\{x = 1.650888583, y = -1.912273880\}, \{x = -1.650888583, y = -1.912273880\}\,]$$

Note that there are nine critical points. We will now find the discriminant and then use the second partial derivative test to find the local extrema and the saddle points.

$$disc := D[1,1](f)(x, y) \cdot D[2,2](f)(x, y) - D[1,2](f)(x, y)^2$$
$$\big(30\, x^4 - 8 - 10.0\, y^2 \big) \big(24\, y^2 - 2 - 10.0\, x^2 \big) - 400.00\, x^2\, y^2 \qquad (14.52)$$
$$array \left(\left[seq \big(eval \big([\, D[1,1](f)(x, y), disc\,], cpr[j] \big), j = 1..9 \big) \right] \right)$$

$$\begin{bmatrix} -8. & 16. \\ 32.00000002 & -433.5041728 \\ 32.00000002 & -433.5041728 \\ -10.50000000 & -42.00000000 \\ -10.50000000 & -42.00000000 \\ 178.2716558 & 6443.899961 \\ 178.2716558 & 6443.899961 \\ 178.2716558 & 6443.899961 \\ 178.2716558 & 6443.899961 \end{bmatrix}$$

(14.53)

Hence, by the second derivative test, there is a local maximum when $(x,y)=(0.,0.)$, since the discriminant is positive and $f_{xx}(x_1,y_1)<0$; there are saddle points when (x,y) = $(1.074569932,0.)$, (x,y) = $(-1.074569932,0.)$, (x,y) = $(0.,0.5000000000)$, and (x,y) = $(0.,-0.5000000000)$, since the discriminant is negative; and there are local minima when (x,y) = $(1.650888583,1.912273880)$, (x,y) = $(-1.650888583,1.912273880)$, (x,y) = $(1.650888583,-1.912273880)$, and (x,y) = $(-1.650888583,-1.912273880)$, since the discriminant is positive and $f_{xx}(x_1,y_1)>0$. We will now plot this function and its contour graph on separate axes.

$plot3d\left(f(x,y), x=-2.0..2.0, y=-2.0..2.0, axes=box\right)$

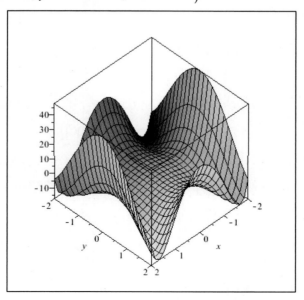

$with\left(plots\right):$

$contourplot\left(f\left(x,y\right),x=-1.9\,..\,1.95,y=-1.9\,..\,1.95,contours=50,grid=\left[45,45\right]\right)$

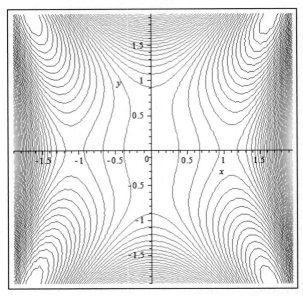

From the contour plot above, one can estimate that the four local minima occur approximately at the diagonal points of the above points. There must be one local maxima in the center, and the four saddle points, two each on the horizontal and vertical axes.

Remark: We can use the *SecondDerivativeTest* command to check if the critical points are local maxima, local minima or saddle points.

$f := \left(x,y\right) \rightarrow x^6 + 2\cdot y^4 - 4\cdot x^2 - y^2 - 5.0\cdot x^2\cdot y^2$

$$\left(x,y\right)\rightarrow x^6 + 2\,y^4 - 4\,x^2 - y^2 - 5.0\,x^2\,y^2 \qquad (14.54)$$

$with\left(Student\left[MultivariateCalculus\right]\right):$

$SecondDerivativeTest\left(f\left(x,y\right),\left[x,y\right]=\left[0.,0.\right]\right)$

$$LocalMin=\left[\ \right],LocalMax=\left[\left[0.,0.\right]\right],Saddle=\left[\ \right] \qquad (14.55)$$

$SecondDerivativeTest\left(f\left(x,y\right),\left[x,y\right]=\left[0.,-0.5000000000\right]\right)$

$$LocalMin=\left[\ \right],LocalMax=\left[\ \right],Saddle=\left[\left[0.,-0.5000000000\right]\right] \quad (14.56)$$

$SecondDerivativeTest\left(f\left(x,y\right),\left[x,y\right]=\left[-1.650888583,1.912273880\right]\right)$

$$LocalMin=\left[\left[-1.650888583,1.912273880\right]\right],LocalMax=\left[\ \right],Saddle=\left[\ \right] (14.57)$$

Hence $\left(0.,0.\right)$ is a local maximum $\left(0.,-0.5000000000\right)$ is a saddle point, and $\left(-1.650888583,1.912273880\right)$ is a local minimum.

Note: Let f be a real valued function of two variables with continuous first partial derivatives. Assume further that g be a real valued function of two variables with continuous first partial derivatives such that $g(x,y)=0$, $g_x(x,y)\neq 0$, and $g_y(x,y)\neq 0$. In order to maximize or minimize the function f subject to the constraint $g(x,y)=0$, one needs to solve the equations $f_x(x,y)=\lambda g_x(x,y)$, $f_y(x,y)=\lambda g_y(x,y)$, and $g(x,y)=0$ simultaneously. All solutions to this system of equations are critical points for the constrained problem. The constant λ is called *the Lagrange multiplier.*

Example 7: Find the minimum value of the function $f(x,y)=x^2+y^2$ subject to the constraint $g(x,y)=3x+2y-6=0$. Find the local extrema and saddle points of this function. Verify your answer by constructing a contour plot.

Solution: We will enter f as a *Maple* function and find the first and second partial derivatives. We enter the coefficient -5 of x^2y^2 as -5.0 to get floating-decimal answers.

$f:=(x,y)\rightarrow x^2+y^2$

$$(x,y)\rightarrow x^2+y^2 \tag{14.58}$$

$g:=(x,y)\rightarrow 3\cdot x+2\cdot y-6$

$$(x,y)\rightarrow 3\cdot x+2\cdot y-6 \tag{14.59}$$

We will now solve the equations $f_x(x,y)=\lambda g_x(x,y)$, $f_y(x,y)=\lambda g_y(x,y)$, and $g(x,y)=0$ simultaneously.

$cp:=\left[solve\left(\{D[1](f)(x,y)=\lambda\cdot D[1](g)(x,y),D[2](f)(x,y)=\lambda\cdot D[2](g)(x,y),g(x,y)=0\}\right)\right]$

$$\left\{x=\frac{18}{13},y=\frac{12}{13},\lambda=\frac{12}{13}\right\} \tag{14.60}$$

We will now evaluate $f\left(\frac{18}{13},\frac{12}{13}\right)$.

$f\left(\frac{18}{13},\frac{12}{13}\right)$

$$\frac{36}{13} \tag{14.61}$$

In order to confirm that the minimum occurs when $(x, y) = \left(\dfrac{18}{13}, \dfrac{12}{13}\right)$, we pick a point on

$3x + 2y - 6 = 0$, say $(x, y) = \left(1, \dfrac{3}{2}\right)$, and find $f\left(1, \dfrac{3}{2}\right)$.

$f\left(1, \dfrac{3}{2}\right)$

$$\dfrac{13}{4} \qquad\qquad\qquad \textbf{(14.62)}$$

Since $f\left(1, \dfrac{3}{2}\right) > f\left(\dfrac{18}{13}, \dfrac{12}{13}\right)$, f has a minimum when $(x, y) = \left(\dfrac{18}{13}, \dfrac{12}{13}\right)$. We will see this

graphically by plotting the contours of $f(x, y)$, $g(x, y) = 0$, and the point $\left(\dfrac{18}{13}, \dfrac{12}{13}\right)$ on

the same set of axes. Do not forget to activate the $with(plots)$ package.

$with(plots):$

$s1 := contourplot\left(f(x, y), x = 3 .. 3, y = -3 .. 3, contours = 50, grid = [100, 100]\right)$

$$PLOT(...) \qquad\qquad \textbf{(14.63)}$$

$s2 := pointplot\left(\left[\dfrac{18}{13}, \dfrac{12}{13}\right], color = blue, thickness = 2, symbol = circle, symbolsize = 15\right)$

$$PLOT(...) \qquad\qquad \textbf{(14.64)}$$

$s3 := implicitplot\left(g(x, y) = 0, x = 3 .. 3, y = -3 .. 3, thickness = 2\right)$

$$PLOT(...) \qquad\qquad \textbf{(14.65)}$$

$display(s1, s2, s3)$

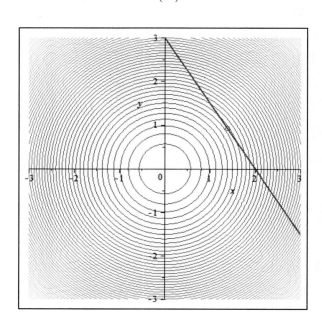

Remark: We can use the *with*(*Student*[*MultivariateCalculus*]) to obtain the Lagrange multipliers.

$f := (x, y) \rightarrow x^2 + y^2$

$$(x, y) \rightarrow x^2 + y^2 \qquad \textbf{(14.66)}$$

with(*Student*[*MultivariateCalculus*]):

LagrangeMultipliers$(f(x, y), [3 \cdot x + 2 \cdot y - 6], [x, y])$

$$\left[\frac{18}{13}, \frac{12}{13} \right] \qquad \textbf{(14.67)}$$

Note that if we do not give options for the *output* we get the critical point. However, if we want the value of the Lagrange multiplier and the value of $f\left(\dfrac{18}{13}, \dfrac{12}{13}\right)$, we must add the option *output = detailed* .

LagrangeMultipliers$(f(x, y), [3 \cdot x + 2 \cdot y - 6], [x, y], output = detailed)$

$$\left[x = \frac{18}{13}, y = \frac{12}{13}, \lambda_1 = \frac{12}{13}, x^2 + y^2 = \frac{36}{13} \right] \qquad \textbf{(14.68)}$$

The option of *output = plot* gives us a plot of the function with the point.

LagrangeMultipliers$(f(x, y), [3 \cdot x + 2 \cdot y - 6], [x, y], output = plot)$

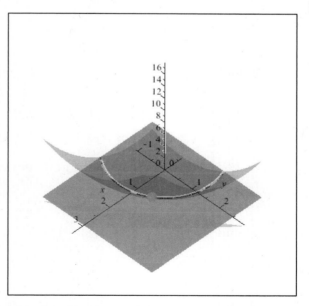

An additional option of *showlevelcurves = false* ensures that the level curves are not plotted.

$LagrangeMultipliers\left(f\left(x,y\right),\left[3\cdot x+2\cdot y-6\right],\left[x,y\right],output=plot,showlevelcurves=false\right)$

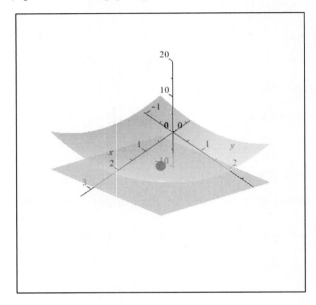

Vector Functions

Definition: Let f, g, and h be three real valued functions. A (*three dimensional*) *vector function* is of the form $v\left(t\right)=\left\langle f\left(t\right),g\left(t\right),h\left(t\right)\right\rangle$.

Definition: Let c be a real number and $v\left(t\right)=\left\langle f\left(t\right),g\left(t\right),h\left(t\right)\right\rangle$ be a vector function. Then the *limit*

$$\lim_{x\to c}v\left(t\right)=\left\langle \lim_{x\to c}f\left(t\right),\lim_{x\to c}g\left(t\right),\lim_{x\to c}h\left(t\right)\right\rangle$$

exists provided the individual limits exist.

Definition: Let c be a real number and $v\left(t\right)=\left\langle f\left(t\right),g\left(t\right),h\left(t\right)\right\rangle$ be a vector function. Then the vector function $v\left(t\right)$ is said to be *continuous at* c if $\lim\limits_{x\to c}v\left(t\right)$ exists and $\lim\limits_{x\to c}v\left(t\right)=v\left(c\right)$. If $v\left(t\right)$ is continuous at every point in its domain then it is said to be a *continuous vector function*.

Remark: Let c be a real number and $v\left(t\right)=\left\langle f\left(t\right),g\left(t\right),h\left(t\right)\right\rangle$ be a vector function. If the derivative of the vector function exists then

$$v'\left(t\right)=\left\langle f'\left(t\right),g'\left(t\right),h'\left(t\right)\right\rangle$$

Definition: Let $v(t) = \langle f(t), g(t), h(t) \rangle$ be a vector function. Then the *unit tangent vector* is defined as $\dfrac{v'(t)}{\|v'(t)\|}$.

Definition: Let $v(t) = \langle f(t), g(t), h(t) \rangle$ be a vector function and $v'(t)$ is continuous on some interval $t_0 \le t \le t_1$, then the *length of the curve* in three dimensions that is depicted by $v(t)$ is given by the formula $\int_{t_0}^{t_1} \sqrt{(f'(t))^2 + (g'(t))^2 + (h'(t))^2}\, dt$.

Example 8: Consider the vector function $v(t) = \langle \sin(t), e^t, t \rangle$. Use *Maple* to graph this function for the values of t in the interval $0 \le t \le 11$ and find its tangent vector. Estimate the length of the curve depicted by the vector function $v(t)$ on the interval $0 \le t \le 11$.

Solution: In order to solve this example, we will activate the $with(Student[VectorCalculus])$ package.

$with(Student[VectorCalculus])$:

$v := t \rightarrow \langle \sin(t), e^t, t \rangle$

$$t \rightarrow Student : -VectorCalculus : -\langle\,,\,\rangle(\sin(t), e^t, t) \qquad (14.69)$$

$v(t)$

$$(\sin(t))e_x + (e^t)e_y + (t)e_z \qquad (14.70)$$

Note that *Maple* expresses the vector as a sum of its coordinate vectors. These vectors e_x, e_y, and e_z are the unit vectors along the *x*-axis, *y*-axis, and *z*-axis, respectively. We will use the *SpaceCurve* command to plot this vector function.

$SpaceCurve(v(t), t = 0..11, axes = normal, numpoints = 100, thickness = 3)$

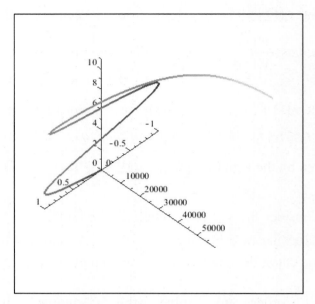

We will use the *TangentVector* command to obtain the tangent vector.

TangentVector$\left(v\left(t\right)\right)$

$$\begin{bmatrix} \cos\left(t\right) \\ \mathrm{e}^{t} \\ 1 \end{bmatrix} \tag{14.71}$$

The built in *ArcLength* command will be used to find the arc length. This command, for vectors, is available in the *with*(*Student*[*VectorCalculus*]) package.

ArcLength$\left(v\left(t\right),t=0..11\right)$

$$\int_{0}^{11} \sqrt{1+\cos\left(t\right)^{2}+\left(\mathrm{e}^{t}\right)^{2}}\ \mathrm{d}t \tag{14.72}$$

Since the value of the integral is not evaluated, we will need to use the *evalf* command to get the floating-point solution to the arc length.

evalf $\left(\left(\mathbf{14.72}\right)\right)$

$$59873.84756 \tag{14.73}$$

Hence the length of the arc is 59873.84756.

<u>Gradient and Directional Derivatives</u>

Definition: The *gradient of* $f(x,y)$ *at the point* (a,b) is the vector $\langle f_x(a,b), f_y(a,b) \rangle$. We will denote the gradient of $f(x,y)$ at the point (a,b) by the symbol $\nabla f(a,b)$. We can therefore write that $\nabla f(a,b) = \langle f_x(a,b), f_y(a,b) \rangle$.

Remark: We can rewrite the formula on page 463 of this chapter for the equation of the tangent plane at the point (x_1, y_1) on the surface of $f(x,y)$ as $z - z_1 = \nabla f(x_1, y_1) \cdot \langle x - x_1, y - y_1 \rangle$.

Definition: The *gradient of* $f(x,y,z)$ *at the point* (a,b,c) is the vector $\langle f_x(a,b,c), f_y(a,b,c), f_z(a,b,c) \rangle$. We will denote the gradient of $f(x,y,z)$ at the point (a,b,c) by the symbol $\nabla f(a,b,c)$. We can therefore write that $\nabla f(a,b,c) = \langle f_x(a,b,c), f_y(a,b,c), f_z(a,b,c) \rangle$.

Remark: ∇f gives the direction of greatest rise in f and $-\nabla f$ gives the direction of the greatest drop in f, provided $\nabla f \neq 0$.

Definition: For a unit vector $v = \langle v_1, v_2 \rangle$ in two dimensions the *directional derivative of* $f(x,y)$ *at the point* (a,b) *in the direction of* v is given by the formula $D_v f(a,b) = f_x(a,b)v_1 + f_y(a,b)v_2$. We can rewrite this formula more compactly as a vector dot product, i.e. $D_v f(a,b) = \langle f_x(a,b), f_y(a,b) \rangle \cdot \langle v_1, v_2 \rangle = \nabla f(a,b) \cdot v$

Definition: For a unit vector $v = \langle v_1, v_2, v_3 \rangle$ in three dimensions *the directional derivative of* $f(x,y,z)$ *at the point* (a,b,c) *in the direction of* v is given by the formula $D_v f(a,b,c) = f_x(a,b,c)v_1 + f_y(a,b,c)v_2 + f_z(a,b,c)v_3$. We can rewrite this formula more compactly as a vector dot product, i.e. $D_v f(a,b,c) = \langle f_x(a,b,c), f_y(a,b,c), f_z(a,b,c) \rangle \cdot \langle v_1, v_2, v_3 \rangle = \nabla f(a,b,c) \cdot v$

Theorem: Let f be a real valued function of several variables with $\nabla f \neq 0$ and let v be a unit vector. Then $\|\nabla f\|$ is the maximum value of the directional derivative $D_v f$, in the direction of the vector v. This maximum value occurs when v is in the same direction as the gradient vector. $-\|\nabla f\|$ is the minimum of the directional derivative $D_v f$, in the direction of the vector v. This minimum value occurs when v is in the opposite direction of the gradient vector.

Example 9: Let $f(x, y) = x^3 y - 3e^{\sin(x)} + 1$.

(a)   Find the directional derivative of the function $f(x, y) = x^3 y - 3e^{\sin(x)} + 1$, at the point $\left(\dfrac{\pi}{4}, -1\right)$, and in the direction of the vector $\langle 2, -1 \rangle$.

(b)   What is the maximum value of the directional derivative $D_v f\left(\dfrac{\pi}{4}, -1\right)$, for any two dimensional vector v ?

Solution:

(a)   Before entering f as a *Maple* function, we will need to execute the *restart* command.

> *restart*
> $f := (x, y) \rightarrow x^3 \cdot y - 3 \cdot e^{\sin(x)} + 1$
$$(x, y) \rightarrow x^3 y - 3e^{\sin(x)} + 1 \tag{14.74}$$

We will use the *DirectionalDerivative* command in the *with(Student[MultivariateCalculus])* package.

> *with(Student[MultivariateCalculus])* :
> $DirectionalDerivative\left(f(x, y), [x, y] = \left[\dfrac{\pi}{4}, -1\right], [2, -1] \right)$
$$\frac{2}{5}\left(-\frac{3}{16}\pi^2 - \frac{3}{2}\sqrt{2}\ e^{\frac{1}{2}\sqrt{2}} \right)\sqrt{5} - \frac{1}{320}\pi^3\sqrt{5} \tag{14.75}$$

> *evalf* ((**14.75**))
$$-5.719923537 \tag{14.76}$$

Hence, an approximate value of the directional derivative is -5.719923537.

(b)   We will first find the gradient g, of $f(x, y)$.

> $g := Gradient\left(f(x, y), [x, y] \right)$
$$\begin{bmatrix} 3x^2 y - 3\cos(x)e^{\sin(x)} \\ x^3 \end{bmatrix} \tag{14.77}$$

Hence the gradient of $f(x, y)$ is $\begin{bmatrix} 3x^2 y - 3\cos(x)e^{\sin(x)} \\ x^3 \end{bmatrix}$. We will now find the magnitude of this gradient vector.

$with(LinearAlgebra):$
$Norm(g, 2)$

$$\sqrt{\left| 3x^2 y - 3\cos(x)e^{\sin(x)} \right|^2 + \left| x \right|^6} \tag{14.78}$$

We will now substitute $\left(\dfrac{\pi}{4}, -1 \right)$ into **(14.78)**.

$$evalf\left(subs\left(x = \frac{\pi}{4}, y = -1, (\textbf{14.78}) \right) \right)$$

$$6.171876592 \tag{14.79}$$

Therefore, by the theorem before this example, the maximum value of $D_v f\left(\dfrac{\pi}{4}, -1 \right)$ is approximately 6.171876592 in the direction of the unit vector

$$v = \frac{\nabla f\left(\dfrac{\pi}{4}, -1 \right)}{\left| \nabla f\left(\dfrac{\pi}{4}, -1 \right) \right|}.$$

Remark: We can plot this directional derivative.

$$f := (x, y) \rightarrow x^3 \cdot y - 3 \cdot e^{\sin(x)} + 1$$

$$(x, y) \rightarrow x^3 y - 3e^{\sin(x)} + 1 \tag{14.80}$$

$with\left(Student\left[MultivariateCalculus \right] \right):$

$DirectionalDerivative\left(f(x, y), [x, y] = \left[\dfrac{\pi}{4}, -1 \right], [2, -1], x = -4 .. 6, y = -6 .. 4, \right.$

$\left. z = -1400 .. 700, output = plot \right)$

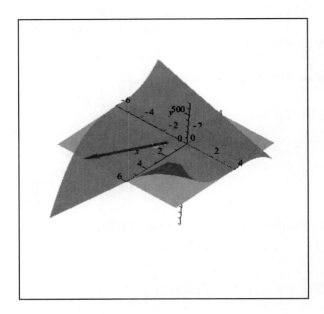

Definition: For any surface defined implicitly as $f(x,y,z)=c$, where c is a constant, the *equation of the tangent plane at the point* (x_0, y_0, z_0), where $f(x_0, y_0, z_0)=c$ is given by the equation $\nabla f(x_0, y_0, z_0) \cdot \langle x-x_0, y-y_0, z-z_0 \rangle = 0$, providing $\nabla f(x_0, y_0, z_0) \neq 0$.

Example 10: Find the equation of the tangent plane to the surface $xy^2 e^z = 4$ at the point $(1,2,0)$. Make an appropriate plot of your surface and its tangent plane on the same axes.

Solution: We will enter f as a *Maple* function.

$f := (x,y,z) \rightarrow x \cdot y^2 \cdot e^z$

$$(x,y,z) \rightarrow x\,y^2\,e^z \tag{14.81}$$

In order to compute the gradient function, we will find the partial derivatives of $f(x,y,z) = xy^2 e^z$ with respect to x, y and z.

$f1 := D[1](f)$

$$(x,y,z) \rightarrow y^2 e^z \tag{14.82}$$

$f2 := D[2](f)$

$$(x,y,z) \rightarrow 2\,x\,y\,e^z \tag{14.83}$$

$f3 := D[3](f)$

$$(x,y,z) \rightarrow x\,y^2\,e^z \tag{14.84}$$

$gradient := \langle f1(x,y,z), f2(x,y,z), f3(x,y,z) \rangle$

$$\begin{bmatrix} y^2 e^z \\ 2xye^z \\ xy^2 e^z \end{bmatrix} \qquad\qquad \textbf{(14.85)}$$

Hence the gradient of f is $\begin{bmatrix} y^2 e^z \\ 2xye^z \\ xy^2 e^z \end{bmatrix}$. We will now evaluate this gradient at the point $(1,2,0)$.

$g := \langle f1(1,2,0), f2(1,2,0), f3(1,2,0)\rangle$

$$\begin{bmatrix} 4 \\ 4 \\ 4 \end{bmatrix} \qquad\qquad \textbf{(14.86)}$$

In order to find the equation of the tangent plane to the surface $xy^2 e^z = 4$ at the point $(1,2,0)$, we will define the vector $p = \langle x-1, y-2, z-0\rangle$ and then find the dot product $p.g$.

$p := \langle x-1, y-2, z-0\rangle$

$$\begin{bmatrix} x-1 \\ y-2 \\ z \end{bmatrix} \qquad\qquad \textbf{(14.87)}$$

$with\left(Student\left[LinearAlgebra\right]\right):$
$p.g$

$$4x-12+4y+4z \qquad\qquad \textbf{(14.88)}$$

Hence the equation of the tangent plane to the surface $xy^2 e^z = 4$ at the point $(1,2,0)$ is $4x-12+4y+4z=0$ or $x+y+z=3$. We will now make a plot of the surface and the plane on the same axes using the *implicitplot3d* command.

$with\left(plots\right):$
$implicitplot3d(\{4\cdot x+4\cdot y+4\cdot z-12=0, f(x,y,z)=4\}, x=-5..5, y=0..5, z=-10..10,$
$\quad grid=[20,20,20], axes=box)$

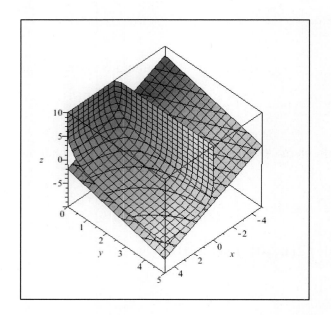

Exercises

1 – 5   Graph the functions $f(x, y)$. Find and graph the first partial derivatives.

1. $f(x, y) = 3x^5 - 5x^4 y - 2x^3 y^4 + 9$

2. $f(x, y) = 3e^{2x} \sin(3xy)$

3. $f(x, y) = \arctan\left(\dfrac{y}{x}\right)$

4. $f(x, y) = \dfrac{\sin\left(\sqrt{x^2 + y^2}\right)}{\sqrt{x^2 + y^2}}$

5. $f(x, y) = xy^2 + x^{\tan(y)}$

6 – 10 Find all the second partial derivatives of $f(x, y)$. Is $f_{xy} = f_{yx}$ for every (x, y)? Give reasons for your answer.

6. $f(x, y) = 3x^5 - 5x^4 y - 2x^3 y^4 + 9$.

7. $f(x, y) = 3e^{2x} \sin(3xy)$.

8. $f(x, y) = \arctan\left(\dfrac{y}{x}\right)$.

9. $f(x, y) = y \ln(x) + \csc\left(xy^2\right)$.

10. $f(x, y) = (3xy)^{x+y}$.

11.   Let $f(x, y) = x^3 y + 2xy + \ln xy$. Find the equation of the tangent plane to the surface of $z = f(x, y)$ when $(x, y) = \left(\sqrt{7}, 2\right)$. Plot the surface f along with the tangent plane.

12.   Let $f(x, y) = 3x^5 - 5x^4 y - 2x^3 y^4 + 9$. Find the equation of the tangent plane to the surface of $z = f(x, y)$ when $(x, y) = (-1, 5)$. Plot the surface f along with the tangent plane.

13. Let $f(x,y) = \sin(x+y) - \sin(x) - \sin(y)$. Find the equation of the tangent plane to the surface of $z = f(x,y)$ when $(x,y) = \left(\dfrac{\pi}{3}, \dfrac{\pi}{2}\right)$. Plot the surface f along with the tangent plane.

14. Find the linearization at $(x,y) = \left(0, \dfrac{1}{2}\right)$ of $f(x,y) = e^{-x}\sin(\pi y)$. Use your answer to approximate $f(-0.001, 0.483)$. Find the relative error of your answer.

15. Find the linearization at $(x,y) = (4,9)$ of $f(x,y) = \sqrt{x} + \sqrt{y} + \sqrt{xy}$. Use your answer to approximate $f(4.002, 8.784)$. Find the relative error of your answer.

16. Find the linearization at $(x,y) = (1, 2\pi)$ of $f(x,y) = x^{\cos(y)}$. Use your answer to approximate $f(0.994, 5.963)$. Find the relative error of your answer.

17. Let $f(x,y) = x^6 + 2y^4 - 4x^2 - y^2 - 5x^2 y^2$. Find the local extrema and saddle points of this function. Verify your answer by graphing f using the *plot3d* command and constructing a contour plot.

18. Let $f(x,y) = 2x^4 - x^2 + 3y^2$. Find the local extrema and saddle points of this function. Verify your answer by graphing f using the *plot3d* command and constructing an appropriate contour plot.

19. Let $f(x,y) = xy + \dfrac{4}{x} + \dfrac{2}{y}$. Find the local extrema and saddle points of this function. Verify your answer by graphing f using the *plot3d* command and constructing a contour plot.

20. Let $f(x,y) = xye^{-\left(x^2 + y^2 - 6x\right)}$. Find the local extrema and saddle points of this function. Verify your answer by graphing f using the *plot3d* command and constructing an appropriate contour plot.

21. Find the minimum/maximum value of the function $f(x,y) = e^{-|y|}\sin(x^2 + y^2)$ $f(x,y) = e^{-|y|}\sin(x^2 + y^2)$ subject to the constraint $4x^2 + y^2 = 1$. Find the local extrema and saddle points of this function. Verify your answer by constructing an appropriate contour plot.

22. Find the minimum/maximum value of the function $f(x, y) = x^3 y^3$ subject to the constraint $x^2 + y^2 = 9$. Find the local extrema and saddle points of this function. Verify your answer by constructing an appropriate contour plot.

23. Find the minimum/maximum value of the function $f(x, y) = x^2 + y^2 - 9$ subject to the constraint $x^3 y^3 = 9$. Find the local extrema and saddle points of this function. Verify your answer by constructing an appropriate contour plot.

24. Find the equation of the tangent plane to the surface $x^3 + 5y^4 + z^4 = 2011$ at the point $\left(10, 1, \sqrt[4]{1006}\right)$. Make an appropriate plot of your surface and its tangent plane on the same axes.

25. Find the equation of the tangent plane to the surface $z^4 = \ln\left(2011 + 5y^2 + x^4\right)$ at the point $(1, 2, 1.6612798)$. Make an appropriate plot of your surface and its tangent plane on the same axes.

26. Find the equation of the tangent plane to the surface $z = \sin(xy) + e^{xy} \cos(xy)$ at the point $\left(\dfrac{\pi}{2}, 1, 1\right)$. Make an appropriate plot of your surface and its tangent plane on the same axes.

27 – 31 For the curves depicted by the vector function $v(t)$ in an appropriate interval, find the tangent vector, and the normalized tangent vector at the given value of t in each case. Make an appropriate space plot of each curve. (Hint: For the normalized tangent vector, use the *Maple* command *TangentVector* $(t \rightarrow v(t), normalized)$.)

27. $v(t) = \left\langle t^3, \sin(t), 8t \right\rangle$ at $t = \dfrac{\pi}{6}$

28. $v(t) = \left\langle e^t \cos(t), e^t \sin(t), t^2 \right\rangle$ at $t = 0$

29. $v(t) = \left\langle \sqrt{t^2 + 5}, \cos(t), 1 \right\rangle$ at $t = \sqrt{3}$

30. $v(t) = \left\langle t^3, e^{2t}, t \ln(t) \right\rangle$ at $t = 9$

31. $v(t) = \left\langle \tan(t), \sin(t), 8t \right\rangle$ at $t = 1.21$

32. Find the length of the curve depicted by the vector function $v(t) = \langle t^3, e^{2t}, t\ln(t) \rangle$ over the interval $1 \le t \le 5$.

33. Find the length of the curve depicted by the vector function $v(t) = \langle e^t \cos(t), e^t \sin(t), t^2 \rangle$ over the interval $0 \le t \le 5$.

34. Find the directional derivative of the function $f(x, y) = e^{xy^2}$, at the point $(-2,1)$, and in the direction of the vector $\langle -2,1 \rangle$.

35. Let $f(x, y) = 2011\sin(3x^2 + y^2)\cos(3x^2 + y^2)$, find its directional derivative at the point $(-1,1)$, and in the direction of the vector $\langle \sqrt{2}, -7 \rangle$.

36. Find the directional derivative of the function $f(x, y) = x^4 y + 65x^2 + \sqrt{x+y}$, at the point $(2,6)$, and in the direction of the vector $\langle -2,1 \rangle$.

37. Find the directional derivative of the function $f(x, y, z) = x^4 y + z^2 + \cos(zy)$, at the point $(1,0,-1)$, and in the direction of the vector $\langle -3,1,9 \rangle$.

38. Find the directional derivative of the function $f(x, y, z) = z^2 \ln(x^2 + 3xy)$, at the point $(1,10,101)$, and in the direction of the vector $\langle -3,1,9 \rangle$.

39. What is the maximum value of $D_v f(1,2,-3)$ as v varies, for the function $f(x, y, z) = e^{x+y+x^2 z}$.

40. What is the minimum value of $D_v f(-1,2)$ as v varies, for the function $f(x, y) = e^{3x-9y^4}$.

41. An explorer is standing at the point with coordinates $(1,1,3)$ on a mountain that has the shape of the graph $f(x, y) = 5 - x^2 - y^2$. Plot the surface $f(x, y)$, the point $(1,1,3)$ and the contour line at the level curve $f(x, y) = 3$. In what direction should the explorer move to neither climb nor descend the mountain.

42. Below is a list of some *Maple* commands that were used in this chapter. Describe the significance of each command. Can you find examples where each command is used?

(a) $contourplot\left(f\left(x,y\right),x=a\,..\,b,y=c\,..\,d,contours=k,grid=\left[m,n\right]\right)$

(b) $implicitplot3d\left(implicit\ equation,a\,..\,b,c\,..\,d,p\,..\,q,options\right)$

(c) $LagrangeMultipliers\left(f\left(x,y\right),\left[g\left(x,y\right)\right],\left[x,y\right],output=plot\right)$

(d) $plot3d\left(f\left(x,y\right),x=a\,..\,b,y=c\,..\,d,axes=box\right)$

(e) $with\left(Student\left[MultivariateCalculus\right]\right)$

43. Can you list new *Maple* commands that were used in this chapter but were not listed in exercise 42?

An extension of integration to encompass multiple integrals will be made. *Maple* will be used to convert between different coordinate systems in evaluating n-fold integrals (mostly in $\mathbb{R}^2$ and $\mathbb{R}^3$). Students will be able to evaluate multiple integrals in different coordinate systems, find volumes and center of mass of various regions and have a visual · perspective to give them further insights and understanding of integration.

Double Integrals

We will now consider multiple integrals in this chapter. Recall, from chapter 6, that if f is a function defined for $a \leq x \leq b$, then the definite integral of f from a to b, which is denoted by $\int_a^b f(x)\,dx$, is defined as $\int_a^b f(x)\,dx = \lim\limits_{n\to\infty} \sum\limits_{i=1}^{n} f(x_i^*)(x_i - x_{i-1})$, provided this limit exists. We will now extend this definition to functions of two and three variables.

Definition: Let f be any real valued function of two variables defined for $a \leq x \leq b$ and $c \leq y \leq d$. Partition the rectangular block $R = [a,b] \times [c,d]$ into mn rectangles of equal sizes (as outlined here). Let the end points of the subintervals of $[a,b]$ be $x_0 (=a), x_1, x_2 .. x_m (=b)$ and the end points of the subintervals of $[c,d]$ be $y_0 (=c), y_1, y_2, .. y_n (=d)$. Then the length of each subinterval $[x_{i-1}, x_i]$ is $\dfrac{b-a}{m}$ for $i = 1$, 2, .. m and the length of each subinterval $[y_{j-1}, y_j]$ is $\dfrac{d-c}{n}$ for $j = 1, 2, .. n$. Let (x_{ij}^*, y_{ij}^*) be any point in the rectangle $[x_{i-1}, x_i] \times [y_{j-1}, y_j]$. The *definite integral of f on the rectangle R*, denoted by $\iint\limits_R f(x, y)\,dA$, is defined as

$$\iint\limits_R f(x, y)\,dA = \lim\limits_{n\to\infty} \lim\limits_{m\to\infty} \sum\limits_{j=1}^{n} \sum\limits_{i=1}^{m} f(x_{ij}^*, y_{ij}^*) \frac{(b-a)}{m} \frac{(d-c)}{n}$$

provided this limit exists. In this case f is said to be integrable over the rectangle R.

Definition: Let f be any real valued function of two variables defined for $a \leq x \leq b$ and $c \leq y \leq d$. Partition the rectangle $R = [a,b] \times [c,d]$ into mn rectangles of equal length as in the previous definition. Let $\overline{x_i} = a + \dfrac{(b-a)(i-0.5)}{m}$ and $\overline{y_j} = c + \dfrac{(d-c)(j-0.5)}{n}$. Then *the midpoint approximation of definite integral of f* is defined as

$$\iint\limits_R f(x, y)\,dA = \lim\limits_{n\to\infty} \lim\limits_{m\to\infty} \sum\limits_{j=1}^{n} \sum\limits_{i=1}^{m} f(\overline{x_{ij}}, \overline{y_{ij}}) \frac{(b-a)}{m} \frac{(d-c)}{n}$$

provided this limit exists.

Example 1: Estimate the value of the double integral $\int_1^9 \int_1^9 \left(x^2 + y^2\right) dx\, dy$ using the midpoint Riemann approximation sum by partitioning the rectangle $[1,9] \times [1,9]$ into 64 unit squares.

Solution: We will enter the integrand $f(x,y) = x^2 + y^2$ as a *Maple* function and graph it.

$f := (x,y) \rightarrow x^2 + y^2$

$$(x,y) \rightarrow x^2 + y^2 \tag{15.1}$$

$plot3d\left(f(x,y), x = 1..9, y = 1..9, axes = box\right)$

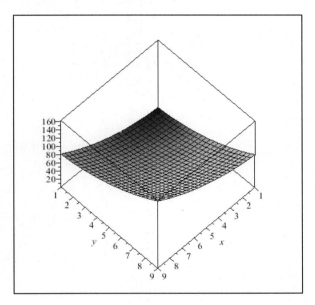

The function $f(x,y)$ is nonnegative in the rectangle $[1,9] \times [1,9]$. In this case the definite integral is the volume of the region bounded by the surface $z = x^2 + y^2$, the planes $z = 0$, $x = 1$, $x = 9$, $y = 1$, and $y = 9$. This volume can be approximated by constructing rectangular blocks that fill the region, then finding the volume of each rectangular block and then taking their sum. In other words, in this case the Riemann sums can be interpreted as the sum of the volumes of these approximating rectangular blocks. The *ApproximateInt* command with the options of *method = midpoint* and *output = plot* sketches rectangular blocks in a manner that gives the height of each block the value of the function at the midpoint of each rectangular block. The default method is the *midpoint* Riemann sum. The number of rectangular blocks (with unit square basis) needs to be specified. The graph below has 64 such rectangular blocks. The default partition has 25 rectangular blocks. We will need to activate the *with*$\left(Student\left[MultivariateCalculus\right]\right)$ package.

$with\big(Student\big[MultivariateCalculus\big]\big):$

$ApproximateInt\big(f(x,y),\,x=1..9,\,y=1..9,\,method=midpoint,\,output=plot,$

$\quad partition=[8,8],\,prismoptions=[color=grey]\big)$

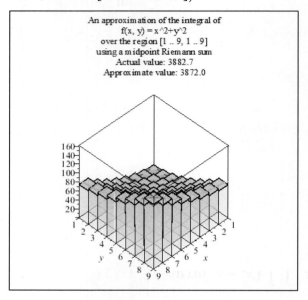

We will now estimate the value of the double integral $\displaystyle\int_1^9\int_1^9\big(x^2+y^2\big)\,dx\,dy$ by finding the volume of these rectangular blocks using the *ApproximateInt* command.

$ApproximateInt\big(f(x,y),\,x=1..9,\,y=1..9,\,method=midpoint,\,partition=[8,8]\big)$

$$3872.000000 \hspace{4cm} \textbf{(15.2)}$$

Hence an estimated value of the double integral $\displaystyle\int_1^9\int_1^9\big(x^2+y^2\big)\,dx\,dy$ using the midpoint Riemann approximation sum is 3872.000000.

Remark: Unlike the case of estimating the definite integral in the case of a function of one variable, the options for the *ApproximateInt* command in the multivariable case does not have an option for sum.

Example 2: Compute the general midpoint Riemann approximation to estimate the double integral $\displaystyle\int_1^9\int_1^9\big(x^2+y^2\big)\,dx\,dy$. Use *Maple* commands to evaluate the integral $\displaystyle\int_1^9\int_1^9\big(x^2+y^2\big)\,dx\,dy$.

Solution: We will enter the integrand $f(x,y)=x^2+y^2$ as a *Maple* function.

$$f := (x, y) \rightarrow x^2 + y^2$$

$$(x, y) \rightarrow x^2 + y^2 \tag{15.3}$$

To estimate the sum in the midpoint Riemann approximation, we need to declare m and n as variables in *Maple* since they no longer take on the value of 8.

$$m := 'm' : \ n := 'n' : xibar := 1 + \frac{(9-1)(i-0.5)}{m} : \ yjbar = 1 + \frac{(9-1)(j-0.5)}{n} :$$

We will now evaluate $\displaystyle \lim_{n \to \infty} \lim_{m \to \infty} \sum_{j=1}^{n} \sum_{i=1}^{m} f\left(\overline{x}_i, \overline{y}_j\right) \frac{(b-a)}{m} \frac{(d-c)}{n}$ using *Maple* commands.

$$\lim_{n \to \infty} \left(\lim_{m \to \infty} \sum_{j=1}^{n} \sum_{i=1}^{m} f\left(xibar, yjbar\right) \cdot \frac{(9-1)}{m} \cdot \frac{(9-1)}{n} \right)$$

$$3882.666666 \tag{15.4}$$

An estimate for $\displaystyle \int_{1}^{9} \int_{1}^{9} \left(x^2 + y^2\right) dx\, dy$ is 3882.666666. We will now use the *Maple* command $int\left(f(x, y), [x = a..b, y = c..d]\right)$ to evaluate the integral $\displaystyle \int_{1}^{9} \int_{1}^{9} \left(x^2 + y^2\right) dx\, dy$.

$$int\left(f(x, y), [x = 1..9, y = 1..9]\right)$$

$$\frac{11648}{3} \tag{15.5}$$

$$evalf\left((\mathbf{15.5})\right)$$

$$3882.666667 \tag{15.6}$$

Hence $\displaystyle \int_{1}^{9} \int_{1}^{9} \left(x^2 + y^2\right) dx\, dy = \frac{11648}{3} = 3882.666667$.

Remark: Double integrals can be extended over general regions D. This will be illustrated in the following definitions.

Definition: For the region $D_1 = \left\{ (x, y) \mid x_0 \leq x \leq x_1, f_1(x) \leq y \leq f_2(x) \right\}$, where the functions f_1 and f_2 are continuous, we write the double integral of $f(x, y)$ over that region as $\displaystyle \iint_{D_1} f(x, y)\, dA = \int_{x_0}^{x_1} \int_{f_1(x)}^{f_2(x)} f(x, y)\, dy\, dx$. The region D_1 is called a *type I region*.

Definition: For the region $D_2 = \{(x, y) \mid y_0 \leq y \leq y_1, g_1(y) \leq x \leq g_2(y)\}$, where the functions g_1 and g_2 are continuous, we write the double integral of $f(x, y)$ over that region as $\iint_{D_2} f(x, y) \, dA = \int_{y_0}^{y_1} \int_{g_1(y)}^{g_2(y)} f(x, y) \, dx \, dy$. The region D_2 is called a *type II region*.

Definition: Let f be a function of two variables and D a region in $\mathbb{R}^2$. Then $\iint_D f(x, y) \, dA$ represents *the signed volume of the solid* between the surface $z = f(x, y)$ and the xy-plane over the region D. That is, the volume will be positive if $f(x, y) > 0$ and negative if $f(x, y) < 0$.

Remark: We can think of the indefinite integral of a function of a single variable f, $\int f(x) \, dx$, as finding the definite integral $\int_a^x f(t) \, dt$ over an arbitrary interval $[a, x]$. We can extend this notion to *indefinite double integrals* to consider them as being integrated over an arbitrary closed region. In addition, the order in which the integral is evaluated is irrelevant and the results may not be equal. It is important to note that the order is extremely important when we are dealing with definite double integrals. This notion is also valid for triple integrals and n-fold integrals in general.

Fubini's Theorem: Let f be a continuous function on the rectangle $R = [a, b] \times [c, d]$. Then $\iint_R f(x, y) \, dA = \int_c^d \int_a^b f(x, y) \, dx \, dy = \int_a^b \int_c^d f(x, y) \, dy \, dx$.

Remark: The two integrals $\int_c^d \int_a^b f(x, y) \, dx \, dy$ and $\int_a^b \int_c^d f(x, y) \, dy \, dx$ are called iterated integrals.

<u>Properties of the definite double integral:</u>

I. Suppose g and h are two real valued functions of one variable. Then
$$\int_c^d \int_a^b g(x) h(y) \, dx \, dy = \int_a^b g(x) \, dx \int_c^d h(y) \, dy$$

II. Let f and g be continuous on the region D and let k be any constant. Then

1. $\iint_D k \, dA = kA$, where A is the area of the region D.

2. $\iint_D k f(x, y) \, dA = k \iint_D f(x, y) \, dA$

3. $\iint_D [f(x, y) + g(x, y)] \, dA = \iint_D f(x, y) \, dA + \iint_D g(x, y) \, dA$

4. $$\iint_D \left[f(x,y) - g(x,y) \right] dA = \iint_D f(x,y) dA - \iint_D g(x,y) dA$$

5. $$\iint_D f(x,y) dA = \iint_{D_1} f(x,y) dA + \iint_{D_2} f(x,y) dA, \text{ where } D = D_1 \cup D_2 \text{ and } D_1 \text{ and } D_2$$
   have no common elements.

6. If $f(x,y) \geq 0$ for every value of (x,y) in the region D, then $\iint_D f(x,y) dA \geq 0$.

7. If $f(x,y) \geq g(x,y)$ for every value of (x,y) in the region D, then
   $$\iint_D f(x,y) dA \geq \iint_D g(x,y) dA.$$

8. If $m \leq f(x,y) \leq M$ for every value of (x,y) in the region D, then
   $$mA \leq \iint_D f(x,y) dA \leq MA, \text{ where } A \text{ is the area of the region } D.$$

Example 3: Evaluate the integrals

(a) $\iint \left(x^5 y^2 + 2x^3 y^8 - 7 \right) dx\, dy$

(b) $\iint \dfrac{y}{x^2 + y^2} dx\, dy$

(c) $\int_1^4 \int_1^2 \ln(x + 3y) dx\, dy$

(d) $\int_{\frac{\pi}{6}}^{\frac{\pi}{3}} \int_0^{\frac{\pi}{2}} \sin(2x - 5y) dx\, dy$

Solution: We will evaluate the integrals using *int* command.

(a) To evaluate $\iint \left(x^5 y^2 + 2x^3 y^8 - 7 \right) dx\, dy$ we will use the $int\left(f(x,y), [x,y] \right)$ command. Written this way we will first integrate with respect to x and then with respect to y.

$$int\left(x^5 \cdot y^2 + 2 \cdot x^3 \cdot y^8 - 7, [x,y] \right)$$

$$\frac{1}{18} x^6 y^3 + \frac{1}{18} x^4 y^9 - 7xy \qquad\qquad (15.7)$$

Hence $\iint \left(x^5 y^2 + 2x^3 y^8 - 7 \right) dx\, dy = \dfrac{1}{18} x^6 y^3 + \dfrac{1}{18} x^4 y^9 - 7xy$. Observe that when evaluating an indefinite integral, *Maple* treats the constant of integration as zero.

(b) To evaluate the integral $\iint \dfrac{y}{x^2 + y^2} dx\, dy$ use the same steps as in part (a).

$$int\left(\frac{y}{x^2 + y^2}, [x,y] \right)$$

$$y \arctan\left(\frac{x}{y}\right) - x \ln\left(\frac{x}{y}\right) + \frac{1}{2} x \ln\left(1 + \frac{x^2}{y^2}\right) \qquad \textbf{(15.8)}$$

Therefore, $\displaystyle\iint \frac{y}{x^2 + y^2} dx\,dy = y \arctan\left(\frac{x}{y}\right) - x \ln\left(\frac{x}{y}\right) + \frac{1}{2} x \ln\left(1 + \frac{x^2}{y^2}\right).$

(c)   To evaluate the integral $\displaystyle\int_1^4 \int_1^2 \ln(x + 3y)\,dx\,dy$, we will use the $int\left(f(x,y), [x = a..b, y = c..d]\right)$ command.

$$int\left(\ln(x + 3 \cdot y), [x = 1..2, y = 1..4]\right)$$

$$-\frac{9}{2} + 38 \ln(2) - \frac{25}{6} \ln(5) - \frac{169}{6} \ln(13) + \frac{98}{3} \ln(7) \qquad \textbf{(15.9)}$$

In order to get a floating-point approximation we will use the command *evalf* .

$$evalf\left((\textbf{15.9})\right)$$

$$6.45392621 \qquad \textbf{(15.10)}$$

Hence $\displaystyle\int_1^4 \int_1^2 \ln(x + 3y)\,dx\,dy = -\frac{9}{2} + 38 \ln(2) - \frac{25}{6} \ln(5) - \frac{169}{6} \ln(13) + \frac{98}{3} \ln(7).$
This integral is approximately equal to 6.45392621.

(d)   To evaluate the integral $\displaystyle\int_{\frac{\pi}{6}}^{\frac{\pi}{3}} \int_0^{\frac{\pi}{2}} \sin(2x - 5y)\,dx\,dy$ use the same steps as in part (c).

$$int\left(\sin(2 \cdot x - 5 \cdot y), \left[x = 0..\frac{\pi}{2}, y = \frac{\pi}{6}..\frac{\pi}{3}\right]\right)$$

$$-\frac{1}{10} - \frac{1}{10}\sqrt{3} \qquad \textbf{(15.11)}$$

Thus $\displaystyle\int_{\frac{\pi}{6}}^{\frac{\pi}{3}} \int_0^{\frac{\pi}{2}} \sin(2x - 5y)\,dx\,dy = -\frac{1}{10} - \frac{1}{10}\sqrt{3}$.

Remarks:
1.   Based on the remark before example 3 on indefinite integrals, we can also evaluate the indefinite integral $\displaystyle\iint \left(x^5 y^2 + 2x^3 y^8 - 7\right) dy\,dx$ by integrating with respect to y first and then with respect to x. We will use the command

$int\left(f\left(x,y\right),\left[y,x\right]\right)$. Written this way we will first integrate with respect to y first and then with respect to x.

$$int\left(x^5\cdot y^2+2\cdot x^3\cdot y^8-7,\left[y,x\right]\right)$$

$$\frac{1}{18}x^6\,y^3+\frac{1}{18}x^4\,y^9-7\,y\,x \qquad\qquad (15.12)$$

Hence $\iint\left(x^5y^2+2x^3y^8-7\right)dy\,dx=\frac{1}{18}x^6\,y^3+\frac{1}{18}x^4\,y^9-7\,x\,y$. In this instance we get the same result, regardless of the order of integration, since the constant of integration is zero.

2. Notice what happens when we change the order of integration for example 3(b).

$$int\left(\frac{y}{x^2+y^2},\left[y,x\right]\right)$$

$$\frac{1}{2}x\ln\left(x^2+y^2\right)-x+y\arctan\left(\frac{x}{y}\right) \qquad\qquad (15.13)$$

Are **(15.8)** and **(15.13)** the same? What happened to Fubini's Theorem?

3. Notice what happens when we change the order of integration for example 1(d).

$$int\left(\sin\left(2\cdot x-5\cdot y\right),\left[y=\frac{\pi}{6}..\frac{\pi}{3},x=0..\frac{\pi}{2}\right]\right)$$

$$-\frac{1}{10}-\frac{1}{10}\sqrt{3} \qquad\qquad (15.14)$$

We see that Fubini's Theorem works. Why?

4. One could use the *MultiInt* command to evaluate definite double integrals. However, the *with*$\left(Student\left[MultivariateCalculus\right]\right)$ must be activated before using this command.

$$with\left(Student\left[MultivariateCalculus\right]\right):$$

$$MultiInt\left(\sin\left(2\cdot x-5\cdot y\right),x=0..\frac{\pi}{2},y=\frac{\pi}{6}..\frac{\pi}{3}\right)$$

$$-\frac{1}{10}-\frac{1}{10}\sqrt{3} \qquad\qquad (15.15)$$

If we include the option of *output = integral* we will get the unevaluated value of the integral.

$$MultiInt\left(\sin\left(2\cdot x-5\cdot y\right), x=0..\frac{\pi}{2}, y=\frac{\pi}{6}..\frac{\pi}{3}, output = integral\right)$$

$$\int_{\frac{1}{6}\pi}^{\frac{1}{3}\pi}\int_{0}^{\frac{1}{2}\pi}\sin\left(2x-5y\right)dx\,dy \tag{15.16}$$

Just as in the case of the *ShowSteps* command seen in chapter 7 we can add the option of *output = steps* to see the steps taken by *Maple* to solve this integral. However, unlike the *ShowSteps*, this does not tell the steps taken to solve this integral even if the *infolevel*[*Student*[*MultivariateCalculus*]] is added.

$$MultiInt\left(\sin\left(2\cdot x-5\cdot y\right), x=0..\frac{\pi}{2}, y=\frac{\pi}{6}..\frac{\pi}{3}, output = steps\right)$$

$$\int_{\frac{1}{6}\pi}^{\frac{1}{3}\pi}\int_{0}^{\frac{1}{2}\pi}\sin\left(2x-5y\right)dx\,dy$$

$$`=\int_{\frac{1}{6}\pi}^{\frac{1}{3}\pi}\left(-\frac{\cos\left(2x-5y\right)}{2}\bigg|_{x=0..\frac{\pi}{2}}\right)dy$$

$$`=\int_{\frac{1}{6}\pi}^{\frac{1}{3}\pi}\cos\left(5y\right)dy$$

$$`=\frac{\sin\left(5y\right)}{5}\bigg|_{y=\frac{\pi}{6}..\frac{\pi}{3}}$$

$$-\frac{1}{10}-\frac{1}{10}\sqrt{3} \tag{15.17}$$

5.   One can click on $\int_{a}^{b}f\,dx$ in the *Expression* palette twice to integrate.

$$\int_{\frac{\pi}{6}}^{\frac{\pi}{3}}\int_{0}^{\frac{\pi}{2}}\sin\left(2x-5y\right)dx\,dy$$

$$-\frac{1}{10}-\frac{1}{10}\sqrt{3} \tag{15.18}$$

Example 4: Let $f\left(x,y\right)=e^{-\left(x^2+y^2\right)}$. Find the volume of the solid between the surface of $z=f\left(x,y\right)$ and the *xy*-plane over the circular region $x^2+y^2=9$.

Solution: We will begin by entering f as a *Maple* function and graphing it.

$$f := (x, y) \to e^{-(x^2+y^2)}$$

$$(x, y) \to e^{-x^2-y^2} \tag{15.19}$$

$$plot3d\left(f(x, y), x = -5..5, y = -\sqrt{9-x^2} .. \sqrt{9-x^2}, axes = box, grid = [30,30]\right)$$

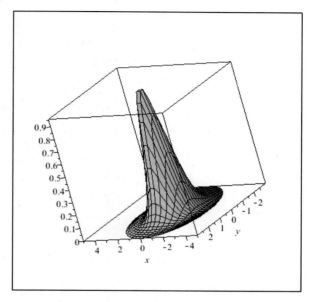

In order to find the volume we will evaluate the integral $\iint_D f(x, y)\,dA$ where D is the circle $x^2 + y^2 = 9$ using *MultiInt* command.

$$with\left(Student[MultivariateCalculus]\right):$$

$$MultiInt\left(f(x, y), y = -\sqrt{9-x^2} .. \sqrt{9-x^2}, x = -3..3\right)$$

$$\int_{-3}^{3} \mathrm{erf}\left(\sqrt{9-x^2}\right)\sqrt{\pi}\, e^{-x^2}\,dx \tag{15.20}$$

Observe that *Maple* did not evaluate this integral as it is currently set up. We will show that the value of this integral is $\pi(1-e^{-9})$ by converting to polar coordinates in example 8. We will find a floating-point approximation for **(15.20)** by using the *evalf* command

$$evalf\left((15.20)\right)$$

$$3.141404950 \tag{15.21}$$

Hence the volume is approximately 3.141404950.

Example 5: Let $f(x,y) = 3x + xy$. Find the volume of the solid formed between the surface of $z = f(x,y)$, and over the triangular region $0 \le x \le 2$ and $0 \le y \le x$ in the xy-plane.

Solution: We will begin by entering f as a *Maple* function and graphing it.

$f := (x,y) \to 3 \cdot x + x \cdot y$

$$(x,y) \to 3x + xy \qquad \textbf{(15.22)}$$

$plot3d\left(f(x,y), x = 0..2, y = 0..x, axes = normal\right)$

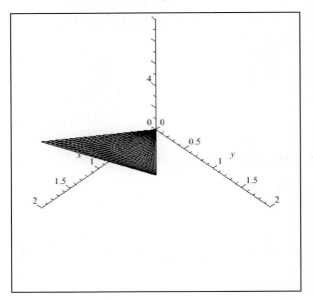

In order to find the volume we will evaluate the integral $\iint\limits_{D} f(x,y)\, dA$ where D is the triangular region $0 \le x \le 2$ and $0 \le y \le x$ using *MultiInt* command.

$with\left(Student\left[MultivariateCalculus\right]\right):$
$MultiInt\left(f(x,y), y = 0..x, x = 0..2\right)$

$$10 \qquad \textbf{(15.23)}$$

Hence the volume is 10.

Remarks:

1. We can add the option *orientation* $= [0,0]$ to the *plot3d* command in example 5 to get another view of the region D in the xy-plane.

$plot3d\left(f(x,y), x = 0..2, y = 0..x, axes = normal, orientation = [0,0], color = khaki\right)$

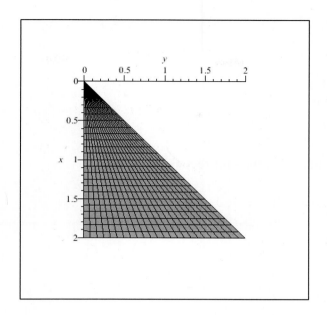

2. We can change the order of integration to evaluate the integral $\iint\limits_D f(x,y)\,dA$ where D is the triangular region $0 \le x \le 2$ and $0 \le y \le x$. The modified command is below.

$with\big(\,Student\,[\,MultivariateCalculus\,]\big):$
$MultiInt\big(f(x,y),x=y\,..\,2,y=0\,..\,2\big)$

$$10 \qquad\qquad\qquad\qquad\qquad\qquad\textbf{(15.24)}$$

Example 6: $f(x,y)=\dfrac{x-y}{(x+y)^3}$ on the unit square $R=\big\{(x,y),0 \le x \le 1 \text{ and } 0 \le x \le 1\big\}.$

Evaluate the integrals $\int_0^1\int_0^1 f(x,y)\,dx\,dy$ and $\int_0^1\int_0^1 f(x,y)\,dy\,dx$. Are the two integrals equal? Does the result contradict Fubini's theorem? Give reasons for your answer.

Solution: We will begin by entering f as a *Maple* function and then evaluating the two integrals. Do not forget to activate the $with\big(\,Student\,[\,MultivariateCalculus\,]\big)$ package.

$$f:=(x,y)\to\frac{x-y}{(x+y)^3}$$

$$(x,y)\to\frac{x-y}{(x+y)^3} \qquad\qquad\qquad\textbf{(15.25)}$$

$with\big(\,Student\,[\,MultivariateCalculus\,]\big):$

$$MultiInt\big(f(x,y),x=0..1,y=0..1\big)$$

$$-\frac{1}{2}$$ (15.26)

$$MultiInt\big(f(x,y),y=0..1,x=0..1\big)$$

$$\frac{1}{2}$$ (15.27)

Let us plot this function in the unit rectangle.

$$plot3d\big(f(x,y),x=0..1,y=0..1,axes=box\big)$$

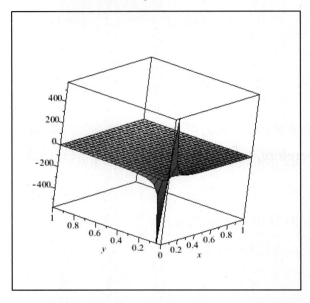

Observe that f is not continuous on the line $y=-x$. We will now graph the region of intersection along with the line $y=-x$.

$$with(plots):$$

$$with(plottools):$$

$$s1:=display\Big(polygon\big(\big[[0,0],[1,0],[0,1],[1,1]\big]\big),color=green,transparency=0.8\Big)$$

$$PLOT(...)$$ (15.28)

$$s2:=plot\big(-x,x=-1..1,thickness=2\big)$$

$$PLOT(...)$$ (15.29)

$$s3:=pointplot\big([0,0],color=blue,symbol=solidcircle,symbolsize=15\big)$$

$$PLOT(...)$$ (15.30)

$$display(s1,s2,s3)$$

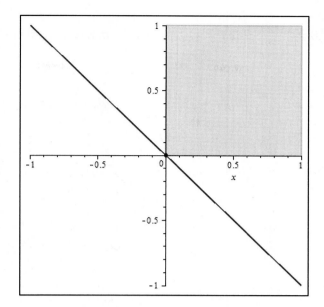

Notice that the region of integration intersects the line $y = -x$ at the origin. Hence our function is not continuous at the origin. Therefore, the hypothesis of Fubini's theorem is not satisfied. Therefore, $\int_0^1 \int_0^1 f(x,y)\,dx\,dy \neq \int_0^1 \int_0^1 f(x,y)\,dy\,dx$ does not contradict Fubini's theorem.

Definition: To convert a double integral into polar coordinates, we make one of the following transformations, called Type 1 and Type 2 respectively:

1. $$\iint_S f(x,y)\,dA = \int_{\theta_1}^{\theta_2} \int_{h_1(\theta)}^{h_2(\theta)} f\left(r\cos(\theta), r\sin(\theta)\right) r\,dr\,d\theta$$

2. $$\iint_S f(x,y)\,dA = \int_{r_1}^{r_2} \int_{g_1(r)}^{g_2(r)} f\left(r\cos(\theta), r\sin(\theta)\right) r\,d\theta\,dr$$

Remark: If $f(x,y) = 1$, then we are basically finding the area enclosed by the region S, or the volume of a solid with height 1 and base S, with appropriate units.

Example 7: Evaluate the double integral $\iint_S f(x,y)\,dA$ where $f(x,y) = \sqrt{x^2 + y^2}$ and

$$S = \left\{ (r,\theta), \frac{r}{2} \leq \theta \leq \frac{\pi}{2}, 1 \leq r \leq 2 \right\}.$$

Solution: We will begin by making appropriate sketches to identify the region S.

$with(plots):$

$$s1 := polarplot\left(1, \theta = 0..\frac{\pi}{2}, filled = true, color = khaki, transparency = 0.2\right)$$

$$PLOT(...) \tag{15.31}$$

$$s2 := polarplot\left(2, \theta = 0..\frac{\pi}{2}, filled = true, color = red, transparency = 0.4\right)$$

$$PLOT(...) \tag{15.32}$$

$$s3 := polarplot\left(2 \cdot \theta, \theta = 0..\frac{\pi}{2}, thickness = 3, color = blue\right)$$

$$PLOT(...) \tag{15.33}$$

$$s4 := textplot\left(\left[0.5, 1.2, `S`\right]\right)$$

$$PLOT(...) \tag{15.34}$$

$$display(s1, s2, s3, s4)$$

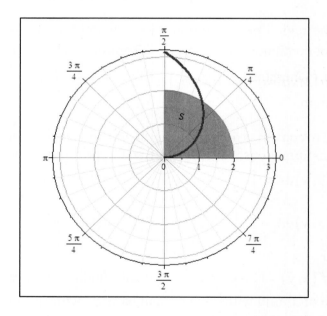

Note that the region S which is red, is the portion that is bounded by the circles $r = 1$, $r = 2$ and the spiral $r = 2\theta$. We will now use the Type 1 formula for polar coordinates to evaluate $\iint_S f(x, y)\,dA$. Let $x = r\cos(\theta)$ and $y = r\sin(\theta)$ then $f(r\cos(\theta), r\sin(\theta))r =$

$r\sqrt{(r\cos(\theta))^2 + (r\sin(\theta))^2} = r\sqrt{r^2(\cos^2(\theta) + \sin^2(\theta))} = r^2$. Using this in Type I formula and the *MultiInt* command we get the following result.

$$f := (x, y) \rightarrow \sqrt{x^2 + y^2}$$

$$(x, y) \rightarrow \sqrt{x^2 + y^2} \tag{15.35}$$

$with\left(Student\left[MultivariateCalculus\right]\right):$

$$MultiInt\left(r\cdot f\left(r\cdot\cos\left(\theta\right),r\cdot\sin\left(\theta\right)\right),\theta=\frac{r}{2}..\frac{\pi}{2},r=1..2\right)$$

$$-\frac{15}{8}+\frac{7\pi}{6}$$

(15.36)

Hence, by using the Type 1 formula, $\displaystyle\iint_{S}\sqrt{x^{2}+y^{2}}\,dA=\int_{r=1}^{r=2}\int_{\theta=r/2}^{\theta=\pi/2}r^{2}\,d\theta\,dr=-\frac{15}{8}+\frac{7\pi}{6}.$

Remark: The region S can also be rewritten as $S=\left\{(r,\theta),1\le\theta\le\frac{\pi}{2},1\le r\le 2\right\}\cup\left\{(r,\theta),\frac{1}{2}\le\theta\le 1,1\le r\le 2\theta\right\}$. Hence we can reverse the order of integration to get:

$$\iint_{S}\sqrt{x^{2}+y^{2}}\,dA=\int_{\theta=1}^{\theta=\pi/2}\int_{r=1}^{r=2}r^{2}\,dr\,d\theta+\int_{\theta=1/2}^{\theta=1}\int_{r=1}^{r=2\cdot\theta}r^{2}\,dr\,d\theta$$

We will evaluate the *MultiInt* command to evaluate the sum of the two integrals.

$with\left(Student\left[MultivariateCalculus\right]\right):$

$$MultiInt\left(r\cdot f\left(r\cdot\cos\left(\theta\right),r\cdot\sin\left(\theta\right)\right),r=1..2,\theta=1..\frac{\pi}{2}\right)+$$

$$MultiInt\left(r\cdot f\left(r\cdot\cos\left(\theta\right),r\cdot\sin\left(\theta\right)\right),r=1..2\cdot\theta,\theta=\frac{1}{2}..1\right)$$

$$-\frac{15}{8}+\frac{7\pi}{6}$$

(15.37)

We have shown that $\displaystyle\iint_{S}\sqrt{x^{2}+y^{2}}\,dA=\int_{\theta=1}^{\theta=\pi/2}\int_{r=1}^{r=2}r^{2}\,dr\,d\theta+\int_{\theta=1/2}^{\theta=1}\int_{r=1}^{r=2\theta}r^{2}\,dr\,d\theta=-\frac{15}{8}+\frac{7\pi}{6}.$

The above integral is a sum of two (Type 2) polar integrals. From a computational perspective, evaluating this integral as a Type 1 polar is simpler.

Example 8: Using polar coordinates, find the volume of the region below the surface $f(x,y)=\sqrt{x^{2}+y^{2}}$ and the xy-plane with the base defined by the inside of the loop S, where $S=\left\{(r,\theta),0\le r\le 2\theta,0\le\theta\le\frac{\pi}{2},\right\}$. The graph of the region of integration is below.

$with\left(plots\right):$

$$s1 := polarplot\left(2\theta, \theta = 0 .. \frac{\pi}{2}, thickness = 3, color = red\right)$$

$$PLOT\,(...) \tag{15.38}$$

$$s2 := implicitplot\,(x = 0, x = 0 .. \pi, y = 0 .. \pi, thickness = 3, color = red)$$

$$PLOT\,(...) \tag{15.39}$$

$$s3 := textplot\left(\left[0.5, 1.2, \text{`}S\text{`}\right]\right)$$

$$PLOT\,(...) \tag{15.40}$$

$$display\,(s1, s2, s3)$$

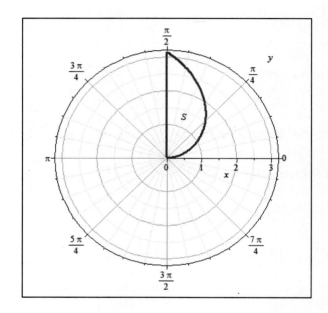

Solution: From example 7 we know that $f\left(r\cos(\theta), r\sin(\theta)\right)r = r^2$. We will use Type 2 formula for polar coordinates to evaluate $\iint_S \sqrt{x^2 + y^2}\, dA$.

$$f := (x, y) \to \sqrt{x^2 + y^2}$$

$$(x, y) \to \sqrt{x^2 + y^2} \tag{15.41}$$

$$with\left(Student\left[MultivariateCalculus\right]\right):$$

$$MultiInt\left(r \cdot f\left(r\cdot\cos(\theta), r\cdot\sin(\theta)\right), r = 0 .. 2\cdot\theta, \theta = 0 .. \frac{\pi}{2}\right)$$

$$\frac{1}{24}\pi^4 \tag{15.42}$$

Therefore, the volume is $\dfrac{1}{24}\pi^4$.

Triple Integrals

Definition: Let f be any real valued function of three variables defined for $a \le x \le b$, $c \le y \le d$ and $p \le z \le q$. Partition the cube $K = [a,b] \times [c,d] \times [p,q]$ into *mnl* rectangular cubes of equal length. Let the end points of the subintervals of $[a,b]$ be $x_0 (= a), x_1, x_2, ... x_m (= b)$, the end points of the subintervals of $[c,d]$ be $y_0 (= c), y_1, y_2, ... y_n (= d)$, and the end points of the subintervals of $[p,q]$ be $z_0 (= c), z_1, z_2, ... z_l (= d)$. Then the length of each subinterval $[x_{i-1}, x_i]$ is $\dfrac{b-a}{m}$ for $i = 1$, 2, .. m, the length of each subinterval $\left[y_{j-1}, y_j \right]$ is $\dfrac{d-c}{n}$ for $j = 1, 2, .. n$, and the length of each subinterval $[z_{k-1}, z_k]$ is $\dfrac{q-p}{l}$ for $k = 1, 2, .. l$. Let $\left(x_{ij}^*, y_{ij}^*, z_{ij}^* \right)$ be any point in the rectangle $[x_{i-1}, x_i] \times \left[y_{j-1}, y_j \right]$. The *definite integral of f on the cube K,* denoted by $\iiint\limits_K f(x, y, z) \, dV$, is defined as

$$\iiint\limits_K f(x,y,z)\,dV = \lim_{l \to \infty} \lim_{n \to \infty} \lim_{m \to \infty} \sum_{k=1}^{l} \sum_{j=1}^{n} \sum_{i=1}^{m} f\left(x_{ij}^*, y_{ij}^*, z_{ij}^* \right)(x_i - x_{i-1})(y_j - y_{j-1})(z_k - z_{k-1})$$

provided this limit exists. In this case f is said to be *integrable over the cube K*.

Definition: If f be any real valued function of three variables and G is any cube in three dimensions, then we say that $\iiint\limits_G f(x,y,z)\,dV$ is an *iterated triple integral over G*.

Fubini's Theorem for Triple Integrals: Let f be a continuous function on a rectangular solid $K = [a,b] \times [c,d] \times [p,q]$. Then $\iiint\limits_K f(x,y,z)\,dV = \int_p^q \int_c^d \int_a^b f(x,y,z)\,dx\,dy\,dz =$

$\int_c^d \int_p^q \int_a^b f(x,y,z)\,dx\,dz\,dy = \int_p^q \int_a^b \int_c^d f(x,y,z)\,dy\,dx\,dz = \int_a^b \int_p^q \int_c^d f(x,y,z)\,dy\,dz\,dx =$

$\int_c^d \int_a^b \int_p^q f(x,y,z)\,dz\,dx\,dy = \int_a^b \int_c^d \int_p^q f(x,y,z)\,dz\,dy\,dx$

Remark: The properties of the triple integral are similar to those listed on pages 490 and 491.

Example 9: Evaluate $\iiint\limits_S f(x,y,z)\,dV$, where $f(x,y,z) = xy^2 + x\sqrt{y} + ze^x$, over the box $S = [0,2] \times [1,6] \times [2,5]$ in the first octant. Verify Fubini's theorem for this example by evaluating the six possible integrals.

Solution: We will begin this problem by generating the solid cube *S* in *Maple*.

$$plot3d\left([2,5], x = 0..2, y = 2..5, axes = box, color = khaki\right)$$

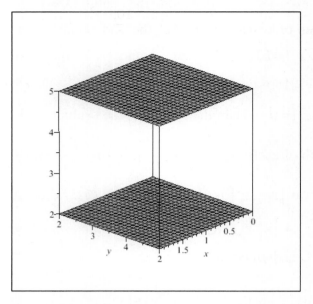

We will now enter *f* as a *Maple* function.

$$f := (x, y, z) \rightarrow x \cdot y^2 + x \cdot \sqrt{y} + z \cdot e^x$$
$$(x, y, z) \rightarrow x\, y^2 + x\, \sqrt{y} + z\, e^x \tag{15.43}$$

The six integrals in Fubini's Theorem, namely, $\int_2^5 \int_1^6 \int_0^2 f(x, y, z)\, dx\, dy\, dz$, $\int_2^6 \int_2^5 \int_1^2 f(x, y, z)\, dx\, dz\, dy$, $\int_2^5 \int_0^2 \int_1^6 f(x, y, z)\, dy\, dx\, dz$, $\int_0^2 \int_2^5 \int_1^6 f(x, y, z)\, dy\, dz\, dx$, $\int_1^6 \int_0^2 \int_2^5 f(x, y, z)\, dz\, dx\, dy$, and $\int_0^2 \int_1^6 \int_2^5 f(x, y, z)\, dz\, dy\, dx$, will be evaluated using the *MultiInt* command.

$with\left(Student\left[MultivariateCalculus\right]\right):$
$MultiInt\left(f(x, y, z), x = 0..2, y = 1..6, z = 2..5\right)$
$$\frac{747}{2} + \frac{105}{2} e^2 + 24\sqrt{6} \tag{15.44}$$
$MultiInt\left(f(x, y, z), x = 0..2, z = 2..5, y = 1..6\right)$
$$\frac{747}{2} + \frac{105}{2} e^2 + 24\sqrt{6} \tag{15.45}$$
$MultiInt\left(f(x, y, z), y = 1..6, x = 0..2, z = 2..5\right)$

$$\frac{747}{2} + \frac{105}{2}\, e^2 + 24\, \sqrt{6} \qquad\qquad \textbf{(15.46)}$$

$MultiInt\left(f\left(x,y,z\right), y = 1..6, z = 2..5, x = 0..2\right)$

$$\frac{747}{2} + \frac{105}{2}\, e^2 + 24\, \sqrt{6} \qquad\qquad \textbf{(15.47)}$$

$MultiInt\left(f\left(x,y,z\right), z = 2..5, x = 0..2, y = 1..6\right)$

$$\frac{747}{2} + \frac{105}{2}\, e^2 + 24\, \sqrt{6} \qquad\qquad \textbf{(15.48)}$$

$MultiInt\left(f\left(x,y,z\right), z = 2..5, y = 1..6, x = 0..2\right)$

$$\frac{747}{2} + \frac{105}{2}\, e^2 + 24\, \sqrt{6} \qquad\qquad \textbf{(15.49)}$$

This verifies Fubini's theorem and $\displaystyle\iiint_S f\left(x,y,z\right) dV = \frac{747}{2} + \frac{105}{2}\, e^2 + 24\, \sqrt{6}$.

Example 10: Evaluate the integral $\displaystyle\int_0^1 \int_{x^4}^x \int_0^{xy} x^2 yz\, dz\, dy\, dx$. Plot the solid region, S, that we are integrating over the xy-plane and reverse the order of integration using $dz\, dx\, dy$.

Solution: We will first plot the solid region S in using the $plot3d$ command.

$plot3d\left(\left[0, x \cdot y\right], y = x^4 .. x, x = 0..1, axes = normal, color = cyan, orientation = [50,50]\right)$

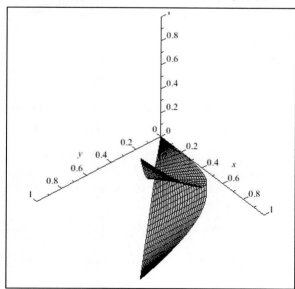

We will use the $MultiInt$ command to evaluate $\displaystyle\int_0^1 \int_{x^4}^x \int_0^{xy} x^2 yz\, dz\, dy\, dx$.

$with\left(Student\left[MultivariateCalculus\right]\right):$

$MultiInt\left(x^2\cdot y\cdot z, z=0..x\cdot y, y=x^4..x, x=0..1\right)$

$$\frac{1}{126}$$

(15.50)

Hence $\displaystyle\int_0^1\int_{x^4}^x\int_0^{xy}x^2yz\,dz\,dy\,dx=\frac{1}{126}$. In order to see the effect of the projection of

$S=\left\{(x,y,z):0\le z\le xy, x^4\le y\le x, 0\le x\le 1\right\}$ onto the xy-plane, we will plot the region in the xy-plane.

$plot3d\left(x\cdot y, y=x^4..x, x=0..1, axes=normal, color=cyan, orientation=[0,0]\right)$

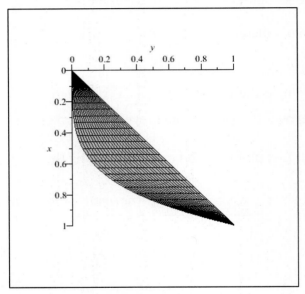

Since we are changing the order of integration from $dz\,dy\,dx$ to $dz\,dx\,dy$, we only need to adjust the limits in the xy-plane. The region in the xy-plane is

$$\left\{(x,y):0\le x\le 1, x^4\le y\le x\right\}=\left\{(x,y):0\le y\le 1, y\le x\le \sqrt[4]{y}\right\}$$

Hence we can rewrite the triple integral to be evaluate as $\displaystyle\int_0^1\int_y^{y^{1/4}}\int_0^{xy}x^2yz\,dz\,dx\,dy$. We

will use the *MultiInt* command to evaluate $\displaystyle\int_0^1\int_y^{y^{1/4}}\int_0^{xy}x^2yz\,dz\,dx\,dy$. We do not need to

activate the $with\left(Student\left[MultivariateCalculus\right]\right)$ at this stage as it was done before evaluating the iterated integral in (15.50).

$MultiInt\left(x^2\cdot y\cdot z, z=0..x\cdot y, x=y..y^{\frac{1}{4}}, y=0..1\right)$

$$\frac{1}{126}$$ **(15.51)**

Thus $\int_0^1 \int_y^{y^{1/4}} \int_0^{xy} x^2 yz\, dz\, dx\, dy = \frac{1}{126}$. This result is expected since Fubini's theorem is valid (why?).

Example 11: Evaluate the integral $\iiint_S xe^y z\, dV$ over the solid region, S, enclosed by the first octant, the surface $z = 5 - x^2$, and the planes $z = 0$, $y = 2x$, and $y = 0$. Plot the solid region S.

Solution: First we will plot the region, S, using *Maple*.

$$plot3d\left(\left[0, 5-x^2\right], y = 0..2\cdot x, x = 0..\sqrt{5}, axes = normal, color = cyan,\right.$$
$$\left. orientation = [30,50]\right)$$

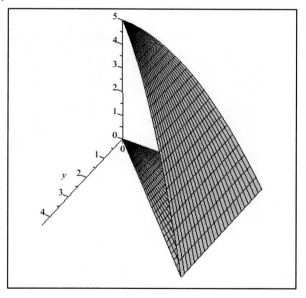

We will now evaluate the integral $\iiint_S xe^y z\, dV = \int_0^{\sqrt{5}} \int_0^{2x} \int_0^{5-x^2} xe^y z\, dz\, dy\, dx$ using *Maple* commands.

$$with\left(Student\left[MultivariateCalculus\right]\right):$$
$$MultiInt\left(x\cdot e^y \cdot z, z = 0..5-x^2, y = 0..2\cdot x, x = 0..\sqrt{5}\right)$$
$$-\frac{395}{48} + \frac{35}{8}\sqrt{5}\, e^{2\sqrt{5}} - \frac{135}{16}e^{2\sqrt{5}}$$ **(15.52)**

We have shown that $\iiint\limits_S xe^y z\, dV = -\dfrac{395}{48} + \dfrac{35}{8}\sqrt{5}\, e^{2\sqrt{5}} - \dfrac{135}{16} e^{2\sqrt{5}}$.

Cylindrical Coordinates

Definition: A point in the Cartesian coordinate system, (x, y, z), gets transformed into cylindrical coordinate system, (r, θ, z), where $r \geq 0$, and (r, θ) are the polar coordinates of (x, y) and z stays the same.

Definition: $\iiint\limits_S f(x, y, z)\, dV = \displaystyle\int_{\theta_1}^{\theta_2} \int_{r_1(\theta)}^{r_2(\theta)} \int_{h_1(r,\theta)}^{h_2(r,\theta)} f\left(r\cos(\theta), r\sin(\theta), z\right) r\, dz\, dr\, d\theta$, is one of

six ways to convert the *triple integral from Cartesian to cylindrical coordinates*. The other forms of the triple integrals can be analogously generated.

Remark: *Maple* has a built in option $coords = cylindrical$ in the $plot3d$ command that allows us to plot graphs in cylindrical coordinates.

Example 12: Find the volume of the solid under the surface $z = x^2 y^2$, above the *xy*-plane, in the first octant, and within the cylinder $x^2 + y^2 - 2y = 0$, using cylindrical coordinates. Graph the solid.

Solution: We will first convert the surface $z = x^2 y^2$ to cylindrical coordinates.

$f := (x, y) \rightarrow x^2 \cdot y^2$

$$(x, y) \rightarrow x^2\, y^2 \tag{15.53}$$

$f\left(r \cdot \cos(\theta), r \cdot \sin(\theta)\right)$

$$r^4 \cos(\theta)^2 \sin(\theta)^2 \tag{15.54}$$

Since z is between the *xy*-plane and $x^2 y^2$, we see that $0 \leq z \leq r^4 \cos(\theta)^2 \sin(\theta)^2$.

Now, we will define the cylinder $h(x, y) = x^2 + y^2 - 2y$ as a *Maple* function and express it in cylindrical coordinates.

$h := (x, y) \rightarrow x^2 + y^2 - 2 \cdot y$

$$(x, y) \rightarrow x^2 + y^2 - 2\, y \tag{15.55}$$

$h1 := h\left(r \cdot \cos(\theta), r \cdot \sin(\theta)\right)$

$$r^2 \cos(\theta)^2 + r^2 \sin(\theta)^2 - 2r\sin(\theta) \tag{15.56}$$

We will now solve this for r in terms of θ and simplify it.

$s := solve(h1, r)$

$$0, \frac{2\sin(\theta)}{\cos(\theta)^2 + \sin(\theta)^2} \qquad\qquad \textbf{(15.57)}$$

$simplify(s[2])$

$$2\sin(\theta) \qquad\qquad \textbf{(15.58)}$$

Therefore, it follows that $0 \le r \le 2\sin(\theta)$. Since the solid is in the first octant we also have $0 \le \theta \le \dfrac{\pi}{2}$. We will now plot the solid region using *Maple* commands by including the option $coords = cylindrical$ in the $plot3d$ command to plot the solid region in cylindrical coordinates.

$$plot3d\left(\left\{\left[r, \theta, r^4\left(\cos(\theta)\right)^2 \cdot \left(\sin(\theta)\right)^2\right], [r, \theta, 0]\right\}, r = 0..2\cdot\sin(\theta), \theta = 0..\frac{\pi}{2}, coords =\right.$$

$$\left. cylindrical, axes = box, orientation = [10, 65]\right)$$

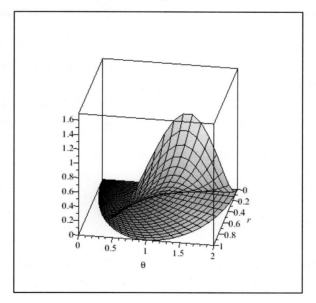

The volume of the solid is found from $\displaystyle\iiint\limits_{S} 1\, dV = \int_0^{\frac{\pi}{2}} \int_0^{2\sin(\theta)} \int_0^{r^4\cos^2(\theta)\sin^2(\theta)} 1\cdot r\, dz\, dr\, d\theta$. We will now evaluate the iterated integral above to get the volume of the solid using the *MultiInt* command.

$with(Student[MultivariateCalculus]):$

$$MultiInt\left(1, z = 0 .. f\left(r \cdot \cos(\theta), r \cdot \sin(\theta)\right), r = 0 .. s[2], \theta = 0 .. \frac{\pi}{2}\right)$$

$$\frac{512}{1575}$$

(15.59)

Therefore, the volume of the solid is $\dfrac{512}{1575}$.

Spherical Coordinates

Definition: A point in the Cartesian coordinate system, (x, y, z), gets transformed into *cylindrical coordinate system,* (ρ, θ, ϕ), where $x = \rho \sin(\phi)\cos(\theta)$, $y = \rho \sin(\phi)\sin(\theta)$, and $z = \rho\cos(\phi)$, for $\rho \geq 0$, $0 \leq \phi \leq \pi$, and $0 \leq \theta \leq 2\pi$.

Definition: One of six ways to convert the triple integral $\iiint\limits_{S} f(x, y, z)\, dV$ from *Cartesian coordinates to spherical coordinates is*

$$\iiint\limits_{S} f(x, y, z)\, dV = \int_{\theta_1}^{\theta_2} \int_{\rho_1(\theta)}^{\rho_2(\theta)} \int_{h_1(\rho,\theta)}^{h_2(\rho,\theta)} f\left(\rho\sin(\phi)\cos(\theta), \rho\sin(\phi)\sin(\theta), \rho\cos(\phi)\right)\rho^2\sin(\phi)\, d\phi\, d\rho\, d\theta$$

The other forms of the triple integrals can be analogously generated.

Remark: As in the case of cylindrical coordinates we can use the built in option *coords = spherical* in the *plot3d* command to plot graphs in spherical coordinates.

Example 13: Let S be the solid ball with radius 2 centered about the origin $(x^2 + y^2 + z^2 \leq 4)$. Plot the solid region S and evaluate the triple integral $\iiint\limits_{S} \dfrac{1}{\sqrt{x^2 + y^2 + (z-3)^2}}\, dV$ using spherical coordinates.

Solution: We will first convert the integrand $\dfrac{1}{\sqrt{x^2 + y^2 + (z-3)^2}}$ and $x^2 + y^2 + z^2$ to spherical coordinates.

$$f := (x, y, z) \rightarrow \frac{1}{\sqrt{x^2 + y^2 + (z-3)^2}}$$

$$(x, y, z) \rightarrow \frac{1}{\sqrt{x^2 + y^2 + (z-3)^2}}$$

(15.60)

$$u := \rho\sin(\phi)\cos(\theta)$$

$$\rho\sin(\phi)\cos(\theta)$$

(15.61)

$v := \rho \sin(\phi) \sin(\theta)$

$$\rho \sin(\phi) \sin(\theta) \qquad\qquad\qquad \textbf{(15.62)}$$

$w := \rho \cos(\phi)$

$$\rho \cos(\phi) \qquad\qquad\qquad \textbf{(15.63)}$$

$simplify\left(f\left(u, v, w\right)\right)$

$$\frac{1}{\sqrt{\rho^2 - 6 \, \rho \cos(\phi) + 9}} \qquad\qquad\qquad \textbf{(15.64)}$$

Since we want to convert S into spherical coordinates we will first define $x^2 + y^2 + z^2$ as a *Maple* function h and express it in cylindrical coordinates.

$h := (x, y, z) \rightarrow x^2 + y^2 + z^2$

$$(x, y, z) \rightarrow x^2 + y^2 + z^2 \qquad\qquad\qquad \textbf{(15.65)}$$

$simplify\left(h\left(u, v, w\right)\right)$

$$\rho^2 \qquad\qquad\qquad \textbf{(15.66)}$$

Therefore, it follows that $S := \left\{ (\rho, \theta, \phi) : 0 \leq \phi \leq \pi, 0 \leq \rho \leq 2, 0 \leq \theta \leq 2\pi \right\}$. Furthermore, our

integral $\displaystyle\iiint_S \frac{1}{\sqrt{x^2 + y^2 + (z - 3)^2}} \, dV$ now becomes $\displaystyle\int_0^{2\pi} \int_0^2 \int_0^\pi \frac{\rho^2 \sin(\phi)}{\sqrt{\rho^2 - 6\rho\cos(\phi) + 9}} \, d\phi \, d\rho \, d\theta$,

using spherical coordinates. We will now plot the solid region using *Maple* commands by including the option *coords = spherical* in the *plot3d* command.

$plot3d\left(2, \theta = 0 .. 2 \cdot \pi, \phi = 0..\pi, coords = spherical, axes = box\right)$

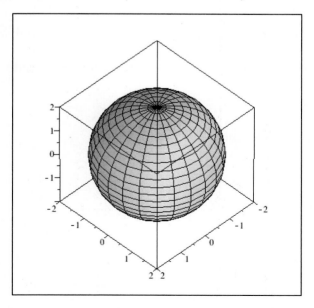

Chapter 15 Multiple Integrals

We will now evaluate the iterated integral above using the *MultiInt* command.

$with\left(Student\left[MultivariateCalculus\right]\right):$

$$MultiInt\left(\frac{\rho^2 \cdot \sin\left(\phi\right)}{\sqrt{\rho^2 - 6\cdot\rho\cdot\cos\left(\phi\right)+9}}, \phi=0..\pi, \rho=0..2, \theta=0..2\cdot\pi\right)$$

$$\frac{32}{9}\pi$$

(15.67)

Hence, $\displaystyle\iiint_{S}\frac{1}{\sqrt{x^2+y^2+\left(z-3\right)^2}}\,dV = \int_0^{2\pi}\int_0^2\int_0^{\pi}\frac{\rho^2\sin\left(\phi\right)}{\sqrt{\rho^2-6\rho\cos\left(\phi\right)+9}}\,d\phi\,d\rho\,d\theta = \frac{32}{9}\pi.$

Center of Mass

Definition: For a lamina (closed flat surface without any holes) described by a region S and density $\delta\left(x,y\right)$ in two dimensions, its *center of mass* $\left(\overline{x},\overline{y}\right)$ is calculated by using the formulas $\overline{x} = \dfrac{\displaystyle\iint_{S}x\delta\left(x,y\right)dA}{\displaystyle\iint_{S}\delta\left(x,y\right)dA}$, and $\overline{y} = \dfrac{\displaystyle\iint_{S}y\delta\left(x,y\right)dA}{\displaystyle\iint_{S}\delta\left(x,y\right)dA}$.

Definition: For a closed solid described by a region S and density $\delta\left(x,y,z\right)$ in three dimensions, its *center of mass* $\left(\overline{x},\overline{y},\overline{z}\right)$ is calculated by using the formulas $\overline{x} = \dfrac{\displaystyle\iiint_{S}x\,\delta\left(x,y,z\right)dV}{\displaystyle\iiint_{S}\delta\left(x,y,z\right)dV}$, $\overline{y} = \dfrac{\displaystyle\iiint_{S}y\,\delta\left(x,y,z\right)dV}{\displaystyle\iiint_{S}\delta\left(x,y,z\right)dV}$, and $\overline{z} = \dfrac{\displaystyle\iiint_{S}z\,\delta\left(x,y,z\right)dV}{\displaystyle\iiint_{S}\delta\left(x,y,z\right)dV}$.

Example 14: Find the center of mass of the lamina L with uniform density $\delta\left(x,y\right)=K$, where K is any constant, that is formed by the region enclosed by the curves $y=\dfrac{1}{x^2}$ and $y=2-x$. Plot this lamina.

Solution: We will first plot the lamina in *Maple*.

$with\left(plots\right):$

$$s1 := implicitplot\left(\left\{y=2-x, y=\frac{1}{x^2}\right\}, x=0.4..2, y=0..1.6\right)$$

$$PLOT\left(...\right)$$

(15.68)

$s2 := textplot\left(\left[1.3, 0.65, \grave{}L\grave{}\right]\right)$

$$PLOT\left(...\right) \qquad\qquad \textbf{(15.69)}$$

$display\left(s1, s2\right)$

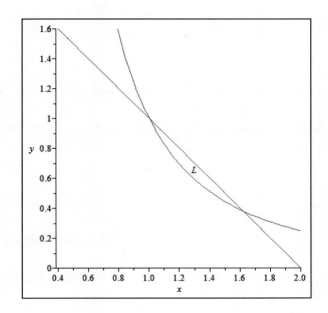

Before finding the center of mass we will need to find the points of intersections of the two curves.

$$solve\left(2 - x = \frac{1}{x^2}, x\right)$$

$$1, \frac{1}{2} - \frac{1}{2}\sqrt{5}, \frac{1}{2} + \frac{1}{2}\sqrt{5} \qquad\qquad \textbf{(15.70)}$$

From the graph and (15.70) the lamina is $L = \left\{(x, y): \dfrac{1}{x^2} \le y \le 2 - x, 1 \le x \le \dfrac{1}{2}\sqrt{5} + \dfrac{1}{2}\right\}$. In order to use the built-in command *CenterOfMass* we will first activate the $with\left(Student\left[MultivariateCalculus\right]\right)$ package.

$with\left(Student\left[MultivariateCalculus\right]\right):$

$$CenterOfMass\left(K, y = \frac{1}{x^2}..2 - x, x = 1..\frac{1}{2}\sqrt{5} + \frac{1}{2}\right)$$

$$-\frac{2\left(\dfrac{1}{6}K + \dfrac{1}{6}K\sqrt{5} + K\ln(2) - K\ln\left(\sqrt{5} + 1\right)\right)\left(\sqrt{5} + 1\right)}{K\left(3\sqrt{5} - 7\right)}, \frac{1}{3}\frac{\left(\sqrt{5} - 3\right)\left(\sqrt{5} + 1\right)}{\left(2 + \sqrt{5}\right)\left(3\sqrt{5} - 7\right)} \qquad \textbf{(15.71)}$$

Right click on (15.71), select *Simplify* and then select *Symbolic* (this option is needed to simplify both the answers in (15.71)) to obtain the following.

$$\xrightarrow{\text{simplify symbolic}}$$

$$-\frac{1}{3}\frac{\left(1+\sqrt{5}+6\ln(2)-6\ln\left(\sqrt{5}+1\right)\right)\left(\sqrt{5}+1\right)}{3\sqrt{5}-7}, \frac{2}{3} \qquad (15.72)$$

To get a floating-point approximation to our answer we will use $y=\dfrac{1.0}{x^2}..2-x$ instead of $y=\dfrac{1}{x^2}..2-x$. Note that when we right click on (15.72) the option of Approximate does not appear.

$$CenterOfMass\left(K, y=\frac{1.0}{x^2}..2-x,\ x=1..\frac{1}{2}\cdot\sqrt{5}+\frac{1}{2}\right)$$

$$1.289406102, 0.6666666668 \qquad (15.73)$$

Hence the center of mass of the lamina is $(\overline{x},\overline{y})=(1.289406102, 0.6666666668)$.

Remark: Note that the default option for the built-in command *CenterOfMass* is value. We may use the option *output = integral* to obtain the inert integral used to calculate the center of mass.

$$with\left(Student\left[MultivariateCalculus\right]\right):$$

$$CenterOfMass\left(K, y=\frac{1}{x^2}..2-x,\ x=1..\frac{1}{2}\sqrt{5}+\frac{1}{2}, output=integral\right)$$

$$\frac{\int_1^{\frac{1}{2}+\frac{1}{2}\sqrt{5}}\int_{\frac{1}{x^2}}^{2-x}xK\,dy\,dx}{\int_1^{\frac{1}{2}+\frac{1}{2}\sqrt{5}}\int_{\frac{1}{x^2}}^{2-x}K\,dy\,dx},\ \frac{\int_1^{\frac{1}{2}+\frac{1}{2}\sqrt{5}}\int_{\frac{1}{x^2}}^{2-x}yK\,dy\,dx}{\int_1^{\frac{1}{2}+\frac{1}{2}\sqrt{5}}\int_{\frac{1}{x^2}}^{2-x}K\,dy\,dx} \qquad (15.74)$$

The option *output = plot* is used to plot the region, density function and the center of mass. In this case we need to give a value of K for the plot. The plot below is given when $K=2$.

$$CenterOfMass\left(2, y=\frac{1}{x^2}..2-x,\ x=1..\frac{1}{2}\sqrt{5}+\frac{1}{2}, output=plot, functionoptions=\right.$$

$$\left[thickness=2\right]\big)$$

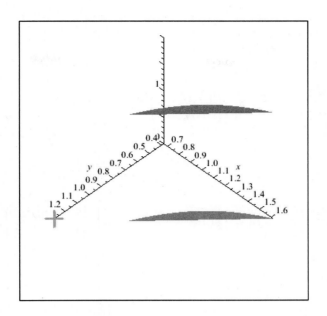

Example 15: Find the center of mass of the solid L with uniform density $\delta(x,y,z) = K$, where K is any constant, bounded by the cylinders $z^2 + x^2 = 1$, $z^2 + y^2 = 1$, and the first octant. Plot this solid using the *plot3d* command.

Solution: The solid can be viewed in the set described by $\left\{(x,y,z): 0 \le x \le \sqrt{1-z^2}, 0 \le z \le \sqrt{1-y^2}, y = 0..1\right\}$. We will plot using the *plot3d* command.

$$plot3d\left(\left[0, \sqrt{1-z^2}\right], z = 0..\sqrt{1-y^2}, y = 0..1, axes = box\right)$$

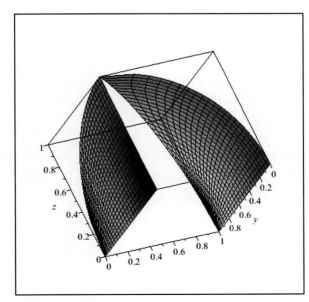

We will use the *CenterOfMass* command.

$with\left(Student\left[MultivariateCalculus\right]\right):$

$CenterOfMass\left(K, x = 0 .. \sqrt{1-z^2}, z = 0 .. \sqrt{1-y^2}, y = 0 .. 1\right)$

$$\frac{9}{64}\pi, \frac{9}{64}\pi, \frac{3}{8}$$

(15.75)

Hence the center of mass of the solid is $\left(\overline{x}, \overline{y}, \overline{z}\right) = \left(\frac{9}{64}\pi, \frac{9}{64}\pi, \frac{3}{8}\right)$.

Jacobian Function

Definition: Let $H(s,t) = H\left(x(s,t), y(s,t)\right)$, then the *Jacobian* of H, denoted by

$J\left(H(s,t)\right)$, is the determinant of the 2×2 matrix $J\left(H(s,t)\right) = \begin{vmatrix} \dfrac{\partial x}{\partial s} & \dfrac{\partial x}{\partial t} \\ \dfrac{\partial y}{\partial s} & \dfrac{\partial y}{\partial t} \end{vmatrix}$.

Definition: Let $H(s,t,r) = H\left(x(s,t,r), y(s,t,r), z(s,t,r)\right)$, then the *Jacobian* of H, denoted by $J\left(H(s,t,r)\right)$, is the determinant of the 3×3 matrix

$$J\left(H(s,t,r)\right) = \begin{vmatrix} \dfrac{\partial x}{\partial s} & \dfrac{\partial x}{\partial t} & \dfrac{\partial x}{\partial r} \\ \dfrac{\partial y}{\partial s} & \dfrac{\partial y}{\partial t} & \dfrac{\partial y}{\partial r} \\ \dfrac{\partial z}{\partial s} & \dfrac{\partial z}{\partial t} & \dfrac{\partial z}{\partial r} \end{vmatrix}.$$

Remark: The Jacobian function is used to make multidimensional u-substitutions for multi-integrals. The definition of the Jacobian function can be extended to higher dimensions.

Note: Let $H(s,t)$ be a mapping of a closed region S_1 with continuous partial derivatives and maps the interior of S_1 in a one to one manner into a region S_2, then we can transform the double integral by the change of variable using the Jacobian as shown below:

$$\iint_{S_2} f(x,y)\,dxdy = \iint_{S_1} f\left(x(s,t), y(s,t)\right)\left|J\left(H(s,t)\right)\right|\,dsdt$$

Similarly, we can extend the formula above to transform a three dimensional integral:

$$\iiint_{S_2} f(x,y,z) \, dx \, dy \, dz = \iiint_{S_1} f\left(x(s,t,r), y(s,t,r), z(s,t,r)\right) \left|J\left(H(s,t,r)\right)\right| ds \, dt \, dr$$

Example 16: Let $H(r,\theta) = \left(x(r,\theta), y(r,\theta)\right) = \left(r\cos(\theta), r\sin(\theta)\right)$. Find $J\left(H(r,\theta)\right)$.

Solution: We will use the *Jacobian* command in *Maple*. This command gives the $n \times n$ matrix and the determinant of this matrix. We will need to activate the $with\left(Student\left[VectorCalculus\right])\right])$ package.

$with\left(Student\left[VectorCalculus\right])\right])$:

$Jacobian\left(\left[r \cdot \cos(\theta), r \cdot \sin(\theta)\right], \,'determinant'\right)$

$$\begin{bmatrix} \cos(\theta) & -r\sin(\theta) \\ \sin(\theta) & r\cos(\theta) \end{bmatrix}, \cos(\theta)^2 r + r\sin(\theta)^2 \qquad \textbf{(15.76)}$$

Right click on (15.75), select *Simplify* and then select *Trig* to obtain the following.

simplify trig

$=$

$$\begin{bmatrix} \cos(\theta) & -r\sin(\theta) \\ \sin(\theta) & r\cos(\theta) \end{bmatrix}, r \qquad \textbf{(15.77)}$$

Hence the Jacobian is r. In fact this Jacobian is precisely the Jacobian used to convert from Cartesian to polar coordinates in double integrals.

Example 17: Use an appropriate change of variables to compute $\iint_{S} \left(x^2 y^2 + 5\right) dA$, where

S is the region bounded by $xy = 0.5$, $xy = 1$, $xy^2 = 0.5$ and $xy^2 = 1$.

Solution: We will first plot the region bounded by the four curves:

$with\left(plots\right)$:

$s1 := implicitplot\left(\left[x \cdot y = 0.5, x \cdot y = 1, x \cdot y^2 = 0.5, x \cdot y^2 = 1\right], x = 0.2 .. 2.5, y = 0.47 .. 2.5,\right.$

$\quad color = \left[red, blue, green, magenta\right], thickness = 3\right)$

$$PLOT\left(...\right) \qquad \textbf{(15.78)}$$

$s2 := textplot\left(\left[0.8, 0.96, `S`\right]\right)$

$$PLOT\left(...\right) \qquad \textbf{(15.79)}$$

$display\left(s1, s2\right)$

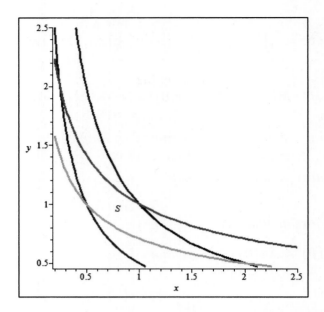

We will use the transformation $s = xy$ and $t = xy^2$ to evaluate the double integral over the region S. We will now use *Maple* to solve for x and y in terms of s and t.

$$solve\left(\left\{s = x \cdot y, t = x \cdot y^2\right\}, [x, y]\right)$$

$$\left[\left[x = \frac{s^2}{t}, y = \frac{t}{s}\right]\right] \tag{15.80}$$

We will use the Jacobian transformation $H(s,t): S_1 \to S$ defined by $H(s,t) = \left(\dfrac{s^2}{t}, \dfrac{t}{s}\right)$

where $S_1 = \left\{(s,t): 0.5 \le s \le 1, 0.5 \le t \le 1\right\}$.

The Jacobian will be found using *Maple*.

$$with\left(Student[VectorCalculus])\right]):$$

$$Jacobian\left(\left[\frac{s^2}{t}, \frac{t}{s}\right], 'determinant'\right)$$

$$\left[\begin{array}{cc} \dfrac{2s}{t} & -\dfrac{s^2}{t^2} \\[3mm] -\dfrac{t}{s^2} & \dfrac{1}{s} \end{array}\right], \frac{1}{t} \tag{15.81}$$

Using the transformation $\iint\limits_S (x^2 y^2 - 5)\, dA = \int_{0.5}^{1} \int_{0.5}^{1} \left(\left(\dfrac{s^2}{t} \right)^2 \left(\dfrac{t}{s} \right)^2 + 5 \right) \left| \dfrac{1}{t} \right| ds\, dt$ we will

evaluate the integral $\iint\limits_S (x^2 y^2 - 5)\, dA$ using the *MultiInt* command.

$with\left(Student \left[MultivariateCalculus \right] \right):$

$MultiInt\left(\left(\left(\dfrac{s}{t} \right)^2 \cdot \left(\dfrac{t}{s} \right)^2 + 5 \right) \cdot \left(\dfrac{1}{t} \right), s = \dfrac{1}{2}\, .. 1, t = \dfrac{1}{2}\, .. 1 \right)$

$$\dfrac{67}{24} \ln(2)$$

(15.82)

Hence $\iint\limits_S (x^2 y^2 + 5)\, dA = \dfrac{67}{24} \ln(2)$.

Exercises

1. Compute the midpoint Riemann approximation sum to estimate the double
    integral $\int_{-5}^{10}\int_{0}^{3}\left(x^2y+e^{xy}\right)dx\,dy$. Sketch using *Maple* commands.
    (a) Use $m=5$ and $n=5$
    (b) Use $m=100$ and $n=150$
    (c) Use $m\to\infty$ and $n\to\infty$.

2. Compute the midpoint Riemann approximation sum to estimate the double
    integral $\int_{0}^{1}\int_{0}^{10}e^{y^2}dx\,dy$. Sketch using *Maple* commands.
    (a) Use $m=10$ and $n=15$
    (b) Use $m=100$ and $n=150$
    (c) Use $m\to\infty$ and $n\to\infty$.

3.* Compute the Riemann approximation sum to estimate $\int_{0}^{5}\int_{0}^{5}\left(xye^{x^2y}\right)dx\,dy$, by
    dividing the region into twenty five unit squares and choosing your sample points
    to be the lower corner of each square. You will need to modify the midpoint
    Riemann approximation sum that is given in the text. Make appropriate sketches.

4. Evaluate the appropriate iterated integral $\iint_{S}\left(xy-y^5\right)dA$ over the regions below:

    Sketch of all the regions using *Maple* commands.
    (a) $S=\left\{(x,y):-5\le x\le 9,\,-3\le y\le 12\right\}$
    (b) S is the region enclosed between the curves $y=x^2$ and $y=x+1$.
    (c) S is the triangular region enclosed by the lines joining the three points
         $(0,0)$, $(1,3)$, and $(-1,5)$.

5. Find the volume of the solid that lies below the surface $f(x,y)=3x+y+y^5$ and
    above the region in the *xy*-plane that is enclosed between the curves $y=x^2$ and
    $y=1-x^2$. Model this solution as in example 3 of this chapter.

6. Find the volume of the solid that is bounded by the coordinate planes and the
    plane $55x+33y-z=2011$. Model this solution as in example 3 of this chapter.

7. Find the volume of the solid that lies below the surface $f(x,y)=xy$, above the
    region in the *xy*-plane that is enclosed between the curves $y=-x+5$ and $y=x^2$.
    Model this solution as in example 3 of this chapter.

8. Let $f(x,y)=e^{-(x^2+y^2)}$. Find the volume of the solid between the surface of $f(x,y)$ and the xy-plane over the circular region: $x^2+y^2=9$ is $\pi(1-e^{-9})$ by transforming to polar coordinates.

9. Find the volume of the solid that lies below the surface $f(x,y)=e^{y^2}$, above the xy-plane and over the region $S=\{(x,y):-1\le x\le y-1, 0\le y\le 2\}$. Use *Maple* commands to sketch of the region of integration. Reverse the order of integration and verify that your solutions are equivalent. Can you do both iterated integrals by hand?

10. Evaluate the integrals by changing to polar coordinates:

    (a) $\iint\limits_{S}(xy)^2 dA$, where S is the region enclosed by the disk $x^2+y^2\le 2011$.

    (b) $\iint\limits_{S}\sqrt[5]{(x^2+y^2)}\, dA$, where S is the region between the disks $x^2+y^2\le 9$ and $x^2+y^2\le 25$.

    (c) $\iint\limits_{S}\sin(x^2+y^2)dA$, where S is the region enclosed by the disk $x^2+y^2\le 9$.

    (d) $\iint\limits_{S}\dfrac{1}{\sqrt{2011-x^2-y^2}}dA$, where S is the region between the disk $x^2+y^2\le 2011$ and the first quadrant.

11. Find the volume of the solid above the cone $z=\sqrt{2x^2+2y^2}$ and below the sphere $x^2+y^2+z^2=9$. Make a *Maple* sketch to illustrate the solid and use polar coordinates to do this problem.

12. Find the volume of the solid below the cone $z=\sqrt{1-x^2-y^2}$ and above the sphere $x^2+y^2+z^2=1$. Make a *Maple* sketch to illustrate the solid and use polar coordinates to do this problem.

13. Evaluate the integral $\iint\limits_{S}\sqrt{(x^2+y^2)}\, dA$, where S is the region inside the circle $x^2+y^2=16$ and outside of the smaller circle $x^2+(y-2)^2=4$. See the picture below.

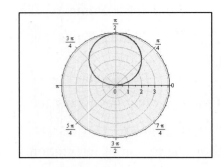

14 - 16. Evaluate the following integrals below by first converting to polar coordinates.

14. $\displaystyle \int_{0}^{0.2}\int_{0}^{\sqrt{2y-y^2}}\left(x^2+y^2\right)^2 dx\,dy$

15. $\displaystyle \int_{-5}^{5}\int_{-\sqrt{25-x^2}}^{\sqrt{25-x^2}}\left(x^2+y^2\right)dy\,dx$

16. $\displaystyle \int_{0}^{5}\int_{0}^{\sqrt{25-x^2}}\tan^{-1}\left(\frac{y}{x}\right)dy\,dx$

17. For example 11 in the text find the iterated integral along the remaining five directions and evaluate them with *Maple*.

18. Evaluate the iterated integrals:

(a) $\displaystyle \int_{2}^{5}\int_{1}^{11x}\int_{3y}^{2x^3+5x}1\,dz\,dy\,dx$

(b) $\displaystyle \int_{0}^{3}\int_{0}^{3}\int_{0}^{3}\frac{1}{\sqrt{x+y+z+2011}}\,dz\,dy\,dx$

(c) $\displaystyle \iiint_{S}\sin\left(x^2+y^2\right)dV$, where $S=\left\{(x,y,z):x^2+y^2\le 9,\,0\le z\le 5\right\}$.

19. Evaluate the integral $\displaystyle \iiint_{S}\frac{1}{\left(x+y+z+2011\right)^5}dV$, where S is the solid enclosed by the coordinate planes and the plane: $x+2y+3z=2011$.

20. Find the volume of the solid under the surface $x^2+y^2+z^2=36$, above the xy-plane and within the cylinder $x^2+y^2-6x=0$, using cylindrical coordinates. Make a sketch of the solid with *Maple*.

21. Find the volume of the solid region between the surfaces $z = 2(x^2 + y^2)$ and $z = 27 - x^2 - y^2$. Carefully sketch the solid with *Maple*.

22. Find the volume of the solid between the cone $z = \sqrt{x^2 + y^2}$ and the sphere $x^2 + y^2 + z^2 = 9$ by using spherical coordinates. Make a sketch of the solid with *Maple*.

23. Evaluate $\iiint\limits_{S} \dfrac{1}{\sqrt{x^2 + y^2 + z^2}}\, dV$ by spherical coordinates, where S is the unit sphere centered at the origin.

24. Redo example 13 of this chapter by using cylindrical coordinates.

25. Evaluate the integral $\displaystyle\int_0^{2\pi} \int_0^2 \int_0^{\pi} \dfrac{\rho^2 \sin(\phi)}{\sqrt{\rho^2 - 6\rho\cos(\phi) + 9}}\, d\phi\, d\rho\, d\theta$, by changing the order of integration to $d\rho\, d\phi\, d\theta$.

26. Find the center of mass of the flat region with constant density and bounded by the curves $y = x^2$ and $y = \sqrt{x}$.

27. Find the center of mass of the flat region with constant density and bounded by the curves $y = x^2$ and $y = 9 - x^2$.

28. Find the center of mass of a solid with constant density and bounded by the surface $z = \sqrt{9 - x^2 - y^2}$ and the plane $x + y + z = 1$.

29. Find the center of mass of a solid with constant density and bounded by the surfaces $z = 1 - x^2 - y^2$ and the plane $z = \sqrt{x^2 + y^2}$.

30. Find the center of mass of a solid with constant density and bounded by the surfaces $z = \sqrt{9 - x^2 - y^2}$ and the plane $z = \sqrt{x^2 + y^2}$.

31. Find the Jacobian of the following:

     (a) $H(s,t) = H(x(s,t), y(s,t))$, where $x = se^t$ and $y = e^t$.

     (b) $H(\rho,\theta,\phi) \quad = \quad H(x(\rho,\theta,\phi), y(\rho,\theta,\phi), z(\rho,\theta,\phi))$, where $x = \rho\sin(\phi)\cos(\theta)$, $y = \rho\sin(\phi)\sin(\theta)$ and $z = \rho\cos(\phi)$.

32. Use an appropriate change of variables to compute $\iint_S (x^3 + y^2) \, dA$, where S is the region bounded by $y = -2x$, $y = -2x + 3$, $y = x$, and $y = x + 5$.

33. Use an appropriate change of variables to compute $\iint_S \cos\left(\dfrac{y-x}{y+x}\right) dA$, where S is the region enclosed by the trapezoid formed by the line segments joining the points $\left(0, \dfrac{1}{2}\right)$, $(0,1)$, $(1,0)$, and $\left(\dfrac{1}{2}, 0\right)$.

34. Use an appropriate change of variables to compute $\iint_S \cos(x+y)\sin(x-y) \, dA$, where $S = \{(x,y) : |x| + |y| \le 2\}$.

35. Below is a list of some *Maple* commands used in this chapter. Describe the significance of each command. Can you find examples where each command is used?

    (a) $CenterOfMass\left(\rho, z = f_1(x,y) .. f_2(x,y), y = h_1(x) .. h_2(x), x = a .. b\right)$

    (b) $int\left(f(x,y), [x, y]\right)$

    (c) $Jacobian\left(\left[x(s,t,r), y(s,t,r), z(s,t,r)\right], 'determinant'\right)$

    (d) $MultiInt\left(f(x,y), x = a .. b, y = c .. d\right)$

    (e) $with\left(Student\left[MultivariateCalculus\right]\right)$

    (f) $with\left(Student\left[VectorCalculus\right]\right)\right)$

36. Can you list new *Maple* commands that were used in this chapter but were not listed in exercise 35?

Generation of vector fields, line integrals, Green's Theorem, Stokes Theorem, surface integrals and the Divergence Theorem will be illustrated using *Maple*.

Line Integrals and Vector Fields

Definition: A *vector field in* $\mathbb{R}^2$, **F**, is a function whose domain is a subset of $\mathbb{R}^2$ and range is a set of two-dimensional vectors. Such a vector is denoted by $\mathbf{F}(x,y) = P(x,y)\,\mathbf{i} + Q(x,y)\,\mathbf{j} = P\,\mathbf{i} + Q\,\mathbf{j}$.

Definition: A *vector field in* $\mathbb{R}^3$, **F**, is a function whose domain is a subset of $\mathbb{R}^3$ and range is a set of three-dimensional vectors. Such a vector is denoted by $\mathbf{F}(x,y,z) = P(x,y,z)\,\mathbf{i} + Q(x,y,z)\,\mathbf{j} + R(x,y,z)\,\mathbf{k} = P\,\mathbf{i} + Q\,\mathbf{j} + R\,\mathbf{k}$.

Definition: Let C be a smooth curve described by the equations $x = x(t)$ and $y = y(t)$ for $a \le t \le b$, and let f be a real valued function of two variables defined on the curve C. Partition the curve C into n arcs of equal length. Let the end points of these arcs be $C_0, C_1, C_2, \ldots C_n$ and let the length of each arc be Δs_i for $i = 1, 2, \ldots n$. Let $\left(x_i^*, y_i^*\right)$ be any point in the arc $\left[C_{i-1}, C_i\right]$. The *line integral of* f *along* C, $\int_C f(x,y)\,ds$, is defined as the limit $\int_C f(x,y)\,ds = \lim_{n\to\infty} \sum_{i=1}^{n} f\left(x_i^*, y_i^*\right) \Delta s_i$, if this limit exists.

Theorem: Suppose that the line integral $\int_C f(x,y)\,ds$ for a function f of two variables exists. Then

1. $\int_C f(x,y)\,ds = \int_a^b f\left(x(t), y(t)\right) \sqrt{\left(\dfrac{dx}{dt}\right)^2 + \left(\dfrac{dy}{dt}\right)^2}\; dt$.

2. Let the parameterized of the curve C be given by $r(t) = \langle r_1(t), r_2(t) \rangle$. Then

    $$\int_C f(x,y)\,ds = \int_a^b f\left(r(t)\right) \sqrt{r'(t) \cdot r'(t)}\; dt\,.$$

    This can easily be extended to three dimensions.

3. $\int_C f(x,y)\,ds = \int_{C_1} f(x,y)\,ds + \int_{C_2} f(x,y)\,ds + \ldots + \int_{C_n} f(x,y)\,ds$, where the curve C is made up of n smooth curves $C_1, C_2, \ldots C_n$.

Definition: Let C be a smooth curve described by the equations $x = f(t)$ and $y = g(t)$ and let f be a real valued function of two variables defined on the curve C. Partition the curve C into n arcs of equal length. Let the end points of these arcs be $C_0, C_1, C_2, \ldots C_n$ and let the coordinates of each of these end points be (x_i, y_i) for $i = 0, 1, 2, \ldots n$. Let

$\left(x_i^*, y_i^*\right)$ be any point in the arc $\left[C_{i-1}, C_i\right]$. The *line integral of* f *along* C *with respect to*

x, is defined as $\displaystyle\int_C f(x,y)\,dx = \lim_{n\to\infty}\sum_{i=1}^{n} f\left(x_i^*, y_i^*\right)\Delta x_i$, provided this limit exists. The *line*

integral of f *along* C *with respect to* y, *is defined as*

$\displaystyle\int_C f(x,y)\,dy = \lim_{n\to\infty}\sum_{i=1}^{n} f\left(x_i^*, y_i^*\right)\Delta y_i$, if this limit exists.

Theorem: Suppose that the line integrals with respect to x and y for a function f exist.
Then

1. $\displaystyle\int_C f(x,y)\,dx = \int_a^b f\left(x(t), y(t)\right) x'(t)\,dt$.

2. $\displaystyle\int_C f(x,y)\,dy = \int_a^b f\left(x(t), y(t)\right) y'(t)\,dt$.

Notation: Suppose that the line integral with respect to x for a function f_1 and the line
integral with respect to y for a function f_2 exist. Then we write

$$\int_C f_1(x,y)\,dx + \int_C f_2(x,y)\,dy = \int_C f_1(x,y)\,dx + f_2(x,y)\,dy.$$

Remark: The definition of the line integral can be extended in $\mathbb{R}^3$.

Theorem: Let C be a smooth curve. Suppose that the line integral $\displaystyle\int_C f(x,y,z)\,ds$ for a
function f three variables exists. Then

$$\int_C f(x,y,z)\,ds = \int_a^b f\left(x(t), y(t), z(t)\right)\sqrt{\left(\frac{dx}{dt}\right)^2 + \left(\frac{dy}{dt}\right)^2 + \left(\frac{dz}{dt}\right)^2}\,dt.$$

Definition: Let $\mathbf{F}$ be a continuous vector field defined on a smooth curve C given by a
vector function $\mathbf{r}(t)$, $a \le t \le b$. Then *the line integral of* $\mathbf{F}$ *along* C, $\displaystyle\int_C \mathbf{F}\cdot d\mathbf{r}$, is defined

as $\displaystyle\int_C \mathbf{F}\cdot d\mathbf{r} = \int_a^b \mathbf{F}\left(\mathbf{r}(t)\right)\cdot\mathbf{r}'(t)\,dt = \int_C \mathbf{F}\cdot\mathbf{T}\,ds$.

Example 1: Evaluate the line integral $\displaystyle\int_C f\,ds$ using *Maple* commands, where

$f(x,y) = \left(x^2 - y^2\right)e^{xy}$ and C is the oriented arc $\mathbf{r}(t) = \left\langle\cos(t), \sin(t)\right\rangle$ for $0 \le t \le \dfrac{\pi}{4}$.

Solution: We will enter f as a *Maple* function and $\mathbf{r}(t)$ as a vector in *Maple*.

restart
$f := (x,y) = \left(x^2 - y^2\right)\cdot e^{xy}$

$$(x,y) \to \left(x^2 - y^2\right)e^{xy} \tag{16.1}$$

$with\left(Student\left[VectorCalculus\right]\right):$

$r:=t\rightarrow\left\langle\cos\left(t\right),\sin\left(t\right)\right\rangle$

$$t\rightarrow Student:-VectorCalculus:-<,>\left(\cos\left(t\right),\sin\left(t\right)\right) \tag{16.2}$$

We will use the formula $\int_{C}f\left(x,y\right)ds=\int_{0}^{\frac{\pi}{4}}f\left(r\left(t\right)\right)\sqrt{r'\left(t\right)\cdot r'\left(t\right)}\ dt$.

$\int_{0}^{\frac{\pi}{4}}f\left(\cos\left(t\right),\sin\left(t\right)\right)\sqrt{r'\left(t\right)\cdot r'\left(t\right)}\ dt$

$$-1+e^{\frac{1}{2}} \tag{16.3}$$

Therefore, $\int_{C}f\left(x,y\right)ds=-1+e^{\frac{1}{2}}$.

Example 2: Let $\mathbf{F}=\left\langle xe^{x},ye^{-y}\right\rangle$ and C be the path defined by the quarter of the unit circle $\mathbf{r}\left(t\right)=\left\langle\cos\left(t\right),\sin\left(t\right)\right\rangle$ and $0\leq t\leq\dfrac{\pi}{2}$ in a counter clockwise direction. Use *Maple* commands to evaluate the line integral $\int_{C}\mathbf{F}\cdot\mathbf{dr}$. Sketch the graph of the vector field $\mathbf{F}$ and the path C .

Solution: We will begin by defining the vector field $\mathbf{F}$ and the curve C . Do not forget to activate the $with\left(Student\left[VectorCalculus\right]\right)$ package.

$with\left(Student\left[VectorCalculus\right]\right):$

$F:=VectorField\left(\left\langle x\cdot e^{x},y\cdot e^{-y}\right\rangle\right)$

$$\left(xe^{x}\right)\overline{e}_{x}+\left(ye^{-y}\right)\overline{e}_{y} \tag{16.4}$$

Here $\overline{e}_{x}$ and $\overline{e}_{y}$ are the basis elements for two dimensional space. The bar is used to indicate that we are working with a vector field.

$r:=\left\langle\cos\left(t\right),\sin\left(t\right)\right\rangle$

$$\left(\cos\left(t\right)\right)e_{x}+\left(\sin\left(t\right)\right)e_{y} \tag{16.5}$$

$C:=Path\left(r,t=0..\dfrac{\pi}{2}\right)$

$$Path\left(\left(\cos\left(t\right)\right)e_{x}+\left(\sin\left(t\right)\right)e_{y},t=0..\dfrac{1}{2}\pi\right) \tag{16.6}$$

We will now evaluate the line integral $\int_C \mathbf{F} \cdot d\mathbf{r}$ by using the *LineInt* command.

LineInt(F, C)

$$-2\,e^{-1} \qquad\qquad\qquad (16.7)$$

Therefore, $\int_C \mathbf{F} \cdot d\mathbf{r} = -2e^{-1}$. We will now generate a plot of the vector field and the path.

LineInt$(F, C, output = plot)$

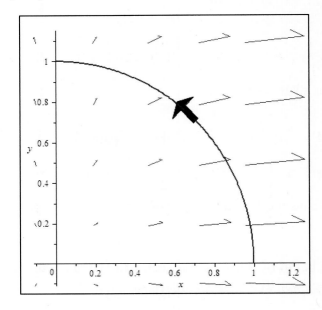

Example 3: Let $\mathbf{F} = \left\langle ye^{xy} + 1, xe^{xy} + 2y \right\rangle$ and C is the path defined by the arc $y = x^4$ for $0 \le x \le 1$ followed by the line segment $y = x$ for $0 \le x \le 1$ in a counterclockwise direction. Use *Maple* commands to evaluate the line integral $\int_C \mathbf{F} \cdot d\mathbf{r}$. Sketch the vector field $\mathbf{F}$ and the path C.

Solution: We will begin by defining the vector field $\mathbf{F}$ and the curve C. Note that for $0 \le t \le 1$, the arc can be parameterized as $\mathbf{r}_1(t) = \left\langle t, t^4 \right\rangle$, which moves from $(0,0)$ to $(1,1)$ and the line segment as $\mathbf{r}_2(t) = \left\langle 1-t, 1-t \right\rangle$, which moves from $(1,1)$ to $(0,0)$.

$with\left(Student\left[VectorCalculus\right]\right):$
$F := VectorField\left(\left\langle y \cdot e^{xy} + 1, x \cdot e^{xy} + 2 \cdot y \right\rangle\right)$

$$\left(y\,e^{xy} + 1\right)\overline{e}_x + \left(x\,e^{xy} + 2\,y\right)\overline{e}_y \qquad\qquad (16.8)$$

$r1 := \left\langle t, t^4 \right\rangle$

$$(t)e_x + (t^4)e_y \qquad \textbf{(16.9)}$$

$C1 := Path(r1, t = 0..1)$

$$Path((t)e_x + (t^4)e_y, t = 0..1) \qquad \textbf{(16.10)}$$

We will now evaluate the line integral $\int_C \mathbf{F} \cdot \mathbf{dr}$ using the same commands as in example 2.

$LineInt(F, C1)$

$$1 + e \qquad \textbf{(16.11)}$$

Hence $\int_C \mathbf{F} \cdot \mathbf{dr} = 1 + e$. We will now generate a plot of the vector field and the path.

$LineInt(F, C1, output = plot)$

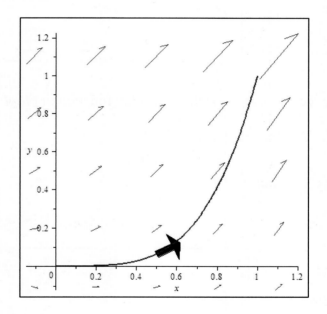

We will now calculate the line integral for the line segment portion.

$r2 := \langle 1 - t, 1 - t \rangle$

$$(1 - t)e_x + (1 - t)e_y \qquad \textbf{(16.12)}$$

$C2 := Path(r2, t = 0..1)$

$$Path((1 - t)e_x + (1 - t)e_y, t = 0..1) \qquad \textbf{(16.13)}$$

$LineInt(F, C2)$

$$-e - 1 \qquad \textbf{(16.14)}$$

Hence $\displaystyle\int_C \mathbf{F}\cdot d\mathbf{r}=-e-1$. We will now generate a plot of the vector field and the path.

$LineInt\left(F,C2,output=plot\right)$

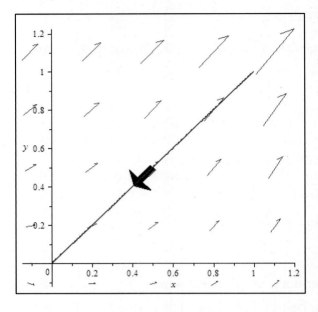

We can combine the plots above with the display function in *Maple*.

$s1:=LineInt(F,C1,output=plot)$

$$PLOT\left(...\right) \tag{16.15}$$

$s2:=LineInt(F,C2,output=plot)$

$$PLOT\left(...\right) \tag{16.16}$$

$display\left(s1,s2\right)$

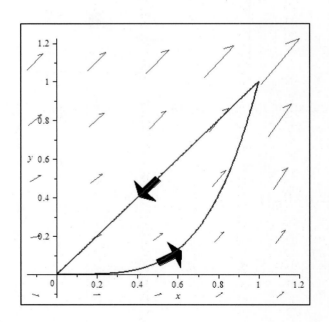

We see that the line integral takes the value of 0 by taking the sum of equations (16.11) and (16.14) above.

Conservative Vector Fields

Definition: The vector field $\mathbf{F}$ that comes from a scalar function f, where $\nabla f = \mathbf{F}$, is called *a conservative field.*

Remarks:
1.   This definition works for both two and three dimensions since we are dealing with vectors. In the case of conservative vector fields the first partial derivatives of $\mathbf{F}$ are continuous in an open connected set.

2.   Line integrals of conservative vector fields over closed paths are zero. It is important to keep in mind that if a line integral is zero, it does not guarantee that the vector field is conservative.

3.   Line integrals of conservative vector fields depend only on the end points of the path. Furthermore, if $\nabla f = \mathbf{F}$, where $f = f(x, y)$ and we have a path C that begins at the point (a_1, b_1) and ends at the point (a_2, b_2), then we can evaluate the line integral as $\int_C \mathbf{F} \cdot d\mathbf{r} = f(a_2, b_2) - f(a_1, b_1)$.

4.   This is easily extended to the case in three dimension with $\nabla f = \mathbf{F}$, where $f = f(x, y, z)$, and we have a path C that begins at the point (a_1, b_1, c_1) and ends at the point (a_2, b_2, c_2). We can evaluate the line integral as $\int_C \mathbf{F} \cdot d\mathbf{r} = f(a_2, b_2, c_2) - f(a_1, b_1, c_1)$.

5.   If $\mathbf{F}$ is conservative and $\mathbf{F} = \langle F_1, F_2 \rangle$, then $\dfrac{\partial F_1}{\partial y} = \dfrac{\partial F_2}{\partial x}$.

6.   If $\mathbf{F}$ is conservative and $\mathbf{F} = \langle F_1, F_2, F_3 \rangle$, then $\dfrac{\partial F_1}{\partial y} = \dfrac{\partial F_2}{\partial x}$, $\dfrac{\partial F_2}{\partial z} = \dfrac{\partial F_3}{\partial y}$, and $\dfrac{\partial F_1}{\partial z} = \dfrac{\partial F_3}{\partial x}$.

Theorem: Let $\mathbf{F} = \langle F_1, F_2 \rangle$ be a conservative vector field. Then the scalar function f for $\mathbf{F}$ can be found from $f(x, y) = \int_{x_0}^x F_1(t, y)\, dt + \int_{y_0}^y F_2(x_0, t)\, dt + c_0$, where (x_0, y_0) is any point in the interior of the simply connected region and c_0 is some constant.

Theorem: Let $\mathbf{F} = \langle F_1, F_2, F_3 \rangle$ be a conservative vector field. Then the scalar function f for the vector field $\mathbf{F}$ can be found using the formula

$$f(x,y,z) = \int_{x_0}^{x} F_1(t,y,z)\,dt + \int_{y_0}^{y} F_2(x_0,t,z)\,dt + \int_{z_0}^{z} F_3(x_0,y_0,t)\,dt + c_0, \quad \text{where} \quad (x_0, y_0, z_0)$$

is any point in the interior of the simply connected region and c_0 is some constant.

Example 4: Let $\mathbf{F} = \langle 2xye^{x^2 y}, x^2 e^{x^2 y}, 2z \rangle$ and C be the helix parameterized by $r(t) = \langle \cos(t), \sin(t), t \rangle$, where $0 \le t \le 3\pi$, oriented in the counterclockwise direction.

(a)   Sketch the vector field $\mathbf{F}$ and the path C.
(b)   Show that the vector field $\mathbf{F}$ is conservative.
(c)   Find the function f such that $\nabla f = \mathbf{F}$.
(d)   Use *Maple* to evaluate the line integral $\int_C \mathbf{F} \cdot d\mathbf{r}$.

Solution:
(a)   We will plot the vector field $\mathbf{F}$ and the path C using *Maple* commands.

$with\left(Student\left[VectorCalculus \right] \right):$

$with(plots):$

$s1 := fieldplot3d\left(\left\langle 2 \cdot x \cdot y \cdot e^{x^2 \cdot y}, x^2 \cdot e^{x^2 \cdot y}, 2 \cdot z \right\rangle, x = -1..1, y = -1..1, z = 0..8, \right.$

$\left. grid = [10,10,10], color = blue, axes = box \right)$

$$PLOT3D(...) \tag{16.17}$$

$s2 := spacecurve\left(\left[\cos(t), \sin(t), t, t = 0..3 \cdot \pi \right], axes = box, thickness = 4 \right)$

$$PLOT3D(...) \tag{16.18}$$

$display(s1, s2)$

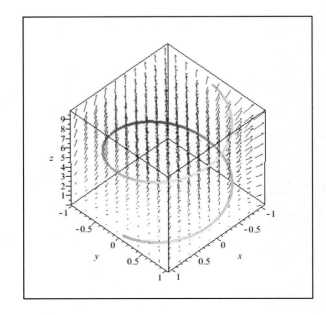

(b)   We will use *Maple* commands to show that **F** is a conservative vector field.

$$F1 := (x, y, z) \rightarrow 2 \cdot x \cdot y \cdot e^{x^2 \cdot y}$$

$$(x, y, z) \rightarrow 2 x y e^{x^2 y} \qquad \textbf{(16.19)}$$

$$F2 := (x, y, z) \rightarrow x^2 \cdot e^{x^2 \cdot y}$$

$$(x, y, z) \rightarrow x^2 e^{x^2 y} \qquad \textbf{(16.20)}$$

$$F3 := (x, y, z) \rightarrow 2 \cdot z$$

$$(x, y, z) \rightarrow 2 z \qquad \textbf{(16.21)}$$

$$\frac{\partial}{\partial y} (F1(x, y, z))$$

$$2 x e^{x^2 y} + 2 x^3 y e^{x^2 y} \qquad \textbf{(16.22)}$$

$$\frac{\partial}{\partial x} (F2(x, y, z))$$

$$2 x e^{x^2 y} + 2 x^3 y e^{x^2 y} \qquad \textbf{(16.23)}$$

$$\frac{\partial}{\partial z} (F2(x, y, z))$$

$$0 \qquad \textbf{(16.24)}$$

$$\frac{\partial}{\partial y} (F3(x, y, z))$$

$$0 \qquad \textbf{(16.25)}$$

$$\frac{\partial}{\partial z} (F1(x, y, z))$$

$$0 \qquad \textbf{(16.26)}$$

$$\frac{\partial}{\partial x} (F3(x, y, z))$$

$$0 \qquad \textbf{(16.27)}$$

Therefore $\dfrac{\partial F_1}{\partial y} = 2xe^{x^2 y} + 2x^3 ye^{x^2 y} = \dfrac{\partial F_2}{\partial x}$, $\dfrac{\partial F_2}{\partial z} = 0 = \dfrac{\partial F_3}{\partial y}$ and $\dfrac{\partial F_1}{\partial z} = 0 = \dfrac{\partial F_3}{\partial x}$.

Furthermore, since the partial derivatives are all continuous, our vector field is clearly conservative.

(c)   We will use the formula from the theorem before this example to find the scalar function f. We will choose the point $(x_0, y_0, z_0) = (0, 0, 0)$ and let $c_0 = 0$. In

other words $f(x, y, z) = \displaystyle\int_0^x F_1(t, y, z)\, dt + \int_0^y F_2(0, t, z)\, dt + \int_0^z F_3(0, 0, t)\, dt$.

$$\int_0^x F1(t,y,z)\,dt + \int_0^y F2(0,t,z)\,dt + \int_0^z F3(0,0,t)\,dt$$
$$-1+e^{x^2 y}+z^2 \qquad\qquad\qquad\qquad (16.28)$$

Hence $f(x,y,z)=-1+e^{x^2 y}+z^2$.

(d) To get the value of the line integral, we will evaluate the scalar function at the points $\left(\cos(0),\sin(0),0\right)$ and $\left(\cos(3\pi),\sin(3\pi),3\pi\right)$. We will enter $\left(\cos(t),\sin(t),t\right)$ as a vector function.

$$f:=(x,y,z)\rightarrow -1+e^{x^2\cdot y}+z^2$$
$$(x,y,z)\rightarrow -1+e^{x^2 y}+z^2 \qquad\qquad (16.29)$$

$with\left(Student\left[VectorCalculus\right]\right):$

$r:=t\rightarrow\left\langle\cos(t),\sin(t)\right\rangle$
$$t\rightarrow Student:-VectorCalculus:-<,>\left(\cos(t),\sin(t),t\right) \qquad (16.30)$$

$r(t)$
$$\left(\cos(t)\right)e_x+\left(\sin(t)\right)e_y+(t)e_z \qquad\qquad (16.31)$$

$r(0)$
$$e_x \qquad\qquad\qquad\qquad (16.32)$$

$r(3\cdot\pi)$
$$-e_x+3\,\pi\,e_z \qquad\qquad\qquad (16.33)$$

Therefore, $\left(\cos(0),\sin(0),0\right)=(1,0,0)$ and $\left(\cos(3\pi),\sin(3\pi),3\pi\right)=(-1,0,3\pi)$.
We will now use the formula $\int_C \mathbf{F}\cdot d\mathbf{r}=f(-1,0,3\cdot\pi)-f(1,0,0)$ (see remark 4 before this example).

$f(-1,0,3\cdot\pi)-f(1,0,0)$
$$9\,\pi^2 \qquad\qquad\qquad\qquad (16.34)$$

Thus $\int_C \mathbf{F}\cdot d\mathbf{r}=9\,\pi^2$.

Example 5: Evaluate the line integral in example 4 above, $\int_C \mathbf{F}\cdot d\mathbf{r}$ directly using *Maple* commands (as was done in the latter part of example 2), where $\mathbf{F}=\left\langle 2xye^{x^2 y},x^2 e^{x^2 y},2z\right\rangle$

and C is the helix parameterized by $r(t) = \langle \cos(t), \sin(t), t \rangle$, where $0 \le t \le 3\pi$, oriented in the counterclockwise direction. Sketch the vector field $\mathbf{F}$ and the path C.

Solution: We will use *Maple* commands to get the same result as in example 4.

$with\left(Student\left[VectorCalculus\right]\right):$

$F := VectorField\left(\left\langle 2 \cdot x \cdot y \cdot e^{x^2 \cdot y}, x^2 \cdot e^{x^2 \cdot y}, 2 \cdot z \right\rangle\right)$

$$\left(2\,x\,y\,e^{x^2 y}\right)\overline{e}_x + \left(x^2\,e^{x^2 y}\right)\overline{e}_y + \left(2\,z\right)\overline{e}_z \qquad \textbf{(16.35)}$$

$r := \langle \cos(t), \sin(t), t \rangle$

$$\left(\cos(t)\right)e_x + \left(\sin(t)\right)e_y + (t)e_z \qquad \textbf{(16.36)}$$

$C := Path\left(r, t = 0..3 \cdot \pi\right)$

$$Path\left(\left(\cos(t)\right)e_x + \left(\sin(t)\right)e_y + (t)e_z, t = 0..3\pi\right) \qquad \textbf{(16.37)}$$

$LineInt\left(F, C\right)$

$$9\pi^2 \qquad \textbf{(16.38)}$$

Hence $\int_C \mathbf{F} \cdot d\mathbf{r} = 9\pi^2$. We will now generate a plot of the vector field and the path.

$LineInt\left(F, C, output = plot, axes = box\right)$

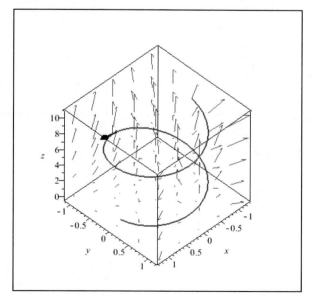

Green's Theorem

Green's Theorem: Let C be a positively oriented, piece-wise smooth, simple closed curve in $\mathbb{R}^2$ and let D be a region bounded by C. Suppose that $\mathbf{F} = \langle F_1, F_2 \rangle$, where the functions F_1 and F_2 have continuous partial derivatives on an open region that contains D. Then

$$\int_C F_1\,dx + F_2\,dy = \iint_D \left(\frac{\partial F_2}{\partial x} - \frac{\partial F_1}{\partial y} \right) dA \ \text{ or } \ \int_C \mathbf{F} \cdot d\mathbf{r} = \iint_D \left(\frac{\partial F_2}{\partial x} - \frac{\partial F_1}{\partial y} \right) dA .$$

Example 6: Use *Maple* commands and Green's Theorem to evaluate the line integral $\int_C \mathbf{F} \cdot d\mathbf{r}$, where $\mathbf{F} = \langle ye^{xy} + 1, xe^{xy} + 2y \rangle$ and C is the path defined by the arc $y = x^4$ for $0 \le x \le 1$ followed by the straight line segment $y = x$ for $0 \le x \le 1$ in a counterclockwise direction. The graph below is from example 3.

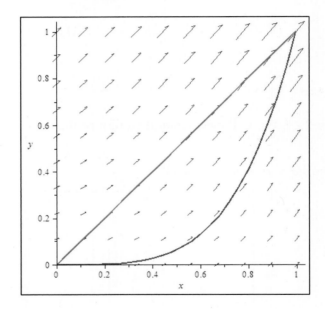

Solution: We will begin by entering $F1(x, y) = ye^{xy} + 1$ and $F2(x, y) = xe^{xy} + 2y$ as *Maple* functions.

restart
$F1 := (x, y) \rightarrow y \cdot e^{x \cdot y} + 1$

$$(x, y) \rightarrow y\,e^{xy} + 1 \tag{16.39}$$

$F2 := (x, y) \rightarrow x \cdot e^{x \cdot y} + 2 \cdot y$

$$(x, y) \rightarrow x\,e^{xy} + 2\,y \tag{16.40}$$

We will now use the formula $\int_C \mathbf{F} \cdot d\mathbf{r} = \iint_D \left(\dfrac{\partial F_2}{\partial x} - \dfrac{\partial F_1}{\partial y} \right) dA$ obtained from Green's theorem to evaluate the integral $\int_C \mathbf{F} \cdot d\mathbf{r}$. From the graph above we see that the region $D = \left\{ (x, y) : x^4 \leq y \leq x, 0 \leq x \leq 1 \right\}$.

$with \left(Student \left[MultivariateCalculus \right] \right):$

$$MultiInt \left(\frac{\partial}{\partial x} F2(x, y) - \frac{\partial}{\partial y} F1(x, y),\, y = x^4 \,..\, x,\, x = 0 \,..\, 1 \right)$$

$$0$$

(16.41)

Hence, $\int_C \mathbf{F} \cdot d\mathbf{r} = 0$.

Definition: The *divergence of a vector field* $\mathbf{F}$ in two dimensions is defined $\operatorname{div} \mathbf{F} = \nabla \cdot \mathbf{F} = \dfrac{\partial F_1}{\partial x} + \dfrac{\partial F_2}{\partial y}$, where $\mathbf{F} = \langle F_1, F_2 \rangle$. This is easily extended to higher dimensions.

Definition: The *curl of a three dimensional vector field* $\mathbf{F} = \langle F_1, F_2, F_3 \rangle$ is defined as

$$\operatorname{curl} \mathbf{F} = \nabla \times \mathbf{F} = \begin{vmatrix} i & j & k \\ \dfrac{\partial}{\partial x} & \dfrac{\partial}{\partial y} & \dfrac{\partial}{\partial z} \\ F_1 & F_2 & F_3 \end{vmatrix}.$$

Stokes Theorem

Stokes Theorem: Let S be an oriented piece-wise smooth surface that is bounded by a piece-wise smooth, simple closed boundary curve C with positive orientation. Let $\mathbf{F}$ be a continuous vector field whose components have smooth partial derivatives on an open region in $\mathbb{R}^3$ that contains S. Then $\int_C \mathbf{F} \cdot d\mathbf{r} = \iint_S \operatorname{curl} \mathbf{F} \cdot d\mathbf{S}$.

Example 7: Use *Maple* commands to show that Stokes theorem holds for the vector field $\mathbf{F} = \langle y, -x, z \rangle$ and the surface S given by $z = x^2 + y^2$ with the circle $x^2 + y^2 = 4$ at $z = 4$ as its boundary and oriented counterclockwise as seen from above.

Solution: To show that Stokes theorem works, we need to prove that $\int_C \mathbf{F} \cdot d\mathbf{r}$ and $\iint_S \operatorname{curl} \mathbf{F} \cdot d\mathbf{S}$ are the same. We will parameterize the curve C as $r(t) = \langle 2\cos(t), 2\sin(t), 4 \rangle$, where $0 \leq t \leq 2\pi$ and evaluate the line integral $\int_C \mathbf{F} \cdot d\mathbf{r}$.

$with\left(Student\left[VectorCalculus\right]\right):$

$F := VectorField\left(\langle y, -x, z \rangle\right)$

$$\left(y\right)\overline{e}_x + \left(-x\right)\overline{e}_y + \left(z\right)\overline{e}_z \qquad \textbf{(16.42)}$$

$r := \langle 2 \cdot \cos\left(t\right), 2 \cdot \sin\left(t\right), 4 \rangle$

$$2\cos\left(t\right)e_x + 2\sin\left(t\right)e_y + 4e_z \qquad \textbf{(16.43)}$$

$C := Path\left(r, t = 0..2 \cdot \pi\right)$

$$Path\left(2\cos\left(t\right)e_x + 2\sin\left(t\right)e_y + 4e_z, t = 0..2\,\pi\right) \qquad \textbf{(16.44)}$$

$LineInt\left(F, C\right)$

$$-8\,\pi \qquad \textbf{(16.45)}$$

Therefore, $\int_C \mathbf{F} \cdot d\mathbf{r} = -8\pi$. We will now generate a plot of the vector field and the path.

$LineInt\left(F, C, output = plot, axes = box\right)$

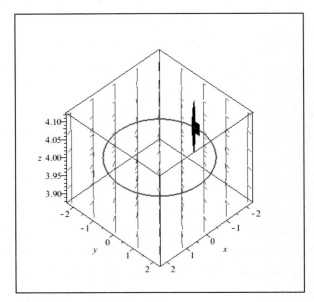

In order to evaluate $\iint_S \mathrm{curl}\,\mathbf{F} \cdot d\mathbf{S}$, we will first find the integrand. We need to find the Flux of the curl through the surface. Note that we will obtain the outward pointing normal vector $\mathbf{n}$ by taking the gradient of $f\left(x, y, z\right) = z - x^2 - y^2$.

$f := \left(x, y, z\right) \rightarrow z - x^2 - y^2$

$$\left(x, y, z\right) \rightarrow z + Student:-VectorCalculus:-'-'\left(x^2\right) + Student:-VectorCalculus:-'-'\left(y^2\right) \qquad \textbf{(16.46)}$$

$$Curl(F).Gradient(f(x,y,z))$$

$$-2 \qquad\qquad\qquad\qquad (16.47)$$

The flux of the $\text{curl}(\mathbf{F})$ through S is equal to $\iint_S \text{curl } \mathbf{F} \cdot d\mathbf{S} = \int_0^{2\cdot\pi} \int_0^2 (-2\cdot r)\, dr\, d\theta$, using polar coordinates (see chapter 15).

We will need to activate the $with(Student[MultivariateCalculus])$ package. Activate the $restart$ command beforehand.

$restart$

$with(Student[MultivariateCalculus])$:

$$\int_0^{2\cdot\pi} \int_0^2 (-2\cdot r)\, dr\, d\theta$$

$$-8\,\pi \qquad\qquad\qquad\qquad (16.48)$$

Therefore, we have established that $\int_C \mathbf{F}\cdot d\mathbf{r} = \iint_S \text{curl } \mathbf{F}\cdot d\mathbf{S}$.

Surface Integral

Definition: Let $\mathbf{F}$ be a continuous vector field on an oriented surface S. Let $r(u,v)$ be a continuous and one to one parameterization of an oriented surface S, with domain D, and let $n(u,v)$ be the normal vector defined by r. Then the surface integral of $\mathbf{F}$ over S is $\int_S \mathbf{F}\cdot d\mathbf{S} = \iint_D \mathbf{F}(r(u,v))\cdot n(u,v)\, du\, dv$.

Note: We sometimes refer to $\int_S \mathbf{F}\cdot d\mathbf{S}$ as the *flux* of $\mathbf{F}$ over S .

Divergence Theorem: Let G be a region in three dimensions with boundary S which is a continuous and smooth oriented surface, with the normal vector pointing to the outside of G, and let $\mathbf{F}$ be a vector field with domain containing G (note that all the partial derivatives are continuous for each component of $\mathbf{F}$). Then $\int_S \mathbf{F}\cdot d\mathbf{S} = \iiint_G \text{div}(\mathbf{F})\, dV$.

Example 8: Let $\mathbf{F} = \langle -y, yz, z^4 \rangle$ be a vector field and S the surface given by $x^2 + y^2 = 9$ between $z = 0$ and $z = 10$.

(a)   Evaluate $\int_S \mathbf{F}\cdot d\mathbf{S}$.

(b)   Show that the divergence theorem holds for the vector field $\mathbf{F}$ and the surface S. Use the outward pointing normal vector.

Solution: Note that the surface S is the union of three surfaces: a cylinder $S1$, the top of the cylinder $S2$, and the bottom of the cylinder $S3$.

(a)   We will calculate the surface integral $\int_S \mathbf{F} \cdot d\mathbf{S}$ as $\int_{S1} \mathbf{F} \cdot d\mathbf{S} + \int_{S2} \mathbf{F} \cdot d\mathbf{S} + \int_{S3} \mathbf{F} \cdot d\mathbf{S}$.

with $\left(Student\left[VectorCalculus \right] \right)$:

$F := VectorField\left(\left\langle -y, y \cdot z, z^4 \right\rangle \right)$

$$(-y)\overline{e}_x + (y \cdot z)\overline{e}_y + (z^4)\overline{e}_z \qquad (16.49)$$

Let $s1$ be the parameterization of the surface of the cylinder $S1$, $s1(\theta, z) = \left\langle 3 \cdot \cos(\theta), 3 \cdot \sin(\theta), z \right\rangle$, where $0 \le \theta \le 2\pi$ and $0 \le z \le 10$.

$s1 := (\theta, z) \to \left\langle 3 \cdot \cos(\theta), 3 \cdot \sin(\theta), z \right\rangle$

$(\theta, z) \to Student : -VectorCalculus : - <,> \left(3 \cdot \cos(\theta), 3 \cdot \sin(\theta), z \right) \qquad (16.50)$

$S1 := Surface\left(s1(\theta, z), [\theta, z] = Rectangle(0..2 \cdot \pi, 0..10) \right)$

$Surface\left(3\cos(\theta)e_x + 3\sin(\theta)e_y + (z)e_z, [\theta, z] = Rectangle(0..2\pi, 0..10) \right)$ (16.51)

$Flux(F1, S1)$

$$450\pi \qquad (16.52)$$

Therefore $\int_{S1} \mathbf{F} \cdot d\mathbf{S} = 450\pi$. We will now find the value of the surface integral for the top of the cylinder $S2$. Let $s2$ be the parameterization for the top of the cylinder $S2$, so that $s2(x, y) = \left\langle x, y, 10 \right\rangle$, where $-\sqrt{9 - y^2} \le x \le \sqrt{9 - y^2}$ and $-3 \le y \le 3$.

$s2 := (x, y) \to \left\langle x, y, 10 \right\rangle$

$(x, y) \to Student : -VectorCalculus : - <,> (x, y, 10) \qquad (16.53)$

$S2 := Surface\left(s2(x, y), [x, y] = Rectangle\left(-\sqrt{9 - y^2} .. \sqrt{9 - y^2}, -3..3 \right) \right)$

$Surface\left((x)e_x + (x)e_z + 10e_z, [x, y] = Rectangle\left(-\sqrt{9 - y^2} .. \sqrt{9 - y^2}, -3..3 \right) \right)$

(16.54)

$Flux(F1, S2)$

$$90000\pi \qquad (16.55)$$

Next we will find the value of the surface integral for the bottom of the cylinder $S3$. Let $s3$ be the parameterization of the surface $S3$, $s3(x, y) = \langle x, y, 0 \rangle$, where $-\sqrt{9 - y^2} \leq x \leq \sqrt{9 - y^2}$ and $-3 \leq y \leq 3$.

$s3 := (x, y) \rightarrow \langle x, y, 0 \rangle$

$$(x, y) \rightarrow Student : -VectorCalculus : - <,> (x, y, 0) \tag{16.56}$$

$S3 := Surface\left(s3(x, y), [x, y] = Rectangle\left(-\sqrt{9 - y^2}..\sqrt{9 - y^2}, -3..3\right)\right)$

$$Surface\left((x)e_x + (x)e_z, [x, y] = Rectangle\left(-\sqrt{9 - y^2}..\sqrt{9 - y^2}, -3..3\right)\right) \tag{16.57}$$

$Flux(F1, S3)$

$$0 \tag{16.58}$$

Therefore $\int_{S1} \mathbf{F} \cdot d\mathbf{S} + \int_{S2} \mathbf{F} \cdot d\mathbf{S} + \int_{S3} \mathbf{F} \cdot d\mathbf{S} = 450\pi + 90000\pi - 0 = 90450\pi$. Hence the surface integral $\int_S \mathbf{F} \cdot d\mathbf{S} = 90450\pi$.

(b)   We will then evaluate the triple integral $\iiint_G div(\mathbf{F}) dV$ and show that $\int_S \mathbf{F} \cdot d\mathbf{S} = \iiint_G div(\mathbf{F}) dV$ where $G = \{(z, r, \theta) : 0 \leq z \leq 10, 0 \leq r \leq 3, 0 \leq \theta \leq 2\pi\}$. Cylindrical coordinates will be used to find the triple integral.

$\int_0^{2 \cdot \pi} \int_0^3 \int_0^{10} r.Divergence(F1) dz\, dr\, d\theta$

$$90450\pi \tag{16.59}$$

Hence, we see from above that $\int_S \mathbf{F} \cdot d\mathbf{S} = \iiint_G div(\mathbf{F}) dV = 90450\pi$.

Exercises

1. Let $\mathbf{F} = \left\langle xe^x, ye^{-y^3} \right\rangle$ and C is the path on the boundary of the region enclosed by the curves $y = x^2$ and $y = 1 - x^2$, oriented in a counter clockwise direction. Use *Maple* commands to evaluate the line integral $\int_C \mathbf{F} \cdot d\mathbf{r}$. Sketch a graph of the vector field and the path C.

2. Let $\mathbf{F} = \left\langle 2 - xy^2 +, ye^{-y} \right\rangle$ and C is the path on the boundary of the region enclosed by the curve $|x| + |y| = 3$, oriented in a counter clockwise direction. Use *Maple* commands to evaluate the line integral $\int_C \mathbf{F} \cdot d\mathbf{r}$. Sketch a graph of the vector field and the path C.

3. Let $\mathbf{F} = \left\langle 2 - xy^2 +, ye^{-y} \right\rangle$ and C is the path on the boundary of the region enclosed by the curve $x^6 + y^6 = 1$, oriented in a counter clockwise direction. Use *Maple* commands to evaluate the line integral $\int_C \mathbf{F} \cdot d\mathbf{r}$. Sketch a graph of the vector field and the path C.

4. Let $\mathbf{F} = \left\langle -zy^2, ye^{-y}, z^2 \right\rangle$ and C is the path $\left\langle \sin(t), t, t^2 \right\rangle$ from $t = 0$ to $t = \dfrac{\pi}{3}$. Use *Maple* commands to evaluate the line integral $\int_C \mathbf{F} \cdot d\mathbf{r}$. Sketch a graph of the vector field and the path C.

5. Let $\mathbf{F} = \left\langle z, y\sin(x), x^3 \right\rangle$ and C be the path $\left\langle t, e^t, t^2 \right\rangle$ from $t = 0$ to $t = 8$. Use *Maple* commands to evaluate the line integral $\int_C \mathbf{F} \cdot d\mathbf{r}$. Sketch graph of the vector field and the path C.

6 - 8. Use *Maple* commands to check whether the following vector fields are conservative and find the potential f in such instances. Note that $\nabla f = \mathbf{F}$.

6. $\mathbf{F} = \left\langle ye^{xy} + 1, xe^{xy} + 2y + y^5 \right\rangle$

7. $\mathbf{F} = \left\langle -\dfrac{y}{x^2 + y^2}, \dfrac{x}{x^2 + y^2} \right\rangle$ and $x > 0$.

8. $\mathbf{F} = \left\langle -\dfrac{1}{x}, -\dfrac{1}{y}, -\dfrac{3}{z} \right\rangle$ and $x > 0, y > 0, z > 0$.

9. Verify Green's theorem for exercise 1.

10.   Verify Green's theorem for exercise 2.

11.   Verify Green's theorem for exercise 3.

12.   Use *Maple* commands to evaluate $\int_C \mathbf{F} \cdot d\mathbf{r}$ and show that Stokes theorem holds for the vector field $\mathbf{F} = \langle x^2, y, yz \rangle$, where C is the boundary of the plane S given by $x + y + 3z = 9$ in the first octant and oriented counterclockwise as seen from above.

13.   Use *Maple* commands to evaluate $\int_C \mathbf{F} \cdot d\mathbf{r}$ and show that Stokes theorem holds for the vector field $\mathbf{F} = \langle \sin(y), -z, x \rangle$, where C is the curve of intersection of $x^2 + y^2 + z^2 = 5$ and $z = 2$ oriented counterclockwise as seen from above.

14.   Use *Maple* commands to evaluate $\int_C \mathbf{F} \cdot d\mathbf{r}$ and show that Stokes theorem holds for the vector field $\mathbf{F} = \langle e^x, 2z + 1, y^3 - 2 \rangle$, where C is the curve of intersection of $z = x^2 + y^2$ and $z = 2$ oriented counterclockwise as seen from above.

15.   Use *Maple* commands to evaluate $\int_C \mathbf{F} \cdot d\mathbf{r}$ and show that Stokes theorem holds for the vector field $\mathbf{F} = \langle x^2, 2xy, 2xz \rangle$, where C is the curve of intersection of $z = x^4 + y^4$ and $z = 2$ oriented counterclockwise as seen from above.

16.   Use *Maple* commands to evaluate $\int_S \mathbf{F} \cdot d\mathbf{S}$ and show that the divergence theorem holds for the vector field $\mathbf{F} = \langle y, -x, z \rangle$ and the boundary surface S of the sphere $x^2 + y^2 + z^2 = 9$. Use the outward pointing normal vector for the surface.

17.   Use *Maple* commands to evaluate $\int_S \mathbf{F} \cdot d\mathbf{S}$ and show that the divergence theorem holds for the vector field $\mathbf{F} = \langle y^2, -x, z \rangle$ and the boundary surface S of the solid cube in the first octant with three of its edges lying on the *x*-, *y*-, and *z*- axes. Use the outward pointing normal vector for the six surfaces.

18.   Use *Maple* commands to evaluate $\int_S \mathbf{F} \cdot d\mathbf{S}$ and show that the divergence theorem holds for the vector field $\mathbf{F} = \langle y, -x^3, z \rangle$ and the boundary surface S of the solid cone $z^2 = x^2 + y^2$ and $0 \le z \le 9$. Use the outward pointing normal vector for the surfaces.

19. Use *Maple* commands to evaluate $\int_S \mathbf{F} \cdot d\mathbf{S}$ and show that the divergence theorem holds for the vector field $\mathbf{F} = \langle y, -x, z \rangle$ and the boundary surface S of the solid formed by $x^2 + y^2 + z^2 = 9$ and $z \geq 6$. Use the outward pointing normal vector for the two surfaces.

20. Below is a list of some *Maple* commands used in this chapter. Describe the significance of each command. Can you find examples where each command is used?

    (a) $Curl(F)$

    (b) $Flux(F, C)$

    (c) $fieldplot(F, x = a \mathbin{..} b, y = c \mathbin{..} d)$

    (d) $LineInt(F, C)$

    (e) $Path(r, t = t_0 \mathbin{..} t_1)$

21. Can you list new *Maple* commands that were used in this chapter but were not listed in exercise 20?

We will list the different *with(packages)* and the commands they activate that are used throughout the book.

with(DETools)

[*Are Similar, DEnormal, DEplot, DEplot3d, DEplot_polygon, DFactor, DFactorLCLM. DFactorsols, Dchangevar, FunctionDecomposition, GCRD, Gosper, Heunsols, Homomorphsims, IVPsol, IsHyperexponential, LCLM, MeijerGsols, MultiplicativeDecomposition, ODEInvariants, PDEchangecoords, PolynomialNormalForm, RationalCanonicalForm, ReduceHyperexp, RiemannPsols, Xchange, Xcommutator, Xgauge, Zeilberger, abelsol, adjoint, autonomous, bernoullisol, buildsol, buildsym, canoni, caseplot, casesplit, checkrank, chinisol, clairautsol, constcoeffsols, convertAlg, convertsys, dalembertsol, dcoeffs, de2diffop, dfieldplot, diff_table, diffop2de, dperiodic_sols, dpolyform, dsubs, eigenring, endomorphism_charpoly, equinv, eta_k, eulersols, exactsol, expsols, exterior_power, firint, firtest, formal_sol, gen_exp, generate_ic, genhomosol, gensys, hamilton_eqs, hypergeomsols, hyperode, indicialeq, infgen, initialdata, integrate_sols, intfactor, invariants, kovacicsols, leftdivision, liesol, line_int, linearsol, matrixDE, matrix_ricati, maxdimsystems, moser_reduce, muchange, mult, mutest, newton_polygon, normalG2, ode_int_y, ode_y1, odeadvisor, odepde, parametricsol, particularsol, phaseportrait, poincare, polysols, power_equivalent, rational_equivalent, ratsols, redode, reduceOrder, reduce_order, regular_parts, regularsp, remove_RootOf, ricati_system, ricatisol, rifread, rifsimp, rightdivision, rtaylor, separablesol, singularities, solve_group, super_reduce, symgen, symmetric_power, symmetric_product, symtest, transinv, translate, untranslate, varparam, zoom*]

with(geometry)

[*Apllonius, AreCollinear, AreConcurrent, AreConcylic, AreConjugate, AreHarmonic, AreOrthogonal, AreParallel, ArePerpendicular, AreaSimilar, AreTangent, CircleOfSimilitude, CrossProduct, CrossRatio, DefinedAs, Equation, EulerCircle, EulerLine, ExteriorAngle, ExternalBisector, FindAngle, GergonnePoint, GlideReflection, HorizontalCoord, HorizontalName, InteriorAngle, IsEquilateral, IsOnCircle, IsOnLine, IsRightTriangle, MajorAxis, MakeSquare, MinorAxis, NagelPoint, OnSegment, ParallelLine, PedalTriangle, PerpenBisector, PerpendicularLine, Polar, Pole, RadicalAxis, RadicalCenter, RegularPolygon, RegularStarPolygon, SensedMagnitude, SimsonLine, SpiralRotation, StretchReflection, StretchRotation, TangentLine, VerticalCoord, VerticalName, altitude, apothem, area, asymptotes, bisector, center, centroid, circle, circumcircle, conic, convexhull, coordinates, detail, diagonal, diameter, dilatation, directrix, distance, draw, dsegment, ellipse, excircle, expansion, foci, focus, form, homology, homothety, hyperbola, incircle, inradius, intersection, inversion, line, medial, median, method, midpoint, orthocenter, parabola, perimeter, point, powerpc, projection, radius, randpoint, reciprocation, reflection, rotation, segment, sides, similitude, slope, square, stretch, tangentpc, translation, triangle, vertex, vertices*]

with (*linalg*)

[*BlockDiagonal, GramSchmidt, JordanBlock, LUdecomp, QRdecomp, Wronksian, addcol, addrow, adj, adjoint, angle, augment, backsub, band, basis, bezout, blockmatrix, charmat, charpoly, cholesky, col, coldim, colspace, colspan, companion, concat, cond, copyinto, crossprod, curl, definite, delcols, delrows, det, diag, diverge, dotprod, eigenvals, eigenvalues, eigenvectors, eigenvects, entermatrix, equal, exponential, extend, ffgausselim, fibonacci, forwardsub, frobenius, gausselim, gaussjord, geneqns, genmatrix, grad, hadamard, hermite, hessian, hilbert, htranspose, ihermite, indexfunc, innerprod, intbasis, inverse, ismith, issimilar, iszero, jacobian, jordan, kernel, laplacian, leastsqrs, linsolve, matadd, matrix, minor, minpoly, mulcol, mulrow, multiply, norm, normalize, nullspace, orthog, permanent, pivot, potential, randmatrix, randvector, rank, ratform, row, rowdim, rowspace, rowspan, rref, scalarmul, singularvals, smith, stackmatrix, submatrix, subvector, sumbasis, swapcol, swaprow, sylvester, toeplitz, trace, transpose, vandermonde, vecpotent, vectdim, vector, wronskian*]

with (*LinearAlgebra*)

[*&x, Add, Adjoint, BackwardSubstitute, BandMatrix, Basis, BezoutMatrix, BidiagonalForm, BilinearForm, CARE, CharacteristicMatrix, CharacteristicPolynomial, Column, ColumnDimension, ColumnOperation, ColumnSpace, CompanionMatrix, ConditionNumber, ConstantMatrix, ConstantVector, Copy, CreatePermutation, CrossProduct, DARE, DeleteColumn, DeleteRow, Determinant, Diagonal, DiagonalMatrix, Dimension, Dimensions, DotProduct, EigenConditionNumbers, Eigenvalues, Eigenvectors, Equal, ForwardSubstitute, FrobeniusForm, GaussianElimination, GenerateEquations, GenerateMatrix, Generic, GetResultDataType, GetResultShape, GivensRotationMatrix, GramSchmidt, HankelMatrix, HermiteForm, HermitianTranspose, HessenbergForm, HilbertMatrix, HouseholderMatrix, IdentityMatrix, IntersectionBasis, IsDefinite, IsOrthogonal, IsSimilar, IsUnitary, JordanBlockMatrix, JordanForm, KroneckerProduct, LA_Main, LUDecomposition, LeastSquares, LinearSolve, LyapunovSolve, Map, Map2, MatrixAdd, MatrixExponential, MatrixFunction, MatrixInverse, MatrixMatrixMultiply, MatrixNorm, MatrixPower, MatrixScalarMultiply, MatrixVectorMultiply, MinimalPolynomial, Minor, Modular, Multiply, NoUserValue, Norm, Normalize, NullSpace, OuterProductMatrix, Permanent, Pivot, PopovForm, QRDecompostion, RandomMatrix, RandomVector, Rank, RationalCanonicalForm, ReducedRowEchelonForm, Row, RowDimension, RowOperation, RowSpace, ScalarMatrix, ScalarMultiply, ScalarVector, SchurForm, SingularValues, SmithForm, StronglyConnectedBlocks, SubMatrix, SubVector, SumBasis, SylvesterMatrix, SylvesterSolve, ToeplitzMatrix, Trace, Transpose, TridiagonalForm, UnitVector, VandermondeMatrix, VectorAdd, VectorAngle, VectorMatrixMultiply, VectorNorm, VectorScalarMultiply, ZeroMatrix, ZeroVector, Zip*]

with (*MTM*)

[*ElementwiseAnd, ElementwiseNot, ElementwiseOr, Map, Minus, Mod, Zip, abs, acos, acosh, acot, acoth, acsc, acsch, array_dims, asec, asech, asin, asinh, atan, atanh, besseli, besselj, besselk, bessely, ccode,* ceil, *char, coeffs, collect, colspace, compose, conj,* cos, cosh, *cosint,* cot, coth, csc, csch, *ctranspose, det, diag, diff, digits, dirac, disp, double, dsolve, eig, end, eq*, erf, exp, *expand, expm, ezcontour, ezcontourf, ezmesh, ezmeshc, ezplot, ezplot3, ezpolar, ezsurf, ezsurfc, factor, findsym, finverse, fix,* floor, *fortran, fourier,* frac, γ, *gcd, ge, gt, heaviside, horner, horzcat,* hypergeom, *ifourier, ilaplace, imag, int, int16, int32, int64, int8, inv, isreal, iztrans, jacobian, jordan, lambertw, laplace, latex, lcm, ldivide, le, limit,* log, log10, *log2, lt, mfun, mldivide, mpower, mrdivide, mtimes, ne, null, numden, numel, plus, poly, poly2sym, power, pretty, procread, prod, quorem, rank, rdivide, real,* round, *rref,* sec, sech, *simple, simplify,* sin, *single,* sinh, *sinint, size, solve, sort,* sqrt, *struct, subs, subsasgn, subsref, sum, svd, sym2poly, symsum,* tan, tanh, *taylor, times, transpose, tril, triu, uint16, uint32, uint64, uint8, vertcat, vpa,* ζ, *ztrans*]

with(*plots*)

[*animate, animate3d, animatecurve, arrow, changecoords, complexplot, complexplot3d, conformal, conformal3d, contourplot, contourplot3d, coordplot, coordplot3d, densityplot, display, dualaxisplot, fieldplot, fieldplot3d, gradplot, gradplot3d, implicitplot, implicitplot3d, inequal, interactive, interactiveparams, intersectplot, listcontplot, listcontplot3d, listdensityplot, listplot, listplot3d, loglogplot, logplot, matrixplot, multiple, odeplot, pareto, plotcompare, pointplot, pointplot3d, polarplot, polygonplot, polygonplot3d, polyhedral_supported, polyhedraplot, rootlocus, semilogplot, setcolors, setoptions, setoptions3d, spacecurve, sparsematrixplot, surfdata, textplot, textplot3d, tubeplot*]

with(*plottools*)

[*arc, arrow, circle, cone, cuboid, curve, cutin, cutout, cylinder, disk, dodecahedron, ellipse, ellipticArc, hemisphere, hexahedron, homothety, hyperbola, icosahedron, line, octahedron, parallelepiped, pieslice, point, polygon, project, rectangle, reflect, rotate, scale, semitorus, sphere, stellate, tetrahedron, torus, transform, translate*]

with(*RealDomain*)

[$\Im$, $\Re$, `^`, arccos, arccosh, arccot, arccoth, arccsc, arccsch, arcsec, arcsech, arcsin, arcsinh, arctan, arctanh, cos, cosh, cot, coth, csc, csch, *eval,* exp, *expand, limit,* ln, log, sec, sech, signum, *simplify,* sin, sinh, *solve,* sqrt, surd, tan, tanh]

with(*Student*)

[*Calculus1, LinearAlgebra, MultivariateCalculus, NumericalAnalysis, Precalculus, SetColors, VectorCalculus*]

with(*Student*[*Calculus1*])

[*AntiderivativePlot, AntiderivativeTutor, ApproximateInt, ApproximateIntTutor, ArcLength, ArcLengthTutor, Asympototes, Clear, CriticalPoints, CurveAnalysisTutor, DerivativePlot, DerivativeTutor, DiffTutor, ExtremePoints, FunctionAverage, FunctionAverageTutor, FunctionChart, FunctionPlot, GetMessage, GetNumProblems, GetProblem, Hint, InflectionPoints, IntTutor, Integrand, InversePlot, InverseTutor, LimitTutor, MeanValueTheorem, MeanValueTheoremTutor, NewtonQuotient, NewtonsMethod, NewtonsMethodTutor, PointInterpolation, RiemannSum, RollesTheorem, Roots, Rule, Show, ShowIncomplete, ShowSolution, ShowSteps, Summand, SurfaceOfRevolution,*

SurfaceOfRevolutionTutor, Tangent, TangentSecantTutor, TangentTutor,
TaylorApproximation, TaylorApproximationTutor, Understand, Undo,
VolumeOfRevolution, VolumeOfRevolutionTutor, WhatProblem]

with(*Student*[*LinearAlgebra*])

[&x, `.`, AddRow, AddRows, Adjoint, ApplyLinearTransformPlot, BackwardSubstitute,
BandMatrix, Basis, BilinearForm, CharacteristicMatrix, CharacteristicPolynomial,
ColumnDimension, ColumnSpace, CompanionMatrix, ConstantMatrix, ConstantVector,
CrossProduct, CrossProductPlot, Determinant, Diagonal, DiagonalMatrix, Dimension,
Dimensions, EigenPlot, EigenPlotTutor, Eigenvalues EigenvaluesTutor, Eigenvectors
EigenvectorsTutor, Equal, GaussJordanEliminationTutor, GaussianElimination,
GaussianEliminationTutor, GenerateEquations, GenerateMatrix, GramSchmidt,
HermitianTranspose, Id, IdentityMatrix, IntersectionBasis, InverseTutor, IsDefinite,
IsOrthogonal, IsSimilar, IsUnitary, JordanBlockMatrix, JordanForm, LUDecomposition,
LeastSquares, LeastSquaresPlot, LinearSolve, LinearSolveTutor, LinearSystemPlot,
LinearSystemPlotTutor, LinearTransformPlot, LinearTransformPlotTutor, MatrixBuilder,
MinimalPolynomial, Minor, MultiplyRow, Norm, Normalize, NullSpace, Pivot, PlanePlot,
ProjectionPlot, QRDecompostion, RandomMatrix, RandomVector, Rank,
ReducedRowEchelonForm, ReflectionMatrix, RotationMartix, RowDimension, RowSpace,
SetDefault, SetDefaults, SumBasis, SwapRow, Trace, Transpose, UnitVector, VectorAngle,
VectorSumPlot, ZeroMatrix, ZeroVector]

with(*Student*[*MultivariateCalculus*])

[ApproximateInt, ApproximateIntTutor, CenterOfMass, ChangeOfVariables, CrossSection,
CrossSectionTutor, Del, DirectionalDerivative, DirectionalDerivativeTutor,
FunctionAverage, Gradient, GradientTutor, Jacobian, LagrangeMultipliers, MultiInt,
Nabla, Revert, SecondDerivativeTest, SurfaceArea, TaylorApproximation,
TaylorApproximationTutor]

with(*Student*[*Precalculus*])

[CenterOfMass, CompleteSquare, CompositionPlot, CompositionTutor, ConicsTutor, Distance,
FunctionSlopePoint, FunctionSlopeTutor, LimitPlot, LimitTutor, Line, LineTutor,
LinearInequalitiesTutor, Midpoint, PolynomialTutor, RationalFunctionPlot,
RationalFunctionTutor, Slope, StandardFunctionsTutor]

with(*Student*[*VectorCalculus*])

[&x, `*`, `+`, `-`, `.`, < , >, < | >, About, ArcLength, BasisFormat, Binormal, ConvertVector,
CrossProduct, Curl, Curvature, D, Del, DirectionalDiff, Divergence, DotProduct,
FlowLine, Flux, GetCoordinates, GetPVDescription, GetRootPoint, GetSpace, Gradient,
Hessian, IsPositionVector, IsRootedVector, IsVectorField, Jacobian, Laplacian, LineInt,
MapToBasis, Nabla, Norm, Normalize, PathInt, PlotPositionVector, PlotVector,
PositionVector, PrincipalNormal, RadiusOfCurvature, RootedVector, ScalarPotential,
SetCoordinates, SpaceCurve, SpaceCurveTutor, SurfaceInt, TNBFrame, Tangent,

TangentLine, TangentPlane, TangentVector, Torsion, Vector, VectorField, VectorFieldTutor, VectorPotential, VectorSpace, diff, evalVF, int, limit, series]

with(*student*)

[*D, Diff, Doubleint, Int, Limit, Lineint, Product, Sum, Tripleint, changevar, completesquare, distance, equate, integrand, intercept, intparts, leftbox, leftsum, makeproc, middlebox, middlesum, midpoint, powsubs, rightbox, rightsum, showtangent, simpson, slope, summand, trapezoid*]

with(*VectorCalculus*)

[&*x*, `*`, `+`, `-`, `.`, <, >, < | >, *About, AddCoordinates, ArcLength, BasisFormat, Binormal, Compatibility, ConvertVector, CrossProd, CrossProduct, Curl, Curvature, D, Del, DirectionalDiff, Divergence, DotProd, DotProduct, Flux, GetCoordinateParameters, GetCoordinates, GetNames, GetPVDescription, GetRootPoint, GetSpace, Gradient, Hessian, IsPositionVector, IsRootedVector, IsVectorField, Jacobian, Laplacian, LineInt, MapToBasis, Nabla, Norm, Normalize, PathInt, PlotPositionVector, PlotVector, PositionVector, PrincipalNormal, RadiusOfCurvature, RootedVector, ScalarPotential, SetCoordinateParameters, SetCoordinates, SpaceCurve, SurfaceInt, TNBFrame, Tangent, TangentLine, TangentPlane, TangentVector, Torsion, Vector, VectorField, VectorPotential, VectorSpace, Wronskian, diff, eval, evalVF, int, limit, series*]

$|a|$

$\sqrt{a}$

$k \cdot A + l \cdot B$

$A[i, j]$

A^{-1}

$A^{\%T}$

$ApproximateInt\left(f\left(x \right), x = a\,..\,b, method = left,\ partition = n \right)$

$ApproximateInt\left(f\left(x \right), x = a\,..\,b, method = left,\ output = plot,\ partition = n \right)$

$ApproximateInt\left(f\left(x \right), x = a\,..\,b, method = left,\ output = integral,\ partition = n \right)$

$ApproximateInt\left(f\left(x \right), x = a\,..\,b, method = midpoint,\ partition = n \right)$

$ApproximateInt\left(f\left(x \right), x = a\,..\,b, method = midpoint,\ output = plot,\ partition = n \right)$

$ApproximateInt\left(f\left(x \right), x = a\,..\,b, method = midpoint,\ output = integral,\ partition = n \right)$

$ApproximateInt\left(f\left(x \right), x = a\,..\,b, method = random,\ partition = n \right)$

$ApproximateInt\left(f\left(x \right), x = a\,..\,b, method = random,\ output = animation,\ partition = n \right)$

$ApproximateInt\left(f\left(x \right), x = a\,..\,b, method = random,\ output = animation,\ partiton = n, \right.$
$\left. iterations = m,\ refinement = k \right)$

$ApproximateInt\left(f\left(x \right), x = a\,..\,b, method = random,\ output = plot,\ partition = n \right)$

$ApproximateInt\left(f\left(x \right), x = a\,..\,b, method = random,\ output = integral,\ partition = n \right)$

$ApproximateInt\left(f\left(x\right),\, x=a\mathbin{..}b,\, method=right,\, partition=n\right)$

$ApproximateInt\left(f\left(x\right),\, x=a\mathbin{..}b,\, method=right,\, output=plot,\, partition=n\right)$

$ApproximateInt\left(f\left(x\right),\, x=a\mathbin{..}b,\, method=right,\, output=integral,\, partition=n\right)$

$ApproximateInt\left(f\left(x\right),\, x=a\mathbin{..}b,\, method=simpson,\, partition=n\right)$

$ApproximateInt\left(f\left(x\right),\, x=a\mathbin{..}b,\, method=trapezoid,\, partition=n\right)$

$ApproximateInt\left(f\left(x,y\right),\, x=a\mathbin{..}b,\, y=c\mathbin{..}d,\, method=lower,\, partition=\left[n,n\right]\right)$

$ApproximateInt\left(f\left(x,y\right),\, x=a\mathbin{..}b,\, y=c\mathbin{..}d,\, method=lower,\, partition=\left[n,n\right],\right.$
$\quad\left. output=animation\right)$

$ApproximateInt\left(f\left(x,y\right),\, x=a\mathbin{..}b,\, y=c\mathbin{..}d,\, method=lower,\, partition=\left[n,n\right],\, output=plot\right)$

$ApproximateInt\left(f\left(x,y\right),\, x=a\mathbin{..}b,\, y=c\mathbin{..}d,\, method=midpoint,\, partition=\left[n,n\right]\right)$

$ApproximateInt\left(f\left(x,y\right),\, x=a\mathbin{..}b,\, y=c\mathbin{..}d,\, method=midpoint,\, partition=\left[n,n\right],\right.$
$\quad\left. output=animation\right)$

$ApproximateInt\left(f\left(x,y\right),\, x=a\mathbin{..}b,\, y=c\mathbin{..}d,\, method=midpoint,\, partition=\left[n,n\right],\, output=plot\right)$

$ApproximateInt\left(f\left(x,y\right),\, x=a\mathbin{..}b,\, y=c\mathbin{..}d,\, method=random,\, partition=\left[n,n\right]\right)$

$ApproximateInt\left(f\left(x,y\right),\, x=a\mathbin{..}b,\, y=c\mathbin{..}d,\, method=random,\, partition=\left[n,n\right],\right.$
$\quad\left. output=animation\right)$

$ApproximateInt\left(f\left(x,y\right),\, x=a\mathbin{..}b,\, y=c\mathbin{..}d,\, method=random,\, partition=\left[n,n\right],\, output=plot\right)$

$ApproximateInt\left(f\left(x,y\right),\, x=a\mathbin{..}b,\, y=c\mathbin{..}d,\, method=upper,\, partition=\left[n,n\right]\right)$

$ApproximateInt\left(f\left(x,y\right),x=a\mathbin{..}b,\ y=c\mathbin{..}d,method=upper,\ partition=\left[n,n\right],\right.$
$\left.output=animation\right)$

$ApproximateInt\left(f\left(x,y\right),\ x=a\mathbin{..}b,\ y=c\mathbin{..}d,method=upper,\ partition=\left[n,n\right],output=plot\right)$

$ArcLength\left(f\left(x\right),x=a\mathbin{..}b\right)$

$ArcLength\left(f\left(x\right),\ x=a\mathbin{..}b,output=integral\right)$

$ArcLength\left(f\left(x\right),\ x=a\mathbin{..}b,output=plot\right)$

$ArcLength\left(r\left(\theta\right),\theta=a\mathbin{..}b,coordinates=\text{polar}\right)$

$ArcLength\left(v\left(t\right),t=a\mathbin{..}b\right)$

$ArcLength\left(\left[x\left(t\right),y\left(t\right)\right],t=a\mathbin{..}b\right)$

$ArcLength\left(\left[x\left(t\right),y\left(t\right)\right],t=a\mathbin{..}b,\ output=integral\right)$

$ArcLength\left(\left[x\left(t\right),y\left(t\right)\right],t=a\mathbin{..}b,\ output=plot\right)$

$Asymptotes\left(f\left(x\right),x\right)$

$Asymptotes\left(f\left(x\right),x=a\mathbin{..}b\right)$

$Asymptotes\left(f\left(x\right),x=a\mathbin{..}b,numeric\right)$

$\text{abs}\left(a\right)$

$angle\left(v,w\right)$

$animate\left(r\left(t,\theta\right),\theta=a\mathbin{..}b,t=c\mathbin{..}d,coordinates=\text{polar},frames=n\right)$

$array\left(1\mathbin{..}m,1\mathbin{..}n\right)$

$CenterOfMass\left(\rho, z = f_1(x, y) .. f_2(x, y), y = h_2(x) .. h_2(x), x = a .. b\right)$

$CenterOfMass\left(\rho, z = f_1(x, y) .. f_2(x, y), y = h_2(x) .. h_2(x), x = a .. b, output = integral\right)$

$CenterOfMass\left(\rho, z = f_1(x, y) .. f_2(x, y), y = h_2(x) .. h_2(x), x = a .. b, output = plot\right)$

$CompleteSquare(c, x)$

$CriticalPoints(f(x))$

$CriticalPoints(f(x), numeric)$

$CrossProduct(v, w)$

$Curl(F)$

$conic(c, equation, [x, y])$

$contourplot\left(f(x, y), x = a .. b, y = c .. d, contours = k, grid = [m, n]\right)$

$contourplot3d\left(f(x, y), x = a .. b, y = c .. d, contours = k, axes = box\right)$

$convert(t, polynom)$

$crossprod(v, w)$

$\dfrac{\mathrm{d}}{\mathrm{d}x} f$

$\dfrac{\partial}{\partial x} f$

$D(f)$

$D(f)(x)$

$D[i]$

$D[i, j]$

$DEplot\left(ode,\left[y(x)\right], x = a \mathinner{..} b, y = c \mathinner{..} d, initialcondition\right)$

$DerivativePlot\left(f(x), x = a \mathinner{..} b, order = 1 \mathinner{..} n\right)$

$DerivativePlot\left(f(x), x = a \mathinner{..} b, order = 1 \mathinner{..} 2, functionoptions = \left[color = color1\right],\right.$
$\quad derivativeoptions\left[1\right] = \left[color = color2, linestyle = linestyle1, thickness = n\right],$
$\quad \left.derivativeoptions\left[2\right] = \left[color = color3, linestyle = linestyle2, thickness = m\right]\right)$

$Diff\left(f(x), x\right)$

$DirectionalDerivative\left(f(x, y), \left[x, y\right] = \left[a, b\right], \left[c, d\right]\right)$

$DirectionalDerivative\left(f(x, y), \left[x, y\right] = \left[a, b\right], \left[c, d\right], x = a \mathinner{..} b, y = c \mathinner{..} d, z = p \mathinner{..} q, output = plot\right)$

$DotProduct(v, w)$

$dfieldplot\left(\dfrac{d}{dx} y(x) = f(x), y(x), x = a \mathinner{..} b, y = c \mathinner{..} d, title = \,'directionfield\ plot\,'\right.$
$\qquad \left.arrows = SLIM, dirfield = \left[m, n\right]\right)$

$denom\left(f(x)\right)$

$detail(c)$

$diag\left(a_{11}, a_{22}, \ldots, a_{nn}\right)$

$diff\left(f(x), x\right)$

$diff\left(f(x, y), y\right)$

$diff \left(f \left(x, y \right), x, x \right)$

$diff \left(f \left(x, y \right), x, y \right)$

$diff \left(f \left(x, y \right), y, x \right)$

$diff \left(f \left(x, y \right), y, y \right)$

$discont \left(f \left(x \right), x \right)$

$display \left(p1, p2, ..., pn \right)$

$dotprod \left(v, w \right)$

$draw \left(c \right)$

$draw \left(c, axes = normal, thickness = n \right)$

$dsolve \left(ode \right)$

$dsolve \left(\left\{ ode, initialvalues \right\} \right)$

$dsolve \left(ode, \left[linear \right], useInt \right)$

$dsolve \left(ode, \left[separable \right] \right)$

$dsolve \left(ode, \left[separable \right], useInt \right)$

$dsolve_{\mathrm{int}\,eractive} \left(\dfrac{\mathrm{d}}{\mathrm{d}x} y \left(x \right) = f \left(x \right) \right)$

$\_ EnvHorizontalName := \,' x \,'$

$\_ EnvVerticalName := \,' y \,'$

$ExtremePoints\big(f(x), x = a..b\big)$

$ExtremePoints\big(f(x), x = a..b, numeric\big)$

elif

end do

end if

$evalf(a)$

$evalf(a, n)$

$evalf\big((a.b), n\big)$

$evalm(A \& \cdot B)$

$evalm(sA + tB)$

$expand\big(f(x)\big)$

$f'(x)$

$f := x \rightarrow y$

$f(a)$

$(f @ g)(x)$

$(f \circ g)(x)$

$Flux(F, C)$

$FunctionAverage\big(f(x), x = a..b\big)$

$FunctionAverage\left(f\left(x\right),x=a\mathbin{..}b,output=integral\right)$

$FunctionAverage\left(f\left(x\right),x=a\mathbin{..}b,output=plot\right)$

$FunctionChart\left(f\left(x\right),x=a\mathbin{..}b\right)$

$FunctionChart\left(f\left(x\right),x=a\mathbin{..}b,pointoptions=\left[symbolsize=n,color=color1\right],\right.$
$\quad sign=\left[thickness\left(m,m\right)\right],slope=color\left(color2,color3\right),$
$\quad \left. concavity=\left[filled\left(color4,color5\right),arrow\right]\right)$

$factor\left(f\left(x\right)\right)$

$fieldplot\left(\left\langle v\left(x,y\right),w\left(x,y\right)\right\rangle,x=a\mathbin{..}b,y=c\mathbin{..}d,arrows=option\right)$

$fieldplot3d\left(\left\langle u\left(x,y,z\right),v\left(x,y,z\right),w\left(x,y,z\right)\right\rangle,x=a\mathbin{..}b,y=c\mathbin{..}d,y=p\mathbin{..}q,grid=\left[m,n,l\right]\right)$

$form\left(c\right)$

$fsolve\left(f\left(x\right)=0,x\right)$

$fsolve\left(f\left(x\right)=a,x\right)$

$fsolve\left(f\left(x\right)=a,x,complex\right)$

$Gradient\left(f\left(x,y\right),[x,y]\right)$

$Gradient\left(f\left(x,y,z\right)\right)$

$\int_{a}^{b}f\,dx$

$InflectionPoints\left(f\left(x\right)\right)$

$InflectionPoints\left(f\left(x\right),numeric\right)$

$InversePlot\left(f(x), x = a..b\right)$

$InversePlot\left(f(x), view = [a..b, c..d], functionoptions = [color = color1, thickness = n1],\right.$
$\left.\qquad inverseoptions = [color = color2, thickness = n2], lineoptions = [color = color3, thickness = n3]\right)$

$implicitdiff\left(f, y, x\right)$

$implicitplot\left(implicit\ equation, x = a..b, y = c..d\right)$

$implicitplot\left([implicit\ equation1, implicit\ equation2], x = a..b, y = c..d, color = [color1, color2]\right)$

$implicitplot\left(implicit\ equation, x = a..b, y = c..d, scaling = constrained\right)$

$implicitplot\left([implicit\ equation1, implicit\ equation2], x = a..b, y = c..d, color = [color1, color2],\right.$
$\left.\qquad grid = [m, n], thickness = t, scaling = constrained\right)$

$implicitplot3d\left(implicit\ equation, a..b, c..d, p..q, options\right)$

$inverse\left(A\right)$

$infolevel\left[dsolve\right] := n$

$infolevel\left[Student\left[Calculus1\right]\right] := n$

$int\left(f(x, y), [x, y]\right)$

$int\left(f(x, y), [x = a..b, y = c..d]\right)$

$Jacobian\left([x(s, t, r), y(s, t, r), z(s, t, r)], 'determinant'\right)$

$\lim\limits_{x \to a} f$

$LagrangeMultipliers\left(f(x, y), [g(x, y)], [x, y], output = detailed\right)$

$LagrangeMultipliers\left(f\left(x,y\right),\left[g\left(x,y\right)\right],\left[x,y\right],output=plot\right)$

$Limit\left(f\left(x\right),x=a\right)$

$Limit\left(f\left(x\right),x=a,left\right)$

$Limit\left(f\left(x\right),x=a,right\right)$

$LineInt\left(F,C\right)$

$LineInt\left(F,C,output=plot\right)$

$limit\left(f\left(x\right),x=a\right)$

$limit\left(f\left(x\right),x=a,left\right)$

$limit\left(f\left(x\right),x=a,right\right)$

$limit\left(f\left(x,y\right),\{x=c,y=d\}\right)$

$limit\left(f\left(x,y\right),\{y=mx+b\}\right)$

$limit\left(limit\left(f\left(x,y\right),\{x=c\}\right),\{y=d\}\right)$

$Matrix\left(\left[a_{11},a_{12},...,a_{1n}\right],\left[a_{21},a_{22},...,a_{2n}\right],...,\left[a_{m1},a_{m2},...,a_{mn}\right]\right)$

$Matrix\left(m,n,\left[a_{11},a_{12},...,a_{mn}\right]\right)$

$Matrix\left(m,n,\left(i,j\right)\rightarrow f\left(i,j\right)\right)$

$MeanValueTheorem\left(f\left(x\right),x=a..b\right)$

$MeanValueTheorem\left(f\left(x\right),x=a..b,output=points\right)$

$MeanValueTheorem\left(f\left(x\right), x = a .. b, output = points, numeric\right)$

$MeanValueTheorem\left(f\left(x\right), x = a .. b, output = plot\right)$

$MultiInt\left(f\left(x, y\right), x = a .. b, y = c .. d\right)$

$MultiInt\left(f\left(x, y\right), x = a .. b, y = c .. d, output = integral\right)$

$MultiInt\left(f\left(x, y\right), x = a .. b, y = c .. d, output = steps\right)$

$MultiInt\left(f\left(x, y\right), x = a .. b, y = c .. d, z = p .. q\right)$

$MultiInt\left(f\left(x, y, z\right), x = a .. b, y = c .. d, z = p .. q, output = integral\right)$

$MultiInt\left(f\left(x, y, z\right), x = a .. b, y = c .. d, z = p .. q, output = steps\right)$

$makeproc\left(\left[x, f\left(x\right)\right], `slope` = m\right)$

$maximize\left(f\left(x\right), x = a .. b\right)$

$minimize\left(f\left(x\right), x = a .. b\right)$

$NewtonsMethod\left(f\left(x\right), c, iterations = n\right)$

$NewtonsMethod\left(f\left(x\right), c, iterations = n, output = animation\right)$

$NewtonsMethod\left(f\left(x\right), c, iterations = n, output = plot\right)$

$NewtonsMethod\left(f\left(x\right), c, iterations = n, output = sequence\right)$

$NewtonsMethod\left(f\left(x\right), c, iterations = n, output = value\right)$

$Norm\left(v, 2\right)$

$normal\left(f\left(x\right)\right)$

$numer\left(f\left(x\right)\right)$

$odeadvisor\left(ode\right)$

$Path\left(r\left(t\right), t = a\,..\,b\right)$

$ProjectionPlot\left(v, w\right)$

$piecewise(\,function)$

$print\left(A\right)$

$plot\left(\left[\,`@`(f, g), `@`(g, f)\right], a\,..\,b, color = \left[color1, color\right], linestyle = \left[linestyle1, linestyle2\right]\right)$

$plot\left(f, a\,..\,b\right)$

$plot\left(f, a\,..\,b, discont = true\right)$

$plot\left(f\left(x\right), x\right)$

$plot\left(f\left(x\right), x = a\,..\,b\right)$

$plot\left(f\left(x\right), x = a\,..\,b, y = c\,..\,d\right)$

$plot\left(f\left(x\right), x = a\,..\,b, y = c\,..\,d, discont = \left[showremovable\right]\right)$

$plot\left(f\left(x\right), x = a\,..\,b, y = c\,..\,d, discont = true\right)$

$plot\left(f\left(x\right), x = a\,..\,b, y = c\,..\,d, numpoints = n\right)$

$plot\left(f\left(x\right), x = a\,..\,b, y = c\,..\,d, thickness = n\right)$

$plot\left(f\left(x\right), x = a\,..\,b, y = c\,..\,d, discont = true\right)$

$$plot\left(\{f(x),g(x)\}, x=a..b, y=c..d\right)$$

$$plot\left(\left[f(x),g(x)\right], x=a..b, y=c..d, color=[color1, color2]\right)$$

$$plot\left(\left[f(x),g(x)\right], x=a..b, y=c..d, color=[color1, color2], linestyle=\right.$$
$$\left.[linestyle1, linestyle2]\right)$$

$$plot\left(\left[x(t), y(t), t=a..b\right], scaling=constrained\right)$$

$$plot\left(\left[seq([n,a(n)], n=1..k)\right], style=style1, thickness=m, symbolsize=l\right)$$

$$plot3d\left(f(x,y), x=a..b, y=c..d\right)$$

$$plot3d\left(f(x,y), x=a..b, y=c..d, axes=box\right)$$

$$plot3d\left(f(x,y), x=a..b, y=c..d, axes=normal\right)$$

$$plot3d\left(f(x,y), x=a..b, y=c..d, orientation=[m,n]\right)$$

$$plot3d\left(f(\theta,z), r=a..b, \theta=c..d, coords=cylindrical, axes=box\right)$$

$$plot3d\left(f(\theta,\phi), \theta=a..b, \phi=c..d, coords=cylindrical, axes=box\right)$$

$$plot3d\left(2, \theta=0..2\cdot\pi, \phi=0..\pi, coords=spherical, axes=box\right)$$

$$pointlist := \left[seq([n,a(n)], n=k..l)\right]$$

$$pointplot\left([x_1, y_1], [x_2, y_2], ..., [x_n, y_n]\right)$$

$$polarplot\left(\{r(\theta)\}, \theta=a..b\right)$$

$$polarplot\left(\{r(\theta)\}, \theta=a..b, scaling=constrained\right)$$

$print(A)$

$proc(x)$

$quo\big(numer\big(f(x)\big),denom\big(f(x)\big),x\big)$

$RationalFunctionPlot\big(f(x),view=[a..b]\big)$

$RollesTheorem\big(f(x),x=a..b\big)$

$RollesTheorem\big(f(x),x=a..b,output=plot\big)$

$RollesTheorem\big(f(x),x=a..b,output=points\big)$

$RollesTheorem\big(f(x),x=a..b,output=points,numeric\big)$

$Roots\big(f(x)=a,x\big)$

$Roots\big(f(x)=a,x,numeric\big)$

$Roots\big(f(x)=g(x),x\big)$

$Roots\big(f(x)=g(x),x,numeric\big)$

$Roots\big(f(x)=g(x),x=a..b,numeric\big)$

$Rule[\ ]\big(Diff\big(f(x),x\big)\big)$

$Rule[\ ]\big(Int\big(f(x),x\big)\big)$

$Rule[\ ]\big(Int\big(f(x),x=a..b\big)\big)$

$Rule[\ ]\big(Limit\big(f(x),x=a\big)\big)$

Rule[*name of rule*]((*label number*))

remove(*has*, *A*, *I*)

restart

round(*a*)

SecondDerivativeTest($f(x, y), [x, y] = [a, b]$)

ShowIncomplete((*label number*))

ShowSolution(*Limit*($f(x), x = a$))

ShowSolution(*Diff*($f(x), x$))

ShowSolution(*Int*($f(x), x$))

ShowSolution(*Int*($f(x), x = a .. b$))

ShowSteps( )

SpaceCurve($v(t), t = a .. b, axes = normal$)

Sum($a(n), n = k .. m$)

SurfaceOfRevolution($f(x), x = a .. b$)

SurfaceOfRevolution($f(x), x = a .. b, axis = vertical$)

SurfaceOfRevolution($f(x), x = a .. b, distancefromaxis = n$)

SurfaceOfRevolution($f(x), x = a .. b, axis = vertical, distancefromaxis = n$)

$SurfaceOfRevolution\left(f\left(x\right), x = a \mathbin{..} b, output = integral\right)$

$SurfaceOfRevolution\left(f\left(x\right), x = a \mathbin{..} b, output = integral, axis = vertical\right)$

$SurfaceOfRevolution\left(f\left(x\right), x = a \mathbin{..} b, output = integral, distancefromaxis = n\right)$

$SurfaceOfRevolution\left(f\left(x\right), x = a \mathbin{..} b, output = integral, axis = vertical, distancefromaxis = n\right)$

$SurfaceOfRevolution\left(f\left(x\right), x = a \mathbin{..} b, output = plot\right)$

$SurfaceOfRevolution\left(f\left(x\right), x = a \mathbin{..} b, output = plot, axis = vertical\right)$

$SurfaceOfRevolution\left(f\left(x\right), x = a \mathbin{..} b, output = plot, distancefromaxis = n\right)$

$SurfaceOfRevolution\left(f\left(x\right), x = a \mathbin{..} b, output = plot, axis = vertical, distancefromaxis = n\right)$

$seq\left(a\left(n\right), n = k \mathbin{..} l\right)$

$showtangent\left(f\left(x\right), x = c, x = a \mathbin{..} b\right)$

$simplify\left(f\left(x\right)\right)$

$slope\left(A, B\right)$

$solve\left(f\left(x\right) = 0, x\right)$

$solve\left(f\left(x\right) = a, x\right)$

$solve\left(f\left(x\right) \geq a, \{x\}\right)$

$solve\left(f\left(x\right) \leq a, \{x\}\right)$

$subs\left(x = a, g\right)$

$sum\left(a(n), n = k \mathbin{..} m\right)$

Appendix B *Maple* Commands

$Tangent\left(f(x), x = c, a \mathinner{..} b\right)$

$Tangent\left(f(x), x = c, a \mathinner{..} b, output = line\right)$

$Tangent\left(f(x), x = c, a \mathinner{..} b, output = plot\right)$

$Tangent\left(f(x), x = c, a \mathinner{..} b, output = slope\right)$

$TangentVector\left(v(t)\right)$

$TaylorApproximation\left(f(x), x = c, order = 1 \mathinner{..} n\right)$

$TaylorApproximation\left(f(x), x = c, a \mathinner{..} b, order = 1 \mathinner{..} n\right)$

$TaylorApproximation\left(f(x), x = c, order = n, view = \left[a \mathinner{..} b, c \mathinner{..} d\right], output = plot\right)$

$taylor\left(f(x), x = c, n\right)$

$transpose\left(A\right)$

$unapply\left(expression, x\right)$

$v \;\&\mathrm{x}\; w$

$v.w$

$VectorAngle\left(v, w\right)$

$VectorField\left(\langle v(x, y), w(x, y)\rangle\right)$

$VolumeOfRevolution\left(f(x), g(x), x = a \mathinner{..} b\right)$

$VolumeOfRevolution\left(f(x), g(x), x = a \mathinner{..} b, axis = vertical\right)$

$VolumeOfRevolution\left(f(x), g(x), x = a \mathinner{..} b, distancefromaxis = n\right)$

$VolumeOfRevolution\left(f(x), g(x), x = a .. b, axis = vertical, distancefromaxis = n\right)$

$VolumeOfRevolution\left(f(x), g(x), x = a .. b, output = integral\right)$

$VolumeOfRevolution\left(f(x), g(x), x = a .. b, output = integral, distancefromaxis = n\right)$

$VolumeOfRevolution\left(f(x), g(x), x = a .. b, output = integral, axis = vertical\right)$

$VolumeOfRevolution\left(f(x), g(x), x = a .. b, output = integral, axis = vertical, distancefromaxis = n\right)$

$VolumeOfRevolution\left(f(x), g(x), x = a .. b, output = plot\right)$

$VolumeOfRevolution\left(f(x), g(x), x = a .. b, output = plot, distancefromaxis = n\right)$

$VolumeOfRevolution\left(f(x), g(x), x = a .. b, output = plot, axis = vertical\right)$

$VolumeOfRevolution\left(f(x), g(x), x = a .. b, output = plot, axis = vertical, distancefromaxis = n\right)$

$with\left(DETools\right)$

$with\left(geometry\right)$

$with\left(linalg\right)$

$with\left(LinearAlgebra\right)$

$with(MTM)$

$with\left(plots\right)$

$with\left(plottools\right)$

$with\left(RealDomain\right)$

$with\left(Student\right)$

with(*Student*[*Calculus1*])

with(*Student*[*LinearAlgebra*])

with(*Student*[*MultivariateCalculus*])

with(*Student*[*Precalculus*])

with(*Student*[*VectorCalculus*])

with(*student*)

Appendix C   <u>References</u>

R.T. Smith and R.B. Minton, *Calculus: Early Transcendental Functions*, 3rd edition (2006), McGraw-Hill.

D. Hughes-Hallett, A. M. Gleason, W. G. McCallum, et al., *Calculus*, 5th edition (2009), John Wiley & Sons, Inc.

J. Rogawski, *Calculus*, (2008), W. H. Freeman and Company, New York.

R. Larson and B. H. Edwards, *Calculus*, 9th edition (2009), Brooks Cole.

J. Stewart, *Calculus: Early Transcendentals*, 6th edition (2008), Cengage Learning USA.

Index

Index

Index

Notes

Notes

Notes

Notes

Notes

Notes

Notes

Notes

Notes